SUSTAINABLE ENERGY LANDSCAPES

Designing, Planning, and Development

Applied Ecology
and Environmental Management

A SERIES

Series Editor
Sven E. Jørgensen
Copenhagen University, Denmark

ADDITIONAL VOLUMES IN PREPARATION

SUSTAINABLE ENERGY LANDSCAPES

Designing, Planning, and Development

Edited by **Sven Stremke and Andy van den Dobbelsteen**

CRC Press
Taylor & Francis Group
Boca Raton   London   New York

CRC Press is an imprint of the
Taylor & Francis Group, an **informa** business

CRC Press
Taylor & Francis Group
6000 Broken Sound Parkway NW, Suite 300
Boca Raton, FL 33487-2742

Printed in the United States of America on acid-free paper
Version Date: 20120622

International Standard Book Number: 978-1-4398-9404-0 (Hardback)

Library of Congress Cataloging-in-Publication Data

Sustainable energy landscapes : designing, planning, and development / edited by Sven Stremke and Andy Van Den Dobbelsteen.
        p. cm. -- (Applied ecology and environmental management)
    Includes bibliographical references and index.
    ISBN 978-1-4398-9404-0 (hardback)
    1. Landscape design. 2. Sustainable design. 3. Renewable energy sources. I. Stremke, Sven. II. Dobbelsteen, Andy van den.

SB472.45.S87 2012
712--dc23   2012021015

Visit the Taylor & Francis Web site at
http://www.taylorandfrancis.com

and the CRC Press Web site at
http://www.crcpress.com

Contents

Part III   Case Studies

Part IV Education

Part V Epilogue

Preface

Dear reader, we hope this book finds you well and tingled with interest in the contents.

The idea for the book was born in August 2010, when we visited the renewable energy island of Samsø in Denmark under the guidance of Professor Sven Erik Jørgensen from Copenhagen University. It was Sven Erik who strongly advocated the necessity for a yet nonexistent book on sustainable energy landscapes. He argued that even though many scholars study sustainable energy landscapes, not a single comprehensible book existed that was devoted to the designing, planning, and development of such landscapes. And he was right. Over the past few years, we had never encountered such a book. When we started discussing a prospective book on sustainable energy landscapes at various international conferences, workshops, and meetings, the idea was very much welcomed by our colleagues, and many promised to share their knowledge and experiences.

Back in 2010, the island of Samsø revealed to us both the possibility and impact of a self-sufficient energy system based on locally available renewable sources, and, not least, we experienced the pride locals take in their achievements. We observed that a renewable energy system does, without a doubt, affect the physical landscape and yet create many synergies and added value for the inhabitants. We were convinced (and still are) that well-designed energy landscapes, realized in close collaboration with (or perhaps even under the lead of) the inhabitants, can strengthen already present landscape qualities, forge a strong community, and substantially improve the regional economy.

We are aware of the fact that Samsø is a special case: an island with a low population density and a mainly agricultural economy that can be sustained by biomass, wind and solar power. Many villages with similar conditions, for example, in Germany and Austria, have followed the same path and developed into self-sufficient renewable (rural) energy landscapes. Many other landscapes, urban landscapes in particular, are confronted with far more challenging assignments when aiming for carbon or energy neutrality. The generation of renewable energy, regardless of location and population density, requires space, and the transition to a sustainable society will affect future landscapes around the world. So Sven Erik was very right in underlining the importance of consolidating the existing knowledge on and experience with sustainable energy landscapes into one book.

We have spent nearly one year of our lives getting this book published, together with the support of 45 experts in the field of sustainable energy landscapes. Yes, it has been hard work, but writing this book has also been very rewarding since *Sustainable Energy Landscapes: Designing, Planning, and Development* provides a comprehensive overview of the state of the art of this exciting new field. We, the editors, and the team of contributing authors hope that we can inspire you; provide you with principles, concepts, theories, and examples; and, last but not least, infect you with the enthusiasm needed to build the energy landscapes of our sustainable future.

Additional material is available from the CRC web site: http://www.crcpress.com/product/isbn/9781439894040.

Sven Stremke
Andy van den Dobbelsteen
Editors

Acknowledgments

We would like to thank all the authors who contributed to this book. It was a daunting task for all of us, and we are very grateful to have been supported by such a strong group of experts. We would also like to thank our colleagues at Taylor & Francis Group, Laurie Schlags and Irma Shagla-Britton, who helped in the book production process. We are also grateful for the helpful and constructive comments we received from the reviewers. Thanks to all of you for helping to realize this important and, not unimportant, beautiful book.

Last but not least we would like to thank our families. Without their understanding and support such an endeavor is impossible. Sven would like to thank his partner Sabrina and his son Luca, who arrived faster than the book. Andy would like to thank his partner Sandra and his lovely daughters Noé and Isha.

Editors

Sven Stremke, Dr. Dipl. Ing., MA (Rostock, Germany, 1976), is an assistant professor of landscape architecture at Wageningen University in the Netherlands. During his undergraduate studies, Sven received a scholarship from the Carl Duisberg trust to move to the United States and gain some international work experience. He received awards for his undergraduate and graduate theses in landscape architecture. In his early career, Sven worked as a designer in Germany, New York City, Amsterdam, and Barcelona.

In 2006, Sven joined Wageningen University to pursue his doctoral thesis entitled *Designing Sustainable Energy Landscapes: Concepts, Principles and Procedures.*

His research commenced with a particular focus on the mitigation of climate change (i.e., sustainable energy landscapes) and has been expanded to also include adaptation to climate change (i.e., climate-robust landscapes). His research and design have been published in scientific journals such as *Environment and Planning B, Landscape Journal, Journal of Landscape Architecture, European Planning Studies,* and *International Journal of Exergy.* Over the past few years, Sven has had the opportunity to present and discuss his work in the Netherlands, Belgium, Germany, Ireland, Czech Republic, Switzerland, South Africa, China, Brazil, and the United States. He has organized and hosted conference sessions devoted to sustainable energy landscapes, for instance, at the 2011 International Federation of Landscape Architects (IFLA) World Congress in Zurich.

Sven combines research and teaching for undergraduate and graduate students of the landscape architecture program at Wageningen University. Several of his thesis students have received awards for their outstanding work, and some projects have resulted in scientific publications.

Each year, Sven and his colleagues conduct commissioned projects on sustainable energy landscapes, for instance, in the Dutch peat colonies, northwest Overijssel, and the Delta region. He was an examiner for the *Mooi Nederland* innovation program on the identity of energy landscapes, funded by the Dutch ministry of spatial planning, and he is a member of the scientific advisory board to the New Development of Methods and Tools for Estimation of Sustainability research project in Denmark. Recently, Sven has initiated the launch of an online platform that provides information on projects, publications, teaching, and other activities with regard to energy landscapes (www.NRGlab.net).

Andy van den Dobbelsteen, PhD, MSc (Tilburg, the Netherlands, 1968), is a full professor of climate design and sustainability at the Faculty of Architecture of the Delft University of Technology, the Netherlands. At the university, he is the coordinator of the Green Building Innovation research program and the built environment theme leader for the Delft Energy Initiative.

Andy lectures and leads research projects in various areas of sustainability in the built environment, most notably on sustainable energy systems for neighborhoods, cities, and regions. He has authored many publications, both scientific and popular, and is one of the expert authors for the Dutch national website on sustainable building (www.duurzaamgebouwd.nl).

Andy has chaired various national and international events, including the award-winning international CIB conference on Smart and Sustainable Built Environments, SASBE2009 (www.sasbe2009.com). He is the joint coordinator of the CIB Working Commission 116 (Smart and Sustainable Built Environments) and a member of several juries, for instance, the Gulden Feniks, the Dutch national renovation prize.

In the past, Andy worked with opMAAT and NIBE, consultants for sustainable building, and he was the chair of the Dutch Green Building Council advisory committee. He was an external examiner for the MA in architecture and urbanism at the Manchester School of Architecture and a visiting fellow at the Melbourne Sustainable Society Institute.

Contributors

Loren Abraham
California State Polytechnic
   University
Pomona, California

and

University of Minnesota
Mankato, Minnesota

Claudia Basta
Department of Environmental
   Sciences
Wageningen University and
   Research Centre
Wageningen, the Netherlands

Simone Bastianoni
Department of Chemistry
Siena University
Siena, Italy

Adri van den Brink
Department of Environmental
   Sciences
Wageningen University and
   Research Centre
Wageningen, the Netherlands

Siebe Broersma
Delft University of Technology
Delft, the Netherlands

Raoul Bunschoten
CHORA
London, United Kingdom
and
Utrecht, the Netherlands

Gerrit J. Carsjens
Department of Environmental
   Sciences
Wageningen University and
   Research Centre
Wageningen, the Netherlands

Daniel Czechowski
Faculty of Architecture
Department of Landscape
Architecture and Regional Open
   Space
Technical University of Munich
Munich, Germany

Andreas R. Dittrich
Faculty of Architecture
Department of Landscape
Architecture and Regional Open
   Space
Technical University of Munich
Munich, Germany

Andy van den Dobbelsteen
Delft University of Technology
Delft, the Netherlands

Machiel van Dorst
Delft University of Technology
Delft, the Netherlands

Michael Dudley
Royal Institute of Art
Stockholm, Sweden

Rudi van Etteger
Department of Environmental
  Sciences
Wageningen University and
  Research Centre
Wageningen, the Netherlands

David Flanders
University of British Columbia
Vancouver, British Columbia,
  Canada

Michiel Fremouw
Delft University of Technology
Delft, the Netherlands

Adrienne Grêt-Regamey
Eidgenössische Technische
  Hochschule Zürich
Zurich, Switzerland

Mary Guzowski
California State Polytechnic
  University
Pomona, California

and

University of Minnesota
Mankato, Minnesota

Sven E. Jørgensen
Copenhagen University
Copenhagen, Denmark

Marianne Karpenstein-Machan
Technical University of Berlin
Berlin, Germany

and

Göttingen University
Göttingen, Germany

Greg Keeffe
School of Planning, Architecture
  and Civil Engineering
Queen's University Belfast
Belfast, United Kingdom

Wim van der Knaap
Department of Environmental
  Sciences
Wageningen University and
  Research Centre
Wageningen, the Netherlands

Barry Lehrman
California State Polytechnic
  University
Pomona, California

and

University of Minnesota
Mankato, Minnesota

Michela Marchi
Department of Chemistry
University of Siena
Siena, Italy

Michael Narodoslawsky
Insitute for Process and Particle
  Engineering
Graz University of Technology
Graz, Austria

Lance Neckar
California State Polytechnic University
Pomona, California

and

University of Minnesota
Mankato, Minnesota

Søren N. Nielsen
Copenhagen University
Copenhagen, Denmark

Henrietta Palmer
Royal Institute of Art
Stockholm, Sweden

Martin J. Pasqualetti
School of Geographical Sciences and
    Urban Planning
and
Global Institue of Sustainability
Arizona State University
Tempe, Arizona

Ellen Pond
University of British Columbia
Vancouver, British Columbia,
    Canada

Riccardo M. Pulselli
Department of Chemistry
University of Siena
Siena, Italy

Rob Roggema
Faculty of Architecture,
Climate Design and Sustainability
Delft University of Technology
Delft, the Netherlands

and

Earth Systems Science Group
Wageningen University and
    Research Centre
Wageningen, the Netherlands

and

RMIT University
Melbourne, Victoria, Australia

Pietro Romano
Department of Chemistry
University of Siena
Siena, Italy

Derek Schilling
California State Polytechnic
    University
Pomona, California

and

University of Minnesota
Mankato, Minnesota

Peter Schmuck
Technical University of Berlin
Berlin, Germany

and

Göttingen University
Göttingen, Germany

Sören Schöbel
Faculty of Architecture
Department of Landscape
Architecture and Regional Open
    Space
Technical University of Munich
Munich, Germany

Olaf Schroth
University of British Columbia
Vancouver, British Columbia,
    Canada

Stephen Sheppard
University of British Columbia
Vancouver, British Columbia,
    Canada

Dirk Sijmons
H+N+S Landscape Architects
Amersfoort, the Netherlands

and

Delft University of Technology
Delft, the Netherlands

Gernot Stoeglehner
Department of Spatial, Landscape
and Infrastructure Sciences
Institute of Spatial Planning and
Rural Development
BOKU—University of Natural
Resources and Life Sciences,
Vienna
Vienna, Austria

Sven Stremke
Department of Environmental
Sciences
Wageningen University and
Research Centre
Wageningen, the Netherlands

Geoffrey Thün
University of Michigan
Ann Arbor, Michigan

and

RVTR
Toronto, Ontario, Canada

Rory Tooke
University of British Columbia
Vancouver, British Columbia, Canada

Elizabeth Turner
California State Polytechnic
University
Pomona, California

and

University of Minnesota
Mankato, Minnesota

Kathy Velikov
University of Michigan
Ann Arbor, Michigan

and

RVTR
Toronto, Ontario, Canada

Renée de Waal
Landscape Architecture Group
Environmental Sciences Group
Wageningen University and
Research Centre
Wageningen, the Netherlands

Ulrike Wissen Hayek
Eidgenössische Technische
Hochschule Zürich
Zurich, Switzerland

André Wüste
Technical University of Berlin
Berlin, Germany

and

Göttingen University
Göttingen, Germany

Part I

Preamble

1

Sustainable Energy Landscapes: An Introduction

Sven Stremke and Andy van den Dobbelsteen

CONTENTS

Much has been said about the introduction of the automobile and the changes in the built environment as a consequence of this unprecedented form of individual and motorized transportation. At present, we are witnessing the emergence of another land use that will affect the appearance and spatial organization of the larger physical environment across the world, including both urban and rural landscapes. Whereas transportation claims no more than 2% of the land surface of most countries, the provision of renewable energy is expected to occupy a much larger and more substantial part of the physical environment. Energy-related land use has already started to compete with food production in some places. There is no doubt that the assimilation, conversion, storage, and transport of renewable energy will be one of the most important land uses of the twenty-first century.

1.1 Problem Statement

Of course, renewable energy is not a new land use; humans have relied on renewable energy sources for 99% of their history. Coal, oil, and gas have only been utilized for the past few centuries. During this recent industrialized period, the world has grown accustomed to an almost decadent exploitation of fossil fuels. In essence, fossil energy resources are dense packages of ancient solar energy, originally captured in biomass and transformed into the highly exergetic oil, natural gas, and coal that we have grown so dependent on. However, the availability of these resources has passed its peak and mankind will have to get used to energy that is acquired mainly from above ground. Because the energetic density of solar, wind, water, and biomass is lower than that of fossil fuel, much larger areas must be allotted to renewable energy. The challenge therefore is to reintegrate renewable energy provision into the existing environment that people have gotten attached to, value, and want to preserve—all in the context of an ever-increasing world population, climate change, and precarious resource depletion.

1.2 Aim of This Book

This book is not about how to add photovoltaic panels to your roof. Nor will we discuss how to select the best exterior color for biogas plants. This book is about energy landscapes—physical environments that provide energy for human use. As with agricultural landscapes or recreation landscapes, energy landscapes are by no means limited to a single land use. The sheer quantity of renewable energy that needs to be generated to sustain humanity may require us to regard, at least conceptually, every landscape as an energy landscape—one of the hypotheses discussed in this book.

In the meantime, environmental designers* should not only participate in the implementation of renewable energy sources but also aim for a socially fair, environmentally sound, and economically feasible transition. Recent trends indicate that the transition to renewable energy is likely to occur in any case, with or without participation of environmental designers. However, ensuring that energy transition takes place in a sustainable manner should be a prerequisite for all of us.

Recent controversies about large-scale hydropower (for instance, in China and Brazil), monoculture production of energy crops (both in Asia and America), and broad opposition to large wind parks (for instance, in Europe) reveal that many of the ongoing renewable energy projects are shortsighted and have purely economic motives. Some may even pose a threat to sustainable development itself. Despite the necessity to transit towards renewable energy sources, several knowledge gaps persist on how to develop sustainable energy landscapes. We define a *sustainable energy landscape* as a physical environment that can evolve on the basis of locally available renewable

---

* *Environmental design* refers to the disciplines concerned with the planning and design of the built environment, such as architecture, urban planning, spatial planning, and landscape architecture.

energy sources without compromising landscape quality, biodiversity, food production, and other life-supporting ecosystem services.

In this book, we do not devote much space to explaining the need for a transition to alternative energy sources. Instead, we aim to contribute to the growing body of knowledge that has been accumulated from people who are committed to exploring a future without fossil fuels and, more specifically, who are eager to discuss the development of a physical environment that can be sustained on the basis of renewable energy sources. The discussion on sustainable energy transition, as confirmed by UNEP and many national policy documents, cannot be limited to the provision of energy alone but must address the reduction of energy use and energy efficiency as well. After all, every joule of energy, regardless of whether it derives from conventional or from renewable sources, consumes resources and takes up space.

1.3 Questions to Be Answered

"Can you help us develop a sustainable energy landscape?" "How can we transform our community so it can rely entirely on locally available, renewable energy sources?" We have been asked to answer these questions many times in recent years. While we remain pleasantly surprised by the large demand for such studies and by the level of commitment from many communities and investors, there remain many challenges for the planner and designer of sustainable energy landscapes. Altogether, they motivated us to compose this book, in which we address the following themes:

1. *Methods.* Many questions that are crucial to the planning and design of sustainable energy landscapes remain to be answered. For example, how can we organize the decision-making process, visualize the future landscapes, and quantify the contribution of energy-conscious interventions to the overall sustainability of a particular landscape, region, or other territory?

2. *Case studies.* Although we have witnessed the emergence of a growing number of renewable energy landscapes across the world, very few best-practice cases have been made accessible to an international audience. To accelerate the transition and to avoid repeating previous mistakes, this book examines some of the most successful projects in a comprehensive manner.

3. *Education.* Energy transition is not only receiving more and more attention in media but has also been included in many higher educational programs—and rightfully so. The pace of innovation

of renewable energy technology, combined with a high level of complexity at the landscape and regional scales, presents a new challenge for the education of young planners and designers. Students and educators, are looking for tools and methodological frameworks that can help them give shape to truly sustainable energy landscapes. A selection of options are presented in this book.

The key question addressed in this book is, how can we design, plan, and develop sustainable energy landscapes? The book draws special attention to the municipal and regional scales—often neglected in the past—where energy-conscious interventions can be implemented effectively.* At the large scale, however, in spite of the many possibilities, complexity increases and continues to challenge conventional approaches to the design, planning, and development of a sustainable environment. This is why the creation of sustainable energy landscapes, among other prerequisites, needs to be approached in a cross and trans-disciplinary manner.

1.4 Bridge between Theory and Practice

In this book, we present a selection of cutting-edge projects from 14 countries: state-of-the-art research, best practices, and education with a focus on sustainable energy landscapes. In doing so, we aim to bridge the gap between fundamental research and theoretical discussion, on the one hand, and planning and design practice, as well as education, on the other. It was our intention from the very beginning to include controversial and at times even contradictory studies. We believe that this strategy can be justified due to the relative novelty of the issues discussed in this book.

Another objective was to facilitate knowledge exchange and discussion between the various disciplines and domains dealing with sustainable energy landscapes. Too many issues are being studied concurrently, but independently, from one another in various countries, while a number of core questions that deserve more attention have escaped national and institutional research agendas. Simultaneously, despite valiant efforts in recent years, the practice of planning and design has not yet fully acknowledged the level of complexity that is inherent to the creation of sustainable energy landscapes. Moreover, some of most breathtakingly innovative research findings have not yet been put into practice.

---

* Many publications are devoted to the scale of the individual building, neighborhood, or small community. There are, in contrast, very few publications that address sustainable energy transition at the larger scale.

This book is intended for (but not limited to) professionals, researchers, educators, students, stakeholders, and decision makers with a background in landscape architecture, landscape planning, land-use planning, urban design, architecture, engineering, or energy management, among other disciplines.

1.5 Outline of the Book

In the second part of this book, experts introduce and discuss innovative methods and approaches to the planning and design of sustainable energy landscapes. The third part illustrates how various approaches have been put into practice in five continents. In the fourth part of the book, educators discuss their perspectives on how to integrate the design, planning, and development of sustainable energy landscapes in higher education. Below we briefly outline the 23 chapters of the book.

The introduction to this book is followed by two chapters that help to establish the larger context of the discussion on energy landscapes. In Chapter 2, Pasqualetti explores how to read the energy landscape around us. His journey around the world and through time is captivating, and many of the photographs are sublime. In Chapter 3, Sijmons and van Dorst discuss why renewable energy technologies face so much opposition. They provide a valuable synthesis of potential harms and trade-offs, with a particular focus on wind turbines—one of the most contested energy technologies.

The second part of the book is devoted to various *methods* for designing, planning, and empirical research on sustainable energy landscapes. In Chapter 4, Dobbelsteen et al. present their approach to the mapping of renewable energy potentials at various spatial scales. This mapping of potentials is doubtlessly one of the initial steps in any meaningful study on sustainable energy landscapes. In Chapter 5, Stremke presents a novel approach to the planning and design of sustainable energy landscapes. The five-step approach can be applied at the municipal, district, and regional scales, where long-term transformation depends on critical uncertainties that go beyond the control of planners and designers. In Chapter 6, Grêt-Regamey and Wissen-Hayek illustrate another method for the planning and design of sustainable energy landscapes. Their multicriteria decision-analysis tool, combined with GIS-based visualizations, can help to integrate renewable energy provision into the landscape, while minimizing adverse effects for other ecosystem services and landscape quality. In Chapter 7, Schöbel et al. discuss the necessity to represent future energy landscapes both realistically and in an ethically correct fashion. They distinguish two fundamentally distinct approaches and enrich the discussion with recent reference studies on the design of energy landscapes.

Starting with Chapter 8, the focus shifts to spatial planning. Roggema classifies energy transition as a wicked problem and subsequently sketches a possible approach that he refers to as swarm planning. Empirical evidence suggests that climate change and energy transition are still approached with more or less conventional planning methods, but he argues that the nature of these issues necessitates a fundamental paradigm shift in planning. In Chapter 9, Basta et al. explore yet another approach to spatial planning that is motivated by the need to change current planning practice. Based on literature research and their experience in education, they call for an advanced approach, described as active planning. In Chapter 10, Stoeglehner and Narodoslawsky propose a systems approach toward integrated spatial and energy planning. They present various typologies for the built environment that can help to operationalize energy-conscious planning at the community scale (see also their case study in Chapter 17).

In the final chapter on methods, Jørgensen describes two quantitative approaches to assess the sustainability of energy landscapes. In Chapter 11, he suggests employing both exergy and carbon models in the assessment of existing and planned energy landscapes (see also the case study in Chapter 18).

The third part of the book is allocated to *case studies*—exemplary and unique projects conducted by planners, designers, and researchers from across the world. In Chapter 12, Bunschoten presents two projects that his UK-based research and design firm CHORA executed in Asia. The proposals for the "Smart City Chengdu" and the "Taiwan Strait Low Carbon Incubator" not only illustrate two fascinating energy landscapes, Bunschoten also discuss how his team encouraged the engagement of citizens and decision makers in the design process. In Chapter 13, Thün and Velikov describe their journey of designing a transnational energy landscape in the Great Lakes megaregion in North America. Based on their research and design, they propose that ecology, economy, and energy should be thought of as unifying agents that can be integrated effectively—and attractively—through new kinds of infrastructure, such as the "post-carbon highway." In Chapter 14, Keeffe introduces yet another aspect of sustainable development in general, and sustainable cities in particular. In this chapter, he first stresses the critical relationships between food production and energy provision and then presents innovative interventions in the built environment of various cities in the United Kingdom that can improve local food production while providing renewable energy. In Chapter 15, Schroth et al. present the work of one of the leading teams of sustainable energy landscape planners in Canada. In recent years, this team from the University of British Columbia has conducted many studies on community energy planning; a selection of projects is presented in this book. In Chapter 16, the focus shifts back to Europe. Schmuck et al., from Göttingen University in Germany, discuss how they initiated, facilitated, and monitored the creation of several "bioenergy villages"—a fascinating journey through more than 10 years of research and planning that has resulted in many zero-carbon municipalities in Germany.

In Chapter 17, Stoeglehner and Narodoslawsky discuss the application of two tools they developed to facilitate the decision-making process in several Austrian municipalities: the GIS-based energy zone mapping and the ELAS calculator. Chapter 18 presents another tool and empirical evidence for the relative sustainability of the Danish island of Samsø. Jørgensen and Nielsen present the first results of their study, in which they examine territorial resource influx and destruction, among other sustainability indicators, by means of yearly exergy balances. In Chapter 19, Pulselli et al. combine two other systemic indicators to inform the planning and design of sustainable energy landscapes. Their case study on the existing energy landscape in the Italian Abruzzo region identifies greenhouse gas source areas and sinks, correlates emissions with the use of environmental resources, and, by doing so, highlights which areas should be focused on when transforming this region into a sustainable energy landscape.

The fourth part of the book focuses on educating energy-conscious planners and designers across the world. In Chapter 20, De Waal et al. reflect on the application of the five-step approach (presented in Chapter 5) in the interdisciplinary training of graduate student at Wageningen University, the Netherlands. They share their experiences, present propositional designs, and discuss the five-step approach that they applied in many studies. In Chapter 21, Lehrman et al. share their perspective on multidisciplinary and energy-conscious design pedagogy in the United States. Examples from courses conducted at the University of Minnesota exemplify how this team of educators integrated energy and water into students' curricula and how the university campus could significantly reduce its greenhouse gas emissions. In Chapter 22, a group of Swedish educators and their students explore sustainable solutions dealing with energy, water, and food in some of the largest metropolitan areas in the world. Palmer and Dudley present a selection of projects from their graduate studios in Shanghai (China), Los Angeles (United States), and Pune (India) and discuss the challenges energy transition poses to all of us, regardless of our professional background and field of activity.

In Chapter 23, the editors attempt to synthesize and conclude the work presented in this book. This was a daunting task, given the many mind-changing ideas, tools, and methods presented in this book. We did our best regardless.

2

Reading the Changing Energy Landscape

Martin J. Pasqualetti

CONTENTS

2.1 Introduction

So dominating a role does energy play in the lives we lead and the land we use that its impacts are everywhere. We see them in the lost forests of England, in the reshaped countryside of the Ruhr, in the oil fields of Azerbaijan, in the abandoned swaths near Chernobyl and Fukushima, and in the vast reservoirs that have formed behind giant dams everywhere from Asia to Zambia. All these are energy landscapes, and once we become attuned to their origins, we begin realizing that our need for energy, in all its many forms, results in much of what awaits us when we venture outdoors. The more capable we become at reading the energy landscape, the more we appreciate how we can use this ability to recall the past, explain the present, and foretell the future.

Reading the energy landscape helps us understand the processes that are involved in changing natural landscapes into human landscapes. We learn, for example, that such conversions have been little noticed and relatively unimportant for most of our time on the planet. Even as recently as 10,000 years ago, at the time of early plant and animal domestication, world population was small but the planet was incomprehensibly big. Even when people occupied every continent, such was their low density that most of the natural landscape remained largely intact.

Once population grew and new tools extended individual power, people started noticing a change in the old patterns. By the fifth century BC, Herodotus was recording how humans were beginning to demonstrate dominion over nature. Before long we were leaving our imprint in deserts, mountains, grasslands, forests, everywhere. We became ingeniously powerful. Soon it was beyond our capability to control either the pace of change or the damage we were causing. Today, with seven billion souls occupying the planet, only a few places with any semblance of Nature's original bequest remain unscathed.

When tools and machines began multiplying in number and diversity, few people cautioned against putting them to their fullest use. Even fewer seemed worried about how their use could drastically change the appearance or use of the land. Instead, most people were disinterested observers, innocently accepting the consequences that the use of such tools could bring. They found at the confluence of their need for energy and their faith in landscape permanence, a helplessness to control a growing ability to change the Earth's surface.

Once machines began augmenting muscle power, deep and lasting harm to the landscape increased to such a degree that it was not only intense and difficult to slow, but also impossible to reverse. It was then that the development and use of energy began pointing us toward our present position, a place where landscapes once thought immutable now commonly lay transformed beyond easy recognition or redemption. More often than we might realize, these "energy landscapes" now appear in our field of view just about everywhere.

Ironically, however, while we might see these landscapes all about us, we often recognize neither their origins nor their value as palimpsests, windows on the past. We might hike across a grassy hillside without realizing that before suffering from the poisoning emissions of power plants, it had been covered with forests. We might stroll along a lovely canal without a thought that it had originally been constructed to transport coal. We might visit an art museum such as London's Tate Modern without recognizing that it once had been an electrical generating station. So common have they become that energy landscapes are now part of our everyday experience. Despite such ubiquity, however, recognition and study of energy landscapes have remained a peripheral theme in the literature, possibly due to indifference or at least helplessness of what might be done to flange up our interest in landscape study with our need to develop energy. As the present book demonstrates, this pattern has obviously changed.

We came to this point gradually, starting in earnest in the mid-1960 with the publication of *The Machine in the Garden* by Leo Marx (1964). Marx' book inspired substantial thought on the relationship of environment and technology, including energy landscapes, by several people including Thayer (1994), Pasqualetti (1997, 2001, 2011), Ghosn (2009), and Burtynsky (2009), as well as a recent collection of several articles published in volume 35, number 2 of the journal *Landscape Research*. This growing cadre is increasingly dedicating itself to understanding the stories that energy landscapes have to tell.

One part of that understanding is to recognize the obvious—that energy landscapes constitute a very large topic. We can help organize our thinking by recognizing that energy landscapes result from a mixture of two groups of influences that might be called "constructs" and "layers." The constructs include the form of the resource, the supply chain that makes these resources available to consumers, as well as the natural and cultural conditions where each step in the supply process is located (Table 2.1). Upon these constructs are imposed several layers (Table 2.2). First, we have the "direct layer," whose changes are expected and whose costs are internalized. Second, we have the "indirect layer," whose landscape changes are more variable and whose costs are externalized. Third, we have the "mitigation layer," whose landscape changes result from attempts to mitigate the changes that fall within the first two categories.

We find that, taken together, these constructs and layers help delimit a wide variety of energy landscapes and that we can best discuss these landscapes within the context of four major economic stages: the organic economy, the mineral economy, the electric economy, and the sustainable economy.

TABLE 2.1

Constructs of Energy Landscapes

Inherent	Solid, liquid, gaseous
Functional	Exploration, extraction, transportation, processing, generation, distribution, disposal, use
Natural	Topography, geology, land cover, climate
Cultural	Economic necessity, standard of living, public perception, technological achievement, habitation, jurisdiction

TABLE 2.2

Layers of Energy Landscapes

Direct	Mining scars, waste piles, rights of way for roads, railroads, canals, and pipelines; storage tanks; derricks; pumping equipment; purpose-built habitation and support structures
Indirect	Subsidence depressions, local pollution impacts on vegetation, sludge ponds, land sterilization, mine fires, town abandonment, oil spills, sludge spills, acid mine drainage, acid rain
Mitigation	Recreational lakes and parks, consolidation drilling, camouflage techniques

2.2 Stage One: Energy Landscapes of the Organic Economy

Looking back 100,000 years, we find humans acquiring all their land-based food through the dual activities of hunting and gathering. This was the period of the "organic economy," when virtually all energy was derived from the capture of solar energy trapped within plants and animals (Wrigley 1988). Any energy landscape that resulted from these activities was minor; the world's population was small, and environmental systems held plenty of excess resiliency. This pattern of hunting and gathering (supplemented by some fishing) persisted until plants and animals fell under human control, and a less nomadic lifestyle became feasible and even preferred. Finally, with the retreat of the great ice sheets, people started spreading poleward and population shot up. The creation of lasting energy landscapes would not be far behind.

2.2.1 Wood as Energy Source

The use of wood as an energy source was easily seen in many places many centuries ago, especially in Europe. Even by the seventeenth century BC, the urbanized areas of Greece were treeless, and by the fifth century BC, the forests over much of Europe were discontinuous. In Britain, large parcels had been removed toward the end of the Middle Iron Age around 250–100 BC, but by the twelfth and thirteenth centuries after Christ the impact on the British landscape was striking. The percentage of forest cover in Ireland dropped from about 34% in 1000 AD to zero by 1800, while that in England dropped from about 24% to about 6% between 1000 AD and 1700 AD (Kaplan et al. 2009). What was left was the first energy landscape, a land stripped of trees. It was the most noticeable energy landscape of the organic economy.

2.2.2 Wind and Water

The major energy landscapes that followed were produced by additions rather than subtractions. Thousands of years before present, energy in wood and human muscles was being meaningfully supplemented with the regular work not only of animals but also of machines. These machines were driven by one or the other forms of solar energy, principally wind and water. It is believed that Heron of Alexandria in the first century AD used a wind-driven wheel to power a machine, although the use of windmills did not become common before the sixth century AD (Drachmann 1961, Lohrmann 1995). Over the years, up to 200,000 windmills were installed in Europe, and they produced a dense energy landscape wherever they were concentrated, becoming so iconic in the Netherlands that some of them

FIGURE 2.1
Example of a well-maintained and highly valued historical windmill "de Gooyer" near the IJ Brewery in Amsterdam, the Netherlands. (Photo by Sven Stremke 2012.)

have been retained as tourist attractions (Figure 2.1). In 1850, there were 9000 windmills in the Netherlands, probably about 5–20 times the number today. Germany in 1895 had over 18,000 windmills.*

Windmills were used for many tasks, including corn milling, irrigation, improved agriculture, hulling barley and rice, saw mills, grinding malt, creating olive oil, and pressing linseed, rapeseed, and hempseed for cooking and lighting. When connected to a waterwheel operating in reverse, they could be effective for draining the polders that have helped create modern Holland (Hoeksema 2006).

In North America, perhaps the most notable energy landscape of the age was created in the Great Plains of the United States. So much did windmills define the landscape there that a traveler could not avoid their distinctive appearance. At one time, hundreds of thousands of wind machines dotted the central part of the country (Righter 1996), and they quickly became beacons of civilization and hope to farmers across the vast lands between the Mississippi and the Rockies. Like many of the wind machines in Holland, most of those on the Great Plains were used to pump water, and they played a critical role in opening the grasslands to permanent habitation.[†]

---

* http://www.lowtechmagazine.com/2009/10/history-of-industrial-windmills.html

[†] The water itself became an early means to store solar energy, an accomplishment that remains an elusive goal of those who promote the use of solar energy for the generation of electricity.

TABLE 2.3

Stage One: Energy Landscapes of the Organic Economy

	Layer One "Direct"	Layer Two "Indirect"	Layer Three "Mitigation"
Wood	Deforestation	Accelerated erosion	n/a
Water	Water wheels (e.g., grinding, lumber, weaving)	Workers' quarters	n/a
Wind	Dutch wind mills Great Plains water pumpers	Reclaimed land Vegetation changes (e.g., native to domesticated grasses with the addition of irrigation)	n/a

The introduction of windmills onto the Great Plains produced one of the most expressive energy landscapes in the world. They became a symbol of the dogged determination of settlers, if not of progress itself. They were welcome neighbors, benign sentinels, visible but clean. Most were eventually displaced by diesel and electric pumps, and the wind energy landscapes that defined several generations of inhabitants reverted to an even more pastoral look (Table 2.3).

2.3 Stage Two: Energy Landscapes of the Mineral Economy

Bending a metaphor, the "Golden Age" of energy landscapes began when the nascent mineral economy grew into the Industrial Revolution. As fossil fuels became more available, the human impact on landscapes became more pronounced and also more destructive. While it was the use of fire and cultivated crops of the organic economy that helped change where and how people could live on Earth, it was the use of fossil fuels of the mineral economy that changed Earth itself. It was a categorical shift; instead of relying on the continuous flow of organic and renewable sources, people started relying on the accumulated stocks of mineral resources. The higher energy densities of fossil fuels also meant that these energy resources could be economically transported longer distances (Jones 2010). Among the starkest reflections of this shift to a mineral economy were the landscapes that it created. Those from coal led the way.

2.3.1 Coal

Many reasons accounted for the shift from renewable resources of the organic economy to the fossil fuels of the mineral economy. Forests were shrinking under the dual demands of a growing population and the rising temperature demands of metallurgy. Eventually, by the latter half of the sixteenth century, wood became hard to come by in places like Great Britain. Eventually, trees were being felled for charcoal faster than they could be regrown, and the scarcity of charcoal began crippling iron production and slowing the momentum of industrialization (Sieferle 2001). Coal was an available substitute, although it held some added cost. For example, burning the coal could release into the atmosphere choking amounts of ash and sulfurous gases. More important to our present purposes, its use would create a new form of energy landscape that would become increasingly more obvious as demand grew.

Coal mining began naturally at surface outcroppings. In Europe, its earliest commercial extraction was at a site near Liège, Belgium in 1113 AD (Nef 1932). A century later, it was being transported to London, and by 1325, coal was first exported from England's Tynemouth region back to France. Common usage of coal did not occur until after the 1750s, and it was not the dominant fuel until 1850 (Debeir et al. 1986).

As demand increased, miners began following surface exposures underground. There they encountered obstacles that would challenge their ingenuity and produce some of the most notorious energy landscapes of the day. Soon England's gracious Midlands became known as the Black Country. Single-purpose coal towns—with their houses, churches, shops, and surrounding agricultural lands—multiplied, creating a settlement pattern that is largely still in place. As populations grew, air pollution and other environmental problems worsened, often reaching notorious levels (Brimblecombe 1987). Coal energy landscapes became common whenever the resource was mined, whether in Europe, the United States, Russia, China, or Australia.

Coal mining and environmental quality were opposing forces. Air quality became sickening in the most literal sense, and slag heaps accumulated wherever coal was mined. With little understanding of how to soften coal's impacts, damaged landscapes were endured as an economic necessity, a curse of progress. If you lived amid the coal measures, there was no escaping the energy landscapes coal extraction produced. Central England, South Wales, eastern Belgium, the Ruhr and Saar regions of Germany, Appalachia in North America, the Donets Basin of Ukraine, and many other places were transformed to sordid, unsafe, and pathetic energy landscapes that would come to include scars, pits, shafts, piles of debris, and dismal assemblages of squalid housing. So notorious did coal landscapes become in many places that they served as the backdrop for novels by Emile Zola, D.H. Lawrence, Richard Llewellyn, and many others (Thesing 2000).

New inventions expanded the range of coal mining. At first, diggings had been in the form of shallow bell mines, named after their profile. Later, when mining equipment became available and water pumping more effective, mines became deeper and more elaborate. While much of the coal mining activities

FIGURE 2.2
Open pit coal mine, south of Wilkes Barre, Pennsylvania. (Photo by author, 1979.)

FIGURE 2.3
Strip mining near Jänschwalde, Germany. May 2011. (Photo by Detlef Stremke [LaNaServ].)

remained out of site underground, landscapes were changed nevertheless from the winding gear, ventilation shafts, and even surface subsidence.

Eventually, surface mining expanded as technology became powerful enough to remove the massive amount of surface materials overlying coal seams. This capability eventually matured into open pit coal mines that became common in some areas where coal seams penetrated the surface, such as in the metamorphosed and steeply inclined anthracite seams of the eastern United States (Figure 2.2).

Further technical capabilities allowed other types of surface mining, including contour mining, augering, and the most extensive form of coal extraction now in practice, strip mining, which produces some of the largest energy landscapes anywhere (Figure 2.3).

Recently, a new form of coal landscape has been emerging, mountain-top removal, and it is producing the most dramatic energy landscapes. No longer must the coal be extracted along the contours, through underground shafts, or by peeling off the overburden; instead, entire mountain tops are removed. While this approach allows retrieval of virtually all the coal underneath, it constitutes an especially vicious insult to the land. Reclamation, of a sort, is practiced, but the landscape is altered to an extent never before witnessed (Figure 2.4).

FIGURE 2.4
Once this area was a quiet rural community, but mining companies can legally come within 100 ft of a family cemetery and 300 ft from a home. The Hobet 21 coal mine owned by Patriot Coal (and before by Arch Coal) looms over one of the few remaining houses in Mud, West Virginia, 2005. (Photo © Melissa Farlow [http://olsonfarlow.com/]. Used with permission.)

Mountains in places like West Virginia are simply disappearing. Some people claim that up to 500 mountains in Appalachia have been lost so far.* In the roughly 12 million acre region of eastern Kentucky, southern West Virginia, western Virginia, and eastern Tennessee where mountaintop removal mining takes place, nearly 7% of the land was disturbed between 1992 and 2012.[†]

Moreover, the new landscapes result not just from the removal of mountains but from the dumping of overburden into nearby valleys. This can double the extent of the energy landscapes produced as well as change the hydrology of the watershed and increase the risk of flooding (Reece 2007). A substantial outcry has accompanied the use of this technique, followed by loud—if ineffective—public attempts to halt its practice.[‡]

Just from the extraction phases, our use of coal produces all the landscapes just described, but additional energy landscapes are created during other phases as well. For example, there are landscapes from transporting coal, storing coal, and even from relocating towns when valuable coal is underneath (Figure 2.5). We have converted mining pits into recreational lakes and turned old mines into tourist destinations. Whatever steps we might wish to employ, coal energy landscapes will have a long half-life.

2.3.2 Oil and Natural Gas

After hundreds of years of coal mining, a new type of energy landscape came on the scene by the mid-nineteenth century. Commercial oil wells were first drilled in Baku, Azerbaijan, and Titusville, Pennsylvania, and this activity—along with similar processes used to develop natural gas several decades later—created a very different form of landscape expression. Since neither oil nor natural gas could be "mined" in the traditional sense, a new infrastructure evolved as techniques were developed to lift oil to the surface and deliver it to consumers. This was, and still is, an often messy undertaking.

Drillers probe everywhere in the hope of finding oil. From the early days, their primary goal was to get the product out of the ground and to do so as quickly as possible. What happened to the surrounding landscapes in consequence was of little concern. Such indifference would produce fields of drilling rigs so dense that they would almost touch one another. It was a chaotic recovery system that not only reflected poor field management, but it also wreaked environmental havoc. Access roads lengthened, derricks proliferated, forests fell, ramshackle towns propagated, and oil was spilled

---

* Mary Anne Hitt. A Bird's Eye View of Mountaintop Destruction with Google Earth, 2007. http://googleblog.blogspot.com/2007/03/birds-eye-view-of-mountaintop.html
† As cited on the following website: http://earthobservatory.nasa.gov/Features/Mountaintop Removal/
‡ For a protest song with video: http://www.youtube.com/watch?v=CddjKEVeos0&feature= player_embedded#!; one of many organizations that is attempting to slow the creation of energy landscapes from mountaintop removal is the National Memorial for the Mountains: http://www.ilovemountains.org/memorial/

with little consideration of the lasting damage to the land. In the United States, such landscapes began in northwestern Pennsylvania and were soon replicated in Texas, California, and many other places (Figure 2.6).

The first commercial oil wells heralded not only the awakening of one of the most prosperous and inventive periods in history, they also inadvertently introduced an intense new agent of landscape change. Areas of oil production quickly became grimy, malodorous landscape disasters. While this pattern of disinterested chaos was not uncommon with the development of many valuable resources of the age, what was different was the scale of damage and the novelty of its appearance. Not only were there tight clusterings of wooden drilling rigs, but there were also fountains, rivers, and lakes of oil when reservoir pressure overcame control efforts. Such events, called "blowouts"—often considered signals of wealth and the entrepreneurial spirit—were more properly considered catastrophic accidents.

Once brought to the surface, the next step was to move the oil to market. This need stimulated the construction of a vast new delivery system of pipelines and shipping facilities. This delivery network grew quickly and with little public attention because, for the most part, it was buried underground,

(a)

FIGURE 2.5
Aerial photographs of the village Heyersdorf/Germany in October 2007 (a) and after demolition in May 2011 (b). (Photos by Detlef Stremke [LaNaServ].)

(continued)

(b)

FIGURE 2.5 (continued)
Aerial photographs of the village Heyersdorf/Germany in October 2007 (a) and after demolition in May 2011 (b). (Photos by Detlef Stremke [LaNaServ].)

out of site, and ignored. In places like Southern California, the oil industry expanded with public encouragement and little competition, generating great wealth and helping finance the quick and prosperous growth of Los Angeles (Williams 1997). From the late 1800s and for several decades afterward, tall drilling rigs multiplied with a conspicuous disregard for how they were affecting the landscape. Such public attitudes were not to change until the energy landscapes of oil became valuable for housing. Nowhere did this occur more noticeably than in southern California, where oil developers and housing developers started vying for the same land. Technological innovations would eventually be developed that would allow a profitable coexistence.

The most important of these innovations was directional drilling, a clever and sophisticated technique that accommodated multiple borings from a single pad, with each well seeking oil at a distinct depth, direction, and distance from a single platform. It was a technology quickly adopted, and it provided several desirable benefits simultaneously; it saved money, increased productivity and spatial efficiency, and freed land for nonenergy uses. It was a three-dimensional solution to a two-dimensional problem, and it would continue to be important when drilling moved to offshore jetties and then completely out of sight of land by 1947.

FIGURE 2.6
Oil field, Saratoga, Texas. May 27, 1908. (From U.S. Library of Congress, Reproduction Number: LC-USZ62-54452 [b&w film copy neg.].)

The benefits of directional drilling were not only financial; they were also cosmetic and political. Because the technique could allow concentration of activities into a relatively small area, operators were better able to hide, muffle, and camouflage their drilling operations and equipment. Oil developers in Long Beach, California, used directional drilling techniques not only to consolidate production wells, but they also integrated them into an entirely new element on the skyline. Starting in April 1965, four elaborate artificial islands were constructed offshore, each to be used as the platform for many directionally drilled wells. The parts of the islands facing the shore were painted with pastel colors and equipped with decorative lighting, palm trees, noise-absorbing barriers, and even waterfalls. They soon became an established part of the landscape of Long Beach harbor, even a tourist attraction (Figure 2.7).

The energy landscapes of oil and natural gas are extensive and wide-ranging in form and function. This means that they are not limited to derricks, oil rivers, subsided lands, pumps, pipelines, or camouflaged equipment. They include harbors scattered all over the world, corporate buildings in key cities, purpose-built towns in remote regions where no one else lives, fields of derelict and disused equipment, and—most noticeably—tank farms and large refineries (Figure 2.8). As a reflection of the importance of oil in the world economy, these energy landscapes are ubiquitous.

FIGURE 2.7
Company boat leaves Grissom Island for White Island, two of the four THUMS islands off Long Beach, California. The artificial islands were constructed as drilling platforms to recover oil from the Wilmington oil field. They are decorated with palm trees, sound and sight barriers, and decorative lighting as a way of camouflaging the recovery operations for those who use the nearby beach for recreation. (Photo by author, 1989.)

2.3.3 Unconventional Fossil Fuels

As we continue consuming conventional oil and natural gas reserves, attention is turning to alternative sources for the same fuels. Developing these unconventional fossil fuels is already creating a new class of energy landscapes, including some of the largest we have ever seen. Heavy oils and oil shale are in this class, but it is the energy landscapes of oil sands and shale gas that are generating the most public attention.

The oil sands of Alberta, Canada, are an enormous reserve, on par with the conventional oil reserves of Saudi Arabia and Iraq. Given their proximity to the largest oil customer in the world, they are considered one of Canada's most important future catalysts of economic development. So tempting is the production of this resource for US markets that much of the future economic health of both countries is being weighed against the startling landscape changes their development is producing; already parts of northeastern Alberta have been transformed into the largest energy landscape in the world (Figure 2.9).*

As energy demand rises and as concerns about environmental quality and energy security continue to make headlines, we are widening our search

---

* The only other energy resource that will likely create landscapes of such scale will be the oil shale reserves that lie beneath a large portion of the western slope of the Rocky Mountains.

FIGURE 2.8
Tank farms and oil refineries are common in oil-rich areas, such as here along the Mississippi River between New Orleans and Baton Rouge, Louisiana. (Photo by author, 2002.)

FIGURE 2.9
Energy landscapes produced from oil sands development are on the grandest. Everything in this scene is part of just one of the several surface oil sands projects north of Fort McMurray, Alberta, Canada. (Photo by author, 2005.)

not just for oil but for natural gas. Two forms are creating new impacts on the landscape: coal-bed methane (CBM) and shale gas. Because of its large internal surface area, coal stores six to seven times more gas than the equivalent rock volume of a conventional gas reservoir. This makes CBM particularly attractive. Its recovery, however, produces energy landscapes that are particularly noticeable because of the tight spacing of the wells and because they are being drilling in areas that have not been developed previously.

These impacts are similar with regard to shale gas, although with a new wrinkle. With improvements in the technique of hydrofracturing (or "fracking"), shale gas has been assuming a strong position in the future supplies of natural gas. Like CBM, surface landscape changes are similar to areas used by rigs drilling for oil and gas anywhere. However, much greater volumes of water are needed for the process of fracking, on the order of millions of gallons per well. This water must be lifted, stored, and then reinjected, activities that all require additional land. Of greater concern in some places, such as those above the Marcellus Formation in New York and Pennsylvania, bucolic farmland is being transformed by the new industrial activities surrounding shale gas recovery much to the consternation and dismay of local residents who never considered their neighborhoods would be attractive to energy developers (Table 2.4).

2.4 Stage Three: Energy Landscapes of the Electricity Economy

The energy landscapes of coal, oil, and natural gas recovery and transport have been accumulating for decades in many places, centuries in others. Created mostly from about the 1930s, however, is another subset of energy landscapes, that of electrical generation.

2.4.1 Electricity Generation and Transmission

The generation and transmission of electricity have produced the most recognizable and public energy landscapes on the planet. Its common feature is a "hub and spoke" signature, the hub being the power plant and the spokes being the pathways of fuel supply entering the plant and the transmission lines that exit. Just in the United States, there are about 17,000 power plants. Tying them with load centers are hundreds of thousands of miles of transmission lines crisscrossing the land. The landscape impacts of the power lines vary according to factors of topography, climate, and vegetation. In forested areas, swaths are cleared for rights of way (Figure 2.10). In deserts, glinting lines are everywhere obvious. Wherever they are, transmission lines have become ubiquitous symbols of progress and civilization. They are energy landscapes in every connotation of the phrase.

TABLE 2.4

Stage Two: Energy Landscapes of the Mineral Economy

Resource	Layer One "Direct"	Layer Two "Indirect"	Layer Three "Mitigation"

Coal	Mining pits
	Surface debris
	Canals
	Railroads
		Subsidence
		Abandoned equipment
		  and buildings
		Waste heaps
			Recreational lakes created
			  out of open cast mining
			Damaged forests by acid
			  rain the result from
			  installation of taller
			  chimneys to mitigate
			  local air pollution
			  problems
			Recreational mounds for
			  discarded waste materials

Oil and
  natural gas
	Drilling derricks
	Well fields
	Distribution networks
	Refineries
	Jack pumps
		Oil spills (on land and
		  water; blowouts)
		Advertising
		New towns
		Shipping facilities
		Abandoned equipment
		Contaminated areas
			Artificial islands
			Directional drilling
			  platforms
			Faux buildings
			  for drilling pads

Oil sands	Roads
	Surface excavations
	Seismic lines
		Sulfur piles
		Downstream changes
		  from water diversions
		Greenhouse gas
		  emissions
		New housing
		  requirements
		Pipeline rights-of-way
		  damage	

Heavy oils	Roads recovery equipment	Forest removal	

Coal-bed
  methane
	Roads recovery
	  equipment
		Changed land use from
		  waste water disposal	

Shale gas	Roads recovery
	  equipment
		Water contamination
		Land use
		  transformation
		Economic development	

Oil shale	Roads recovery
	  equipment
		Additional housing
		  in remote areas	

Although they tend to stand out on the landscape, the impact of power plants varies with their surroundings. For example, in many densely settled industrialized countries, power plants have become so common that they are part of the urbanized infrastructure. In less-populated places, they might be too remote to be seen by many, despite the substantial energy landscape presence they present. When, however, such installations are in picturesque environs, they can be especially intrusive (Figure 2.11).

FIGURE 2.10
Swath of forest is cleared and maintained for transmission lines in Modoc County, California.
(Photo by author, 2005.)

FIGURE 2.11
Navajo Generating Station near Page, Arizona. This power plant is in one of the most remote
and colorful parts of the United States. It is often considered an intrusive element in the otherwise untouched landscape. (Photo by author, 2008.)

Most thermal power plants are equipped with condensers—devices that receive spent steam from turbines and condense it before returning it to the boiler. In many places, cooling towers are necessary to assist in condenser cooling. The cooling towers themselves, and the equipment that sometimes is necessary to support them, are among the largest and most recognizable elements of electricity generation. Their visual profile, particularly the hyperbolic silhouette of the natural-draft variety, has become a symbol of power generation, dominating the energy landscape wherever they are installed (Figure 2.12).

Generating stations and transmission lines allowed the spread of electricity across the land. From the time electrical current was sent down the first wires, consumers at the other end were able to use pure electricity far from the sites of its generation. This allowed consumers to live and work hundreds or even thousands of miles from the impacts of mines, well fields, and power plants. With electrical generation came the spread of electricity's many benefits across the land, even to those not living near any usable energy resource.

FIGURE 2.12
Natural draft cooling towers at the Lippendorf power plant near Leipzig in Germany. (Photo by Detlef Stremke [LaNaServ], 2008.)

2.4.2 Nuclear Power

The introduction of commercial nuclear power created yet another group of energy landscapes. Some of them are similar to fossil fuels, especially in terms of mining operations, transmission lines, and cooling arrangements. Others are unique to nuclear power and result from the radiation hazard and tighter security that accompany development and use.

Every industrial facility produces its own landscape signature, and most people have little difficulty distinguishing power plants from steel mills or car factories. Nuclear power plants are among the most readily recognized energy landscapes in the world. This is especially true in the United States where dome-shaped containment structures have become part of landscape, sometimes producing no apparent public concern (Figure 2.13).

Other parts of the nuclear fuel chain have also produced their own landscape expression, including the landscapes from uranium mills, enrichment, and waste disposal. While the mills and enrichment facilities create energy landscapes that are not too dissimilar to other industrial structures, waste disposal is different. Part of this difference is accounted for by the requirement that they often must carry long-lasting landscape warning markers. In the United States, these markers must endure at least 10,000 years, making them our longest-lasting energy landscape (Pasqualetti 1997) (Table 2.5).

FIGURE 2.13
San Onofre State Beach and San Onofre nuclear power plant near San Clemente, California. (Photo by author, 1981.)

TABLE 2.5

Stage Three: Landscapes of the Electricity Economy

Resource	Layer One "Direct"	Layer Two "Indirect"	Layer Three "Mitigation"
Fossil fuels	Refineries Well fields Fuel storage areas Power plants Cooling towers/lakes Roads Railroads Processing equipment (e.g., mills, enrichment, fabrication)	Greater land needs for emission abatement equipment	Ash disposal areas Disposal areas for sludge from desulfurization
Nuclear fuels	Cooling towers/lakes Power plants	Dead zones Accidental contamination Abandoned areas Radioactive waste heaps (e.g., uranium mills, etc.)	Containment structures Exclusions zones Nuclear waste repositories

2.5 Stage Four: Energy Landscapes of the Sustainable Economy

In most fundamental ways, the "sustainable economy" is based on the same resources of wind, water, and sunlight as the "organic economy." The difference is not *what* resources we are using but *how*. In the past, we used the derivatives of solar energy such as wood for heat, wind for pumping, and water for grinding. Today, while we still use biomass in the form of wood, we also use biomass in the form of corn and sugar cane for the ethanol it can yield. We continue to use wind for water pumping, but we also use it to produce electricity. Water is still used for grinding, but more importantly it is used in hydroelectric dams. While the actual contribution of all these renewables is a small fraction of the commercial energy supplied worldwide, their potential contribution is huge. Experts—many of whom are contributors to this book—believe that we must vastly increase our development of these resources if we are going to create a path to a sustainable energy future. As we do, we will expand the energy landscapes they produce.

2.5.1 Hydropower

Hydropower is a combination of solar energy and gravitational energy. Solar energy evaporates water and places it in the atmosphere as vapor. Gravity pulls it down. An early—and continuing—use of waterpower is to move machines, such as water wheels. These activities continue in traditional societies, but the landscapes they produce are minor intrusions compared to what we now see in the form of today's large hydroelectric dams. Today, there are thousands of hydroelectric dams in the world. Some of them inundated vast areas and displaced millions of people. Among the places where this has occurred are the Three Gorges Dam in China, the Aswan High Dam in Egypt, and the Volta Dam in Ghana, which produced the largest reservoir in the world, covering 8482 km^2 (3275 sq mi.).

In some cases, such as at Lake Powell in Arizona, impoundments promote and ease greater public access to scenic but remote areas, producing support for nearby towns and the establishment of an associated infrastructure. Lake Powell, impounded by Glen Canyon Dam, is about 250 sq mi. in area, and it attracts about four million visitors each year to the picturesque Colorado Plateau (Figure 2.14). Whether this influx of people is a good or bad thing is open to opinion, but there is no question that the creation of the reservoir changed the landscape of northern Arizona.

Pumped water storage is another water-related energy technology employed to store electricity. Commonly, water is pumped up to a storage basin during the night and released during the day to generate electricity

FIGURE 2.14
Starting in 1963, Glen Canyon Dam impounded the Colorado River and transformed an arid remote canyon visited by a few people into recreational Lake Powell visited by millions. (Photo by author, 2007.)

FIGURE 2.15
Goldisthal pumped-storage plant in Thuringia, Germany, was officially commissioned in September 2003 and is, at present, Europe's most advanced pumped-storage plant. Total capacity installed is 1060 MW. (Photo by Detlef Stremke [LaNaServ], 2003.)

to help meet peak demand. This can create a rather odd energy landscape consisting of a reservoir that is not in a flooded canyon but at the top of a mountain (Figure 2.15).

2.5.2 Wind Power

The new era of wind power began in the 1980s, and it immediately attracted attention. By June 2011, the global installed capacity was 215 GW. If the average turbine size was 1 MW, that would mean 215,000 wind turbines would now dot landscapes in more than 100 countries where they are being used. All told, they have created the most extensive modern energy landscape on the planet.

Such modern wind energy landscapes are difficult to hide. This is true for several reasons. First, each turbine has a relatively small generating capacity, meaning that to make an appreciable contribution to the supply of electricity requires hundreds of them spread over a large area, each one typically requiring 0.25 acres. Second, for reasons of aerodynamics and efficiency, the turbines cannot be packed too tightly. Typically, 5–10 turbine diameters of spacing are required between wind turbines. This requirement increases the perimeter of the wind farm appreciably. However, the land

FIGURE 2.16
Turbines rise from a cornfield adjacent to a private residence near Dixon, Illinois. The farmer reports a greater income from wind leases than from corn harvests, an apparent compensation for the wind energy landscape he sees every day. (Photo by author, 2005.)

between the turbines is still usable for its original purpose (Denholm et al. 2009) (Figure 2.16).

The third reason that wind turbines are hard to ignore is that they twist and rotate in the wind, they can glint and flicker, attracting attention. Fourth, wind is a site-specific resource that requires that the turbines be installed where adequate wind resources are attractive. Because they are site specific, they are also not amenable to resiting simply for the aesthetic convenience of observers. Sometimes, this means that turbines are placed on ridge lines, rendering them even more conspicuous. Fifth, the windiest sites often must share the same mountain passes as highways, making them much more apparent than a conventional power plant would be in the same location (Figure 2.17).

Last, although we can paint them, slow them, change out lattice towers for monopoles, quiet them, install systems to protect birds and bats, and make many other adjustments in response to public complaints, we cannot make wind turbines invisible. This has made public complaints a common reaction to these eye-catching landscapes (Figure 2.18). Even admitting this limitation, one could argue that seeing wind turbines has its educational benefits; they remind us that our electricity originates somewhere and not a receptacle in the wall (Pasqualetti 2000). Moreover, the energy landscapes that wind

FIGURE 2.17
Zond victory garden in Tehachapi Pass, California. 1996. Placement of the wind turbines on these steep slopes required substantial earth-moving and produced accelerated erosion downslope. (Photo © Paul Gipe. Used with permission.)

FIGURE 2.18
Wind energy landscape altered the view of Mount San Jacinto and sparked a controversy in the nearby resort community of Palm Springs, California. (Photo by author, 2006.)

FIGURE 2.19
Wind energy landscape merging with a solar energy landscape on a former landfill near Nordhausen in Germany. (Photo by Detlef Stremke [LaNaServ], 2008.)

turbines create can be minimized with care in deployment. One can even double-up these and other energy landscapes in some cases (Figure 2.19).

2.5.3 Geothermal Power

Geothermal energy produces dramatic landscapes in many places, such as in the Yellowstone National Park in the United States, in Iceland, and on the Kamchatka Peninsula in Russia. These are natural energy landscapes, but using geothermal energy to generate electricity creates other energy landscapes that are not natural in origin. One of the factors that contributes to the landscapes that geothermal energy produces is its site-specific nature. This means that developments have to accommodate themselves to the existing landscape conditions, whether these are natural conditions such as steep topography or man-made conditions such as agriculture.

This siting restriction helps recreate various geothermal energy landscapes in different environments. For example, at The Geysers field north of San Francisco, hilly terrain imposes restrictions on road building and construction that do not exist in flatter terrain such as in the Imperial Valley, also in California (Figure 2.20). In addition, the industrial landscapes at The Geysers are not compatible with the recreational landscapes around nearby Clear

FIGURE 2.20
The Geysers geothermal development is located in the Mayacamas Mountains, about 75 miles north of San Francisco, California. Local topography strongly influences the form of energy landscapes produced with development. (Photo by author, 1980.)

Lake a few miles downwind. This and other examples of incompatibility have led to public stances against geothermal energy from time to time and from place to place (Pasqualetti and Dellinger 1988).

In Iceland, at The Geysers and in the Imperial Valley, geothermal operations are easily recognized, but in other locations, development has been much more compatible. Again, the possibility for such design is a function of existing conditions and the potential for adjustments. One such site is located a few miles east of the popular Mammoth Mountain ski area near Bishop, California. Developers of that plant designed it to be as unintrusive as possible, keeping its profile low and using colors that blended in with the surroundings. The local abundance of trees contributes to these efforts to camouflage the equipment (Figure 2.21).

2.5.4 Biomass Energy

While the biomass landscapes in the organic economy were created from its harvesting, those biomass landscapes of the modern era are being created by its planting. In the United States, among other countries, maize is converted to ethanol, predominantly in mid-western states like Minnesota, Iowa, and Illinois. Monetary incentives for various activities that accompany such conversion are made available to reduce national dependence on foreign supplies of oil. At least in terms of volume produced, it has been successful; the 2008 U.S.

FIGURE 2.21
Mammoth-Pacific's geothermal installation in eastern California was designed to have minimal impact on the landscape. The popular Mammoth Mountain ski area is in the background. (Photo by author, 2001.)

Department of Agriculture projects national biofuel production at 13.3 billion gallons of conventional ethanol and 700 million gallons of biodiesel in 2015.

The landscape impact on the land has not, to most people, been very noticeable. Because the areas now cultivated for biofuels have long been cultivated, the resulting energy landscapes that have resulted from the trend toward more biofuel production look no different, only more abundant. However, there have been some additional environmental costs; in particular expanded production on erodible croplands has raised concerns over soil loss both from sheet and rill erosion and by wind (Malcolm et al. 2009).

Maize is not the only feedstock used for ethanol production. In Brazil, the predominant feedstock is sugar cane. Brazil produced 26.2 billion liters (6.92 billion U.S. liquid gallons) in 2010, representing 30.1% of the world's total ethanol used as fuel. It is a well-regarded feedstock, both because it converts sunlight into biomass at a comparatively high efficiency, and because it produces low greenhouse gases in the process (US EPA 2010). Vast areas of Brazil have been converted to sugar cane fields; that area totaled 7.8 million hectares of land in 2008 and mainly used abandoned pasture lands. In other words, the use of sugar cane has turned abandoned pasture into productive land and reduced demand for imports at the same time.

Germany, the Czech Republic, and many other countries use biofuels of another type. In these places, the goal has been to supplant part of the

FIGURE 2.22
Photograph of a biogas installation in one of the first *Bioenergiedorf* in Jühnde (Germany). The installation has been well integrated into the rolling hills of the landscape to the north of the village. (Photo by Sven Stremke, 2010.)

demand for imported natural gases, in this case, by producing biogass from animal manure (Figure 2.22). Again, the landscape impact is not great except, perhaps, when offensive odors affect the types of adjacent land use.

2.5.5 Solar Energy

Modern use of solar energy brings us full circle. We started with the sun and we are returning to the sun. Albeit we still use solar energy in the form of biomass, what is a complete departure is our use of solar energy to generate electricity. These are the operations that are producing the unique solar energy landscapes that are becoming quickly a commonplace landscape feature.

As with all energy resources, solar equipment must be designed to accommodate characteristics inherent of the resource. This means that it takes large installations to generate meaningful amounts of electricity for meaningful commercial sale. Three designs are in use: photovoltaics, parabolic troughs, and central receivers.* Each one creates its own unique solar energy landscape.

---

* Several others are in development, including dish Stirling systems, concentrating photovoltaics, and solar chimneys. They, too, will produce their unique energy landscapes.

Photovoltaic arrays create the simplest energy landscapes. The amount of land required to generate a given amount of electricity depends on latitude, cloud cover, and conversion efficiency of the cells themselves. This means that the scale of the energy landscape for a given amount of generated electricity can fluctuate substantially. This is the form of energy landscape that accounts for the vast majority of installed solar power at present.*

If the entire fuel cycle is included in the calculation of land commitments, the land requirement for a solar plant is not much different than it is for a coal plant, but for a centralized solar facility, the land commitment would all be in one location (Pasqualetti and Miller 1984). Although centralized photovoltaic arrays can require an appreciable area of land locally, the energy landscape they create may actually improve on preexisting conditions, as it has at the former open-cast mines in the Lusatia region of decommissioned military airfields in East Germany (Figure 2.23).

The other major category of utility-scale solar power plants is collectively called concentrating solar power (CSP), and they come in a variety of types. Each type produces its own unique energy landscape. The most common

FIGURE 2.23
Former military airfield near Drewitz in Germany with photovoltaic arrays. For scale, note that the runway has a length of 2.200 m. (Photo by Detlef Stremke [LaNaServ], 2011.)

---

* Centralized solar plants, of course, take no advantage of the ubiquity of solar energy and require large land parcels. Roof top installations require essentially no additional land commitment.

systems use mirrored parabolic troughs to concentrate the sun's rays on a tube filled with a working fluid. These systems track the sun through the sky. The largest group of such devices is near Barstow, California, having a total installed capacity of about 350 MW. Also, a new installation of 280 MW is under construction in Gila Bend, Arizona.

The second type of commercial CSP is the central receiver. It consists of a central receiver 150–200 m tall that receives sunlight reflected from thousands of mirrors called heliostats arranged around the tower. The only operating facilities of this type are in Andalucía, Spain, although 392 MWs are under construction in the Mojave Desert southwest of Las Vegas, Nevada.* All the CSP installations include a turbine, a generator, and some form of cooling apparatus (Table 2.6).

TABLE 2.6

Stage Four: Landscapes of the Sustainable Economy

	Layer One "Direct"	Layer Two "Indirect"	Layer Three "Mitigation"
Modern renewables			
Water	Reservoirs	Altered downstream ecologies, agriculture	
Wind	Roads Turbine installations Movement, flicker	Altered property values (up and down) Night-time impacts (e.g., warning lights) Altered farmland Drained lands	
Geothermal	Visual Roads Drilling pads	Altered vegetation Impacts on surrounding land use Conversion of existing land uses	Disposal sites for sludge from pollution abatement
Solar	Roads Large areas Reflections	Changed appearance of open land Improvements in the impacts of fossil fuels	Distributed deployments (e.g., houses, covered parking) Repositioning to protect endangered species
Biomass	Roads Enlarged agricultural area New and different shaped buildings Conversion of pasture land	Effect on nearby land uses resulting from odors Relocation of farm workers	

---

* http://www.brightsourceenergy.com/projects/ivanpah

2.6 Conclusions

Over the years, energy landscapes have taken the form of assorted strips, piles, trenches, holes, drilling derricks, buildings, pipelines, harbors, and fields of equipment, large and small. Members of this heterogeneous mix are representative of the generic categories of the energy supply chain that includes fuel extraction, processing, transportation, generation, distribution, and disposal. We have not always thought about them, but taken together they create a major influence on what we see, now that they have moved from the shadow of our disinterest into the light of our concern.

Over the years, these landscapes expanded to include unanticipated impacts as well as ones we thought familiar. We began encountering landscapes produced not just from mining but from mine drainage, not just from local air pollution but from climate change, not just from minor spills but from calamitous blowouts, and not just from dangerous uranium mines but from global radioactive contamination.

Energy landscapes became troubling as well as obvious once they no longer could be dismissed as unavoidable ancillaries of progress. The days of simply accepting them as the routine side effects of our lifestyles became a forgotten memory. In trying to keep them at bay, we tried everything: We camouflaged them, grouped them, dispersed them, isolated them, diluted them, and dismissed them. We developed laws to control them and invented techniques to mitigate them. We educated the public and lobbied politicians. And we started to realize that we would need to invest more deliberately in cleaner energy alternatives that could comport with natural landscapes if we were to have any chance of crafting a harmonious home out of the only planet available for us to use.

As surely as energy landscapes project a natural human need for energy, they also reflect political imperatives, public reactions, historical events, technical sophistication, growing population density, and the increasing evidence of Earth's natural limits to the insults we throw in its direction.

Energy landscapes are everywhere, they are varied, and they are instructive. To read these energy landscapes is to trace the history of our changing attitudes and capabilities. We find that we have shifted over time from the simple energy landscapes of the organic economies to the destructive energy landscapes of industrialization and electrification. Now we are on the verge of a new transition, one where we are starting to make choices that will affect where we get our energy in the future. In part because of the increasing stresses on environmental quality, health, safety, and energy security, we seem to be heading back to where we began, to a reliance on the sustainable resources of the organic economy. We have come full circle, having passed from one stage of energy supply and environmental perception to the next. Each of these stages shows up in the landscape, if we but have eyes to see and a willingness to look.

References

Brimblecombe, P. 1987. *The Big Smoke: A History of Air Pollution in London Since Medieval Times*. London, U.K.: Routledge.

Burtynsky, E. 2009. *Oil*. Göttingen, Germany: Steidl Verlag, Göttingen.

Debeir, J.-C., Deléage, J.-P., and D. Hémery. 1986. *In the Servitude of Power: Energy and Civilization through the Ages*. London, U.K.: Zed Books Ltd.

Denholm, P., Hand, M., Jackson, M., and S. Ong. 2009. *Land-Use Requirements of Modern Wind Power Plants in the United States*. US National Renewable Energy Laboratory. Technical Report NREL/TP-6A2-45834.

Drachmann, A.G. 1961. Heron's windmill. *Centaurus* 7: 145–151.

Ghosn, R. (Ed.). 2009. *Landscapes of Energy. New Geographies 02*. Cambridge, U.K.: Harvard University Press. http://www.hup.harvard.edu/catalog.php?recid=29893.

Hoeksema, R.J. 2006. *Designed for Dry Feet: Flood Protection and Land Reclamation in the Netherlands*. Washington, DC: ASCE Press.

Jones, C.F. 2010. A landscape of energy abundance: Anthracite coal canals and the roots of American fossil fuel dependence, 1820–1860. *Environmental History* 15: 449–484, 453.

Kaplan, J.O., Krumhardt, K.M., and N. Zimmermann. 2009. The prehistoric and pre-industrial deforestation of Europe. *Quaternary Science Reviews* 28: 3016–3034.

Lohrmann, D. 1995. Von der östlichen zur westlichen Windmühle. *Archiv für Kulturgeschichte* 77(1): 1–30.

Malcolm, S.A., Marcel, A., and M. Weinberg. 2009. *Ethanol and a Changing Agricultural Landscape*. U.S. Department of Agriculture. Economic Research Report Number 86.

Marks, L. 1964. *The Machine in the Garden: Technology and the Pastoral Ideal in America*. Oxford, U.K.: Oxford University Press.

Nef, J.U. 1932. *The Rise of the British Coal Industry*. London, U.K.: George Routledge & Sons.

Pasqualetti, M.J. 1997. Landscape permanence and nuclear warnings. *The Geographical Review* 7(1): 73–91.

Pasqualetti, M.J. 2000. Morality, space, and the power of wind-energy landscapes. *The Geographical Review* 90(3): 381–394, July issue (published in May 2001).

Pasqualetti, M.J. 2001. Wind energy landscapes: Society and technology in the California desert. *Society and Natural Resources* 14(8): 689–699.

Pasqualetti, M.J. 2011. The next generation of energy landscapes. In *Engineering Earth: The Impacts of Mega-Engineering Projects*, S. Brunn, ed. Dordrecht, the Netherlands: Springer Part 4, pp. 461–482.

Pasqualetti, M.J. and M. Dellinger. 1988. Hazardous waste from geothermal energy: A case study. *The Journal of Energy and Development* 13(2): 275–295.

Pasqualetti, M.J. and B.A. Miller. 1984. Land requirements for the solar and coal options. *The Geographical Journal* 152(2): 192–212.

Reece, E. 2007. *Lost Mountain: A Year in the Vanishing Wilderness Radical Strip Mining and the Devastation of Appalachia*. New York: Riverhead Books.

Righter, R.W. 1996. *Wind Energy in America: A History*. Norman, OK: University of Oklahoma Press.

Sieferle, R.P. 2001 (published in German in 1982). *The Subterranean Forest: Energy Systems and the Industrial Revolution*. Cambridge, U.K.: The White Horse Press.

Thayer, R. 1994. *Grey World, Green Heart: Technology, Nature, and the Sustainable Landscape*. New York: John Wiley & Sons.

Thesing, W.B. (Ed.). 2000. *Caverns of Night: Coal Mines in Art Literature, and Film*. Columbia, SC: University of South Carolina Press.

U.S. Environmental Protection Agency. *Greenhouse Gas Reduction Thresholds*. http://www.epa.gov/OMS/renewablefuels/420f10007.htm#7. Retrieved 2010-02-09.

Williams, J.C. 1997. *Energy and the Making of Modern California*. Akron, OH: The University of Akron Press.

Wrigley, E.A. 1988. *Continuity, Chance and Change: The Character of the Industrial Revolution in England*. Cambridge, U.K.: Cambridge University Press.

3

Strong Feelings: Emotional Landscape of Wind Turbines

Dirk Sijmons and Machiel van Dorst

CONTENTS

3.1 Introduction

3.1.1 Space as the Battlefield of Energy Transition

The spatial consequences of the energy transition to CO_2-free sources will be one of the major challenges for spatial planning and the design community in the coming decades. Spatial planning of "energy" is no longer restricted to the relatively simple task of reserving space and cooling capacity in some remote industrial area. Generating electricity from sustainable sources may have a strongly decentralized character, making "energy" a much more tangible and visible item in everybody's living environment. Energy sources that have long remained "underground" will, literally and figuratively speaking, emerge to the surface. The transition will not only influence the vision of the (urban) landscapes but also have broader impacts on the organization of society and thereby on spatial planning.

Paradoxically, space will, in quantitative terms, not be on the critical path of this transition—with a possible exception of biomass-based energy production. In qualitative terms, the picture changes dramatically for new energy technology—particularly for wind turbines. Policy makers seem to focus strictly on the "measurable" impacts on the landscape and the visibility of the turbines as rational effects to be dealt with in environmental impact statements—however, other connotations seep into the debate. State-of-the-art three-dimensional simulations show what reality will look like after the erection of the wind turbines. This includes parameters such as seasonal shading and curvature of the earth. Universities are even asked to review the validity of such scenic impact assessments.

All these efforts to restrict the playing field to the rational arguments cannot conceal that much of the legislative restrictions, obstacles, political blockades, Not In My BackYard (NIMBY) arguments, and related emotions carry symbolic meanings attached to space. In practice, qualitative connotations overwhelm the technical opportunities; clearly, we are not dealing with just a technical question. We are entering a problem field where debates involve a mix of hard science, extensive amounts of data, opinions, semantics, culture,* private interests, fears, and a range of meanings.

To complicate things even further, one sees a tormenting discourse between two versions of sustainability. First, one involving notions of sustainability that supports "landscape" as the locus of many "sustainable" values. Second, but conversely, sustainability viewed through an energy transition frame introduces radical changes to that very visual landscape. The problem is exacerbated by more efficient yet taller turbines. To avoid this clash, some environmentalists even feel that nuclear power or carbon capture and sequestration (ccs) will be a suitable alternative, thus exorcizing the devil by Beelzebub.

To put it in a nutshell, space and its sociopolitical arena will be the battlefield where the energy transition will be won or lost.

3.1.2 Déjà Vu?

First of all, is the debate apt entering as a designer? Are we suitably trained to fair beyond the realm of design? Should we not restrict ourselves to designing the technical solutions and leave public debate to politicians? Moreover,

---

* This discussion is based on conditions in the Netherlands—which are dissimilar from those in Germany, Denmark, the United States, and Great Britain.

could we not point to earlier versions of this same debate involving scenic impact of high-rise structures and confidently say: "we have been there"?

First, we may refer to the fact that horizon pollution seems to be an endemic Dutch word (*horizonvervuiling*), coined in the 1970s by the environmental movement to summarize their protest against the first high-rise buildings that would affect our Dutch "flat liner" horizon that is thought to be an integral part of our territorial identity. The sacrosanct horizon should, in the conservationist's view, be punctuated only by church towers marking the center of our towns and villages. From this perspective, the panorama of our cities, so carefully staged in the seventeenth century as a typical silhouette of towers as crowns (de Heer, 2000), should remain forever untouched.

> Throughout the country; clumps of trees, town-lands, stumpy towers, churches and elms that contribute to the grand design (Marsman, 1936)

Earlier debates regarding high structures raged about high-rise buildings (mainly housing block slabs in the 1970s until "real high rise" in the present), fodder silo's in the landscape (1980s), high-voltage electricity portals (1980s) (Vrijlandt et al., 1980; Vrijlandt and Kerkstra, 1984), UMTS mast's (Universal Mobile Telecommunications System, 1990–present), and minarets on mosques (2001–present). Looking at these debates, one can conclude that they all resulted in the consensus that they should be carefully designed and positioned. One can also observe that "time" seems to soften these scenic wounds. However; it also shows that the debate on "horizon pollution" is not completely straightforward, and other notions like "identity" and "landscape" pop up immediately. This is only the tip of a deeply layered perception debate, as will now be explained.

3.1.3 Emotional Landscape

Anyone who has had the opportunity to be present at a public meeting on a wind energy project will have noticed the emotional intensity in such debates, to say the least, are remarkable. Rationality seems to be speechless confronted with these emotions. Government officials fail to convince the importance of an energy transition and the pivotal role of wind energy, but contrarily fuel anger. A rationalist would say that it is tragic that emotional utterances without arguments will not make people (and the media!) suspicious and even more important, not let them take sides automatically with those that appeals for reason and comes with verifiable data.* Conversely, accusations are made that proponents of the new technology seem to have a clear lack of morale, are destroying the landscape, are thoughtlessly greedy, and have a reckless belief in progress.

Smits (2002) states that for almost all the technological innovations that appear on the radar of the public discourse, the same patterns can be observed, be it nuclear energy, genetically engineered crops, or cloning of animals. The proponents defend and embrace "their" new technology and argue

---

* Rudy Kousbroek, *De toekomst van het bijgeloof* [The Future of Superstition] De Ingenieur (2001).

that it will solve a multitude of problems of society. The critics on the contrary make statements that some natural boundaries could be overstepped, an argument that mostly will be accompanied by raising alarm over new and unknown dangers. The possibility of a pluralistic and open debate is smothered by these unreconcilable arguments (Smits, 2002). There may be a third path, as we will try to demonstrate in this chapter.

We will take these heated debates on wind turbines as the angle to approach this subject. First, the layered and multifaceted character of this emotional landscape will be explored and the depth of each layer fathomed. We will try to peel away different layers of the debate one by one. This will lead us into fields of psychology, sociology, and cognitive science. Our ambition is to make a brief survey of these layers of perception. We will first explore individual concerns and emotions, including cognition and a psychological approach and the influence of semantics. We move to the social level where the factor of time comes in with demographics to a historical approach to the debate. Finally, a more sociological and philosophical approach is introduced where we try to deal with some obvious aspects connecting the debate with a (changing) worldview or the zeitgeist with a symbolical and phenomenological angle: what do these machines stand for? This section is wrapped up by exploring connections to social science—particularly studies in the relevance of mediation in introducing new techniques that have and will be present throughout cultural history. The chapter will make some concluding remarks regarding how some of these angles might be instrumental for resolving complex debates in energy transitions. The final paragraphs look at the importance of public participation and the role design and designers might play.

3.2 Individual Concerns

3.2.1 Sensory Perception

The appreciation of wind turbines seems to focus mostly on the auditory and the visual qualities. First of all, we can measure sound levels. Sound emitted from wind turbines is calculated per octave band, and the total sound level at the receivers end is the (logarithmic) sum of all octave bands. Turbines are considered to have isotropic point sources and therefore have no directivity (Berg et al., 2008). In total, the exposed levels of wind turbines in the Netherlands are between 24 and 54 dB(A) at a distance between 17 m and 2.1 km. When the background sound level is 30 dB(A) or higher, residents living more than 2.5 km away will hardly be able to perceive wind turbine sound. This distance evidently increases with larger numbers of wind turbines (Berg et al., 2008). Perception is also influenced by a number of variables. First, sound is perceived louder when the wind is blowing from the direction of the noise source.

Sound is also perceived louder at night, with reduced ambient noise. Finally, there is the influence of the landscape on perceived noise levels is considerably bigger in a flat landscape than in a hilly or mountainous environment (Pedersen and Larsman, 2008).

The relation between noise and visual disturbance works conversely—deals with scenic quality and scenic impact. Scenic impact is complex and often subjective (Johansson and Laike, 2007). Assessment criteria include the proximity of views, duration of views, the presence of scenic resources, the size of the wind turbines in relation to the context, and stroboscope effects (NRC, 2007).

Of these, the stroboscopic effect is almost unique to wind turbine technology. It involves first intermittent chopping of the sunlight creating a strobe-like effect and second creating variation in brightness by shadows (NRC, 2007). These problems may be of concern in the direct proximity of the turbine especially in winter (with a lower sun angle). However, flicker frequency of wind turbines is again practically seen as harmless to humans. This is based on the fact that the rotor frequency may be 0.6–1.0 Hz, and epileptic seizures occur at frequencies of 10 Hz and higher (MSU, 2004). Although there is no proven health effect, in some countries turbines are recommended to be set away from roads to avoid distraction for drivers (in Ireland, this is 300 m).

3.2.2 Health and Behavior

Health and health effects could be an objective cornerstone in the debate—but when we peel off this layer, we find complex cause and effect chains. Various studies have reported on a number of health effects resulting from wind turbines, including headaches, sleep problems, learning disabilities for children, ringing in the ears (tinnitus), mood problems (irritability, anxiety), concentration and memory problems, and issues with equilibrium, dizziness, and nausea (Knopper and Ollson, 2011). This is a long list with serious complaints. In analyzing studies, one commonly finds that empirical research is based on popular rather than scientific literature (Knopper and Ollson, 2011). It is a fact that these are self-diagnosed complaints—known as the "wind-turbine syndrome." These (perceived) health problems should be addressed, but the bottom line is that there are no empirical established, proven effects of wind turbines on the physical well-being to the best knowledge of the authors (Pedersen et al., 2007; Berg et al., 2008; Pedersen and Larsman, 2008; Knopper and Ollson, 2011).

We can conclude that the main health-based concern related to wind turbines revolves round the noise. Modern wind turbines produce broadband noise from turbulence produced at trailing edge of the blades (Pedersen and Larsman, 2008). From the environmental point of view, there is only a measurable sound impact, but as noted earlier, this is related to distance and the context. Noise by wind turbines is related to the noises that are produced by the surroundings. There is also a strong relation between the annoyance by noise and the visual cues of the landscape (Pedersen et al., 2007). More extensive research has been done on infrasound (below 20 Hz) produced at high speed. There is no evidence that infrasound from wind turbines has any notable health effects (Berg et al., 2008). Apart from the evidence, there is a growing group of citizens that experience stress through enduring low frequent sound* where sometimes wind turbines are being named as the possible source. Further, the effect of infrasound on sea life (special marine mammals) is unknown and may challenge offshore wind turbines.

For now, there is no empirical evidence for measurable health damage in direct relation with noise or movements of wind turbines. Maybe a general and undefined suspicion explains why a complex (yet harmless) issue such as infrasound levels has been used to argue against wind turbines. This confusion surrounding infrasound helps sway popular opinion, suggesting that there must be something wrong with these relatively new artifacts—in other words, "if it isn't natural, it must be unhealthy." Strikingly, this type of attitude is related to the back to nature lifestyle, rooted in the work of Thoreau (1854)—a starting point for an "eco-based" sustainable lifestyle. This is yet another example of the colliding of two frames of sustainability in the wind turbine debate. In summary, there might be nuisance, but the health impact is not scientifically established, and the perception of wind turbines in this regard places a role in the tolerance toward them.

3.2.3 Resistance against Change

People tend to resist change in their spatial environment. This has to do with the importance of a stable daily environment, be it city or landscape, which is in many ways connected with one's well-being. Not only do people change and influence their environment, but the environment also influences people: "First we shape our buildings, then they shape us" (Winston Churchill, 1943). Hannah Ahrendt even states that people partly derive their identity from the stability of the world that they build, the house that they live in. If this stability is lost, or things quickly change, people are inclined to become very "volatile."

This may be the basis that resistance every (kind of) project has to overcome—and is not restricted to wind-turbine parks. Thus, what is called NYMBY-ism does not primarily or necessarily start from the rejection of the content of a project or its feared disadvantages for the (local) community. It is simply about change itself. There are analogous motives for resistance to change as observed in organizational psychology literature. Like designers, these professionals have daily confrontations with the syrupy nature of beliefs and perceptions. Schuler (2003) identifies the top 10 motives concerning the change persons within an organization might have. The risk of change is seen as greater than the risk of standing still. Making a change requires a kind of

---

* www.laagfrequentgeluid.nl—a Dutch interest group for people that are agonized by all forms of infrasound.

leap of faith: you decide to move in the direction of the unknown on the hope that something will be better for you—but you have no proof. Social motives are also indirect, for instance, "people feel connected to other people who are identified with the old way." They also include personal fears "people anticipate a loss of status or quality of life." Eight out of ten motives, noted by Schuler, relate to these fears. However, only 2 out of 10 motives in his list relate to the content of the proposed change: "people genuinely believe that the proposed change is a bad idea" and also very recognizably "people have a healthy skepticism and want to be sure the new ideas are sound." One could easily jump to the conclusion that only rational objectives are worth being dealt with and leave the eight emotional aspects alone as inevitable collateral damage of progress. This, to our mind, would be a mistake. To have a chance of success in the complex transition trajectory, we have to take the "nonrational," emotional part of the debate very seriously. It is up to us to see what specific fears and emotions concerning wind turbines and wind parks might be identified and discuss if and to what extent they can be dealt with.

3.2.4 Learning Effects

If we take a step beyond this organizational psychology approach, we come to the individual that has the quality to learn. From a cognitive approach, Kaplan and Kaplan's (2005) preference matrix explains the changing appraisal of new artifacts; coherence and legibility on one side and complexity and mystery on the other. First, introducing a new artifact without reference and explanation is like introducing a mysterious object that is not legible or coherent with the context. This creates stress. Confusing instead of informing and supporting the citizen is a mistake that is often made. Professionals who have a background in the issue are inclined to view the conditions rationally and are often blissfully unaware of their head start concerning the level of knowledge. On the other hand, the layperson is likely to be overwhelmed by complexity and mystery—producing very mixed emotions and stress. Thus, from a cognitive perspective, the citizen should be educated or introduced to the issue through experience. Within the second approach, an assumption is made that the layperson will follow the professional—but only if the designer/engineer is taken into the layperson's confidence.

This does not tell the full cognitive story. The legibility and the assessment of the landscape are more than a learning process. There is an almost automatic aspect in the affective responses to the natural environment (Ulrich, 1983), and these are related to the basic needs in a psychoevolutionary framework. So perceptions of wind turbines can be explained as one of the primary responses concerning environment–behavior interaction—that is, control over the living environment (van Dorst, 2005). People may perceive wind turbines as objects that are not in place and cannot be controlled. This perception should be reversed; wind turbines are a symbol of self-sufficient energy supply. To reframe this to the perception that wind turbines are offering control by self sufficiency in energy supply may be more of a challenge than one may initially think—it is emotionally overstepping a natural response, and, moreover, few people understand the significance of the world's energy dependence. An involuntary full blackout would by all means help expose this debate.

   The third layer of cognition might stem from the theory of basic needs, established by Vroon (1990). He finds a relation between wind and stress. This primary reaction to wind (stress) is explained by the fact that in windy environments, one cannot smell a predator. So in windy environments, humans are more inclined to depend on visual perception. Connecting these two issues—wind turbines at high speed give the signal that one is depending on visual perception and that there are these wind turbines in the way. This may explain why there is so much concern regarding the visual impact of wind turbines, although they occupy an average 2% of the space above the horizon. This line of reasoning may also work the other way around; a wind turbine at rest may be perceived as being out of order, which leads to feelings of stress because there is no control over energy supply.

3.2.5 Homo Economicus

The line-up of individual concerns has to be completed by what is called the perception of the "homo economicus" or the "Economic man": an individual ideally acting in a certain setting making rational, economic choices. In this type of rationality, a person is looking at opportunities by balancing benefits against the possible costs (Persky, 1995). In line with what is stated earlier, in reality, the "economical man" does not primarily behave rationally. Benefits may be related to emotions, ethical questions, and social needs. But the theory states that one is trying to gain this against the least possible costs (Persky, 1995). Looking at a case where wind turbines are being introduced, an individual will balance the profits and the disadvantages. The disadvantages are not only formulated in terms of noise effects or possible health implication but particularly in terms of concern that property values will be adversely affected by the wind energy project. Although property value impacts have been found near high voltage transmission lines, the first large scale research on the impact of turbine parks show that the impact is negligible. A research project in the USA collected data from more than seven thousand sales of single family homes situated within ten miles of twenty-four existing wind facilities in nine different U.S. States found that the impacts are either too small and/or too infrequent to result in any widespread, statistically observable impact (Hoen et al., 2009). One may be able to cope with noise and/or the concern about real estate value if there is a direct financial benefit. This effect can be even stronger; the greater the noise tolerated, the greater the profit required. This seems comparable with the false assumption in vehicles that engine noise level is relative to speed; for instance, the noisier vehicle with a disconnected exhaust pipe creates the illusion that it is also faster and more fun.

In the history of autarkic living around the world, the introduction of any form of sustainable energy with an impact on one's behavior or the visual or spatial quality of the living environment, there are hardly any objections. In Mongolia, a spectacular proliferation of (PV) panels and small wind turbines can be observed. The local and personal advantages are understandable and outbalance the disadvantages. The well-known ecovillage, Findhorn, in Scotland, erected four wind turbines, no debate needed. Comparison between the proliferation of wind parks in the Netherlands and neighboring Rheinland-Westfalen shows that the standard procedure in Germany to give price advantages to local communities in Germany really worked in the way that the turbines were seen as 'our' electricity producers (Matton and Wamka, 2004).

In the top-down Dutch policy, one frequently encounters statements like "They put these turbines on our doorstep and we suffer all the disadvantages." Perhaps, the most striking example of the difference between active stakeholders and outsiders is delivered by the Dutch Noordoostpolder (one of the polders of the IJsselmeer). The Noordoostpolder 'being nominated as UNESCO world heritage' was kept relatively free from wind turbines on farms. Farmers and other citizens together with an electricity provider decided on a large-scale 450 MW project with turbines in a triple-line configuration outside the dike in the IJsselmeer. All the inhabitants of the Noordoostpolder were allowed to take a share in the (profit of) project. By any standard, this is an exemplary project in the Dutch context. One crucial social factor was overlooked though—the inhabitants of the old island of Urk. This former fishing community found their island being turned in a nondescript part of an agricultural polder in the late 1940s—without consultation. This old but unforgotten pain held the fishing people from Urk far from their farming neighbors'* energy project and led to a storm of protest by both Urk's inhabitants and its municipality.[†]

3.3 Social Ripples

3.3.1 Zeitgeist

Is the perception of and resistance against change a fact that will always remain the same? No certainly not. Different cultures will judge "change" differently. People in India will perceive new industries different than they

---

* Very illustrative on the bias of "measurable" effects in Environmental Impact reports is that the massive EIS for this project completely missed the social factor of participation in the project (Sijmons, 2010).
[†] The local authorities lost the lawsuit against the state, and the national government lost the chance to greater commitment.

will in the United States. Both ratio and relation will not only differ from culture to culture but also change gradually in time. In different periods people will weight motives differently for the appreciation of the same object and come to different conclusions. So appreciation will express broader societal developments. We are looking differently to the world than our forefathers did in the seventeenth century where our country was an uncompromising modern country with very little interest in cultural heritage.* The generation that rebuilt the country after the war heralded every new reconstruction project and took a completely different position on "change" than our generation does.†

More than 60 years of peace, welfare, and economic growth have changed our worldview radically. After the war, the average worker had long hours and few holidays. The "negotium" (acting) was dominant over the "otium" (rest), like in Roman times where "otium" had a negative connotation. Nowadays, the tables seem to have turned, and the "otium" has gained significance. The free time takes center stage now and penetrates the public realm in the form of all kinds of leisure activities, festivals as well as shopping as the almost last remaining form of public activity (Chuihua et al., 2002). Even "work" has to be challenging and fun.

Masses of people today live in a distorted reality on a rose consumerist cloud on which we are supposedly hovering into a 100% service economy. A world where the only work is typing on our computers and in which "dirty work" is made invisible by "outsourcing" it to the sweat shops of Asia. In this perspective, industrial area, trucks, energy plants, and everything else that is reminding us of the other reality and thus disturbing the dream shall be banned or stopped. NIMBY has a firm base in this new perception. Vuyk (1992) even coined the notion of the "aestheticizing" of our worldview. He observes that science and technique no longer dominate our ways of seeing but have gradually given way to the dominance of culture(s) and "taste."

> The classical instruments of judgment—mind, ratio, conscience and ethics—are not completely eliminated in the formation of judgments (....), but ever more momentous is the contribution on the basis of taste. Conscience is weighed as old fashioned, unless it appeals to reason, or taste. Mind and intelligence still have, in some areas much of their old authority, but even in the most important of these areas, science itself, the influence of the aesthetic judgment is reckoned. Taste is more and more conceded to be the last and most general judgment. (Vuijk, 1992).‡

This aestheticizing of our worldview has produced a structural narcissism and hedonism in Western societies. It has, as we will show, several direct connections with the depth and intensity of the emotions related to wind turbine projects.

---

* Cultural heritage is an invention of the end of the eighteenth and nineteenth centuries.
† Ed Taverne on the role that the notion of renewal ["vernieuwing"] plays in the Netherlands from the 1930s to the 1960s.
‡ Translation of the paragraph by the authors.

3.3.2 Pressure on the Notion of Landscape

First and foremost, this aestheticizing of our worldview makes it harder to accept that the generation of electricity that, not long ago, was no more than a vague plume on the horizon and a world behind it that one should not want to worry about is now entering the living environment of more people. The energy sources that were hidden in the earth are emerging above ground and became decentralized. This is quite a radical shift and is going to change our cityscapes as well as our landscapes. Change is spelled here with a capital C.

Second, when change in the urban environment is, as we saw, not something that will be automatically accepted, change in rural areas may count on strong resistance. "Landscape" as a notion seems pivotal in the debate. The modern perception is that our dynamic urban life should be contrasted with an ever constant and unchanging Arcadian landscape as a backdrop. Almost all forms of modernization of our cultural landscape—like agribusiness centers, new cowsheds, fodder silos, and the scaling up of our landscape—seem to be met with aversion and protests. Always with the best of intensions, but sometimes, a modern form of false romanticism seems to come into play. Most people associate these economical-driven developments with a form of industrializing of the rural areas. Or, to be more precise, seen as an intruder in their landscape (Pedersen et al., 2007). And landscape plays a pivotal role in the well-being of the urban population because it functions for all kinds of imponderable needs apart from simple food production.

It is now almost standard policy that new developments must be consciously "blended" into the landscape [in Dutch *inpassing*, fitting in]. New elements may creep into the landscapes, screened by hedges or trees. Even windturbines seem to try to tip-toeing through the landscape with base of their columns painted in different shades of green. The first generation of wind turbines could be linked to landscape elements, such as roads or dikes, and they could even be said to make the landscape more readable. Configurations of these turbines, if well sited, could connect the observer with the horizon. Possibilities vanish or lessen if many wind-turbine parks are visible at the same time. Some authors think that with the newest generation of large 3–7 MW turbines, with a height of often more than 150 m, these instruments of the landscape architects toolbox lose their grip. Schöne (2007) states that the uniform design of the turbines, combined with the technological aura and the visibility from afar, together bring about a disconnection with the underlying landscape. Fitting the new generation of turbines into the landscape by situating them meaningfully is, in her opinion, out of question.* Here, we encounter the fear that mastodons like that will

---

* Nonetheless, Schöne pleads for a planning policy with the new generation of turbines on the basis of solid experience research.

completely dominate the "landscape experience" and even put pressure on the notion of "landscape" itself. When (at what height) and exactly why this caesura occurs is not made clear. But the arguments are very recognizable and are staged in almost every discussion on wind projects. Landscape is, through its broad range and yet vague definition, also a repository for all kinds of other arguments.

If we move the argumentation to the sea, we might be able to appreciate these arguments in their undiluted state. Zee (1999) summarizes the positive connotation to seascapes in three main motives. First, the experience of the empty coastal seascape offers the possibility to distance oneself from the everyday life and the everyday reality, because elements that would remind you are lacking. By the experience of the infinity of space by the continuous view on the horizon, one can experience the sensation of freedom. The attractive aspect of the seascape is that infinity also has its boundaries. In addition, the borders between land and sea and between land and the sky are sharp. And, third, there is the experience of time in the suggestion of infinity as well as temporality. One can experience the immeasurable time and thereby the transience of life by the natural elements as waves, flow up and down tide, day, and night, with the changing seasons.

You need little imagination to foresee that interference between these three basic experiences and large infrastructural elements, like oilrigs or wind farms, the experience will almost certainly be judged to have a negative impact. But, then again, this is argumented so radically Arcadian that passing ships or wind surfers would also interrupt the sublime experience.

Interesting is the difference between inhabitants of the coast (long stay) and the tourists, sunbathers, and yacht crews (short stay) in judging the effects of near-shore wind parks. The short-stay category tends to be significantly more positive than the long-stay category (Ladenburg and Dubgaard, 2009). The researchers indicate that this might be explained by their findings that the "short-stay" category consider the coastal landscape as a functional landscape fit for different forms of use, both industrial (shipping, wind) and leisure (beach, sunbathing, yachting), while the "long-stay" inhabitants strongly identify with "the ocean" as a pristine resource.

Interim conclusion: one of the reasons for the emotional response is that the new generation of wind turbines introduces a radical change in and puts pressure on the notion of landscape itself. Landscape as a notion and a more general resistance against change are strongly linked to an aestheticized worldview. These categories are historically mediated and will change over time. However, the pace of change fueled by the urgency of an energy transition might be quicker than the gradual shifting of the worldview. We must not rule out, however, that cultural changes might be sped up by real-world incidents in energy security. In the same way, real-world incidents or near disasters in flooding created a rift in the Dutch policy debate on river dike enhancements.

3.3.3 Symbols and Signs

We saw that some people find it hard to combine Arcadian concepts of landscape with modern wind turbines. What do these machines stand for, what associations do they evoke, to make them into anomalies in the eyes of these people? What could be their (hidden) meaning? We are entering the rather fuzzy world of semiotics where signs and symbols and their use or interpretation is being studied. These studies range from charting rather superficial, but well-documented, "associations" to the deeper link symbols that might have interpretations of the unconscious. Out of a myriad of possibilities, we pick some examples that illustrate the relevance of the fields of semiotics to the wind-turbine debate.

The role of symbols and signs can be very dominant in the appraisal of phenomena. It can be an entry to the subjective world. By the same token, let us say a wind-turbine park can be framed and appreciated in very different ways. To some, it stands for government subsidy hungry machines, while the electricity benefits will be even insufficient to recharge the batteries of our electrical toothbrushes. For others, the same technology stands for a real step in the liberation from the carbon era. The first group will also be inclined to be more conservative and rejective while the second group is likely to be more progressive and have a positive appreciation. This example also demonstrates a chicken-and-egg relationship between how things are appreciated and what they signify. Is the positive appreciation a product of the signified or has the sign a positive connotation as a by-product?

Some signs or symbols have a more universal meaning because they are with us for ages. Very relevant to our wind-turbine debate is the symbol of "the tower." In religion, particularly the three major monotheistic religions, the tower signifies the link between heaven and earth and positions the house of the lord, the church, the synagogue, or the mosque in the landscape or cityscape. For centuries, these institutions had an almost monopoly on erecting vertical elements. This strong symbol has a somewhat diluted meaning since the atheist twentieth century rudely broke the spell by introducing profane high-rise buildings and functional infrastructure, like high-voltage power lines. The spectacular contribution to both quality and quantity of the position of towers in the landscape to some appears to be the pagan "coupe-de-grace" to the religious meaning of "the tower."

One can also "read" from the look of things, from their physiognomic. For the technogeeks, admittedly a tiny minority, wind turbines must seem an unlikely candidate for the next generation of electricity production. To these people, these strange and relatively straightforward retromechanical technology—very successful in our seventeenth century—must seem to be a regression phase in technical development, rather than a giant leap forward. Of course, there are high-tech components, but it does not seem to represent the ubiquitous solution for the next energy age, unlike, for example, fusion energy that long dominated the popular science and technology magazines.

Finally, we give an example from the hidden and perhaps deeper layers of consciousness. With a vague connotation to the spiritual role that towers played in the past, wind turbines might be a symbol for some and herald a new age, an uncertain future. This new age, the postenergy transition age, might frighten people or create uncertainty. Will current quality of life change? Will cheap travel come to an end? What will peak oil mean to mobility? Can I keep my house warmed when fuel prices increase? In summary, a deeper layer under the current protest on scenic pollution could, on an unconscious level, resonate with an almost existential fear for the future, the end of what Sloterdijk (2009) calls *the era of Fossil Expressionism*. Denial and wishing away of everything that has to do with the effects of global warming and "peakoil" seem to be one of the strongest emotions in society these days.* People vent their feelings of uncertainty on these symbols of the new energy age.

3.3.4 Mediation

Society is occasionally confronted with mysterious illnesses that seem to spread like bushfire and after some time stabilize on a much lower level, then fading into oblivion or vanish completely. These are real illnesses with very real somatic symptoms, where the medical science classifies and names then, provides diagnostic typologies, and produces formatted treatments. In some cases, these illnesses are linked to the introduction of new technology, where the human body and/or the senses are exposed to new conditions. In a most elegant essay, Kockelkoren shows that in the nineteenth century a number of illnesses were linked to travel by train. The sensation of speed, the unsharp images it produces, the disturbing flicker of foreground and background, and the almost stroboscopic light effects really made some people sick. It was the confusing experience of being disconnected from the unity of senses. Where a person sees, feels the air at walking speed, senses the wind, and smells the surroundings this is regarded as *synesthesia*. The early train traveler had to cope with conflicting and contradictory synesthetic signals, which had no cultural precedent. (Schivelbuch, 1977; Kockelkoren, 2003). These illnesses disappeared without trace after a few decades.

Some less spectacular symptoms were produced later by a few other new modes of transportation such as car and plane travel. This experience, we imagine, was less spectacular, as the human organism and its senses were already "trained" through experiencing train travel. Car sickness and plane sickness are of course related to seasickness and belong to the family of motion sicknesses, where one tries to synchronize the senses with a moving horizon combined with the disruption of the sense of balance. Fatigue caused by excessive or 'unnatural' motion is simply a natural human condition.

---

* Haydn Washington and John Cook, *Climate Change Denial, Heads in Sand*, Taylor & Francis Publishers.

These travel sicknesses are here to stay and every individual has to get over it as a child or keep it lifelong.

In our generation, phenomena like this reemerged in the 1950s when watching (the first) televisions—thought to cause nausea, dizziness, and headaches. In the 1960s, this phenomena emerged in the form of flat neurosis,* where some of the first generation of high-rise-residential residents reacted with physical symptoms, other than fear of heights alone. These symptoms, hyped by the press, were soon all but forgotten. In some ways, flat neurosis (*Slab neurosis*) can be compared with the sick building syndrome of the 1980s where some people reacted with health complaints attributed to controlled humidity from air conditioning and a lack of fresh air due to fixed-shut windows.

In the 1990s, in the wake of the service economy and with the introduction of the personal computer, we could name repetitive strain injury (RSI), as an example where the human body was confronted with a new form of eye, hand, and brain coordination with the long hours behind the screens of our desktops. RSI "infected" our offices and homes and at the height of the hype, an impressive percentage of our computer-using workforce suffered from it. Special workshop, doctors, ergonomists, physiotherapists, and industrial designers helped to gradually discipline the body and mitigate the symptoms. RSI has far from vanished but is very much over its peak as it seems. In our own time, the new frontier of adaptation seems to be the mobile phone, where the "SMS-thumb" produces complaints comparable to RSI. A little bit biased perhaps, researchers from IBM and Microsoft found comparable complaints from the posture over the new tablet computers that they dubbed Tablet-RSI.[†] Perhaps, the wind-turbine syndrome, as sketched out earlier, fits into this row.

Let there be no misunderstanding, none of the phenomena are imaginary diseases. They might be largely *psychosomatic* in the sense that the mind plays a dominant role and also *sociosomatic* in the sense that society and contemporary culture condition public perception—or a complex mix of both.[‡] The symptoms are very real nonetheless. Medical science struggles with these kind of "fuzzy" symptoms. They are assembled under the label of "medically

---

* Possibly an endemic Dutch case, high-rise apartment blocks, really of medium height between 6 and 15 stories high, were introduced relatively late in comparison with other countries in a centuries old low-rise tradition caused by specific foundation problems in our peaty soils.

† Biased perhaps, because IBM and Microsoft do not have tablets on the market and the researched tablet users worked on Samsung and Apple hardware. Justin G. Young et al. Touch-screen tablet user configurations and case-supported tilt affect head and neck flexion angles, *Work*, Volume 41, number 1/2012.

‡ Striking examples of the complex mix between sociosomatic and psychosomatic disorders are obesity and bulimia. One can see eating disorders as a specific sociosomatic reaction to social change, especially to the increased pressure to adapt to achievement and competence in light of observed insecurities and to the public presentations of perfect bodies. This affects adolescents' identity formation and their "body politics." In the context of changed patterns of coping with life, one can see obesity and bulimia as a symbolic "chosen" bodily disorder to reject social demands (in the case of obesity) or to overadjust to the requirements of modern social change (in the case of bulimia) (von Kardorff and Ohlbrecht, 2008).

unexplained symptoms"* or more technically "somatoform and conversion disorders."† Illnesses like this are hardly visible on their radar because medical science generally focuses on the larger research population but mostly over a relative small time frame. Even longitudinal research is measured in years rather than decades or centuries. These long term changes are the domain of sociology, anthropology, and philosophy and make us depart from the realm of the so-called hard sciences‡. As Kockelkoren states:

> One of the striking characteristics of human beings is that their cognition is always mediated through language, technique and art. We are, one could say, equipped with a sort of "natural artificialness" that our observation and even our senses are immersed in and deeply influenced by culture. Unless animals, that completely coincide with themselves, we are in a constant "decentered" state. New techniques that produce, quite literally, new sensations for our senses have the quality to decenter the subject even further. Depending on the specific character of the new invention and its proliferation in the population the "re-centering" process, of getting used to it and to completely integrate in our system, may take years or even decennia. Summarizing one could say that every new technology that puts stress on body or mind might evoke its own specific social diseases. The symptoms tend to disappear when people get gradually accustomed and/ or active mediation. These social diseases also seem differ in their psychosomatic appearance between different countries or cultures. A striking example is the difference between Australia and Europe. In Australia some claim there is proof for a "Wind Turbine Syndrome" while in Europe researchers deny the existence of these symptoms (Barrett, 2012).

3.3.5 Domestication of New Technologies

Dutch philosopher Smits (2002) developed an even more instrumental look on the mediation of new techniques and the recentering of people. She observed that both public and expert reactions to new technologies almost like by a default lead to the same stalemate situations in the public debate. Time after time, public discussion tends to solidify in primarily two worn-out

---

* Patients complain of symptoms that last as long as weeks and where adequate and comprehensive medical research cannot find a somatic illness that can explain the complaint satisfactory.
† Somatization is the propensity to experience a physical disorder and to express it, to attribute or blame it on a physical disease, although a somatic illness that explains the complaints adequately cannot be found. Thus, somatization is the term medical practice uses to denote the process where people keep calling on doctors in the hope that a well-defined illness can be found. Both definitions from: Trimbos Institute: Richtlijn voor de diagnostiek en behandeling van SOLK en Somatoforme Stoornissen, Utrecht, 2010.
‡ The work of Bruno Latour, and especially his contribution to the "Science Studies," makes one wonder how ("hard") science, nature, language, and society are in an interaction where we only start to fathom its complexity. And, sometimes, "hard" medical science proves not to be so hard after all.

grooves: at one extreme that of doom and abhorrence, while at the other end that of salvation and fascination.

The recurring element in this debate is the risk perception of the "naturalness" or "unnaturalness" of the new technology. The critics fear that some "natural" boundaries are to be overstepped with the introduction of a new technique, for example, genetically modified food. The supporters of the new technology will state that we have modified and refined plants for ages and that this is only a turbo on natural selection, thus accelerating evolution and helping humanity to fight hunger.

Smits tried to construct a way out of this unproductive end in which most public debates on technology seem to be smothered. In her research, she looks for mechanisms that can take account of both the fear and the fascination of new technologies. Directly in line with her two case studies on the introduction of plastics in our society and the ineradicable role of pollution, she concludes that these two share the characteristic that they are phenomena that do not fit into definable cultural categories. Waste as "matter-out-of-place," for example, can be classified in the category "cultural" (where it comes from) as well as "natural" (to which it all returns). These cultural categories precondition our perception and appreciation of ambiguous phenomena—provoking fascination and abhorrence—as they cannot be clearly categorized. Smits defines these as "monsters," very analogous to the way Latour speaks of hybrids (Latour, 2005) and Douglas identifies cyborgs (Douglas, 1973).

In the preceding paragraphs, we touched upon a number of arguments regarding why modern wind turbines could be called "monsters"—fitting in more than one cultural category or forming a boundary object between two or more. They are very "urban" but their habitat is both the landscape and the sea. They are towers without a sacred meaning, they are decentralized power plants that should be in one place, they have a retroimage but also high tech, they put the notion of landscape itself under pressure, etc. Not perhaps as scary as hybridization of nature and technology, characterized by "cyborgs," but instigating enough emotion to consider modern wind turbines "monsters" anyway.

Reworking the typology that anthropologist Douglas (1973, 1978) developed on social structure (in terms of the strength of the group and strength of the grid) and combining it with Smits' own typology on worldviews of environmental specialists, she concludes that there are four basic attitudes when it comes to the domestication of "monsters." These four ways in which societies are facing a threatening new technology cover all possible reactions of societies, whether primitive, preindustrial, industrial, or postindustrial: She distinguishes exorcism, adaptation, assimilation, and embrace. *Exorcism* is the very recognizable reaction where the monster is banished completely and kept out of sight. *Adaptation* is the strategy where the monster is adapted and made more acceptable. *Assimilation* is the strategy of gradually changing the cultural categories under which rules one would call something a monster. *Embrace* is the flight forward and in blind admiration welcoming the monster. In an evaluation of these four attitudes, Smits considers the pragmatism of adaptation and *assimilation* as the most promising ones. The first and the last are too dogmatic and completely unable to bend, respectively, the group consensus and/or the cultural categories. *Adaptation* and particularly *assimilation* give way to an active domestication of the "monster" through cultural integration, thus softening both extremes of the public debate stalemate.

3.4 Instrumental Aspects

All these angles exploring the emotional side of the debate in the energy transition can deepen our understanding and allow it to be resolved systematically. But will it deepen our understanding of cultural "deadlocks," and will this knowledge in turn result in better projects? In the following pages, we consider how some of the findings may contribute to design and policy making.

3.4.1 Public Participation

Real public participation takes the form of a dialogue. Such a dialogue cannot start without clearly articulating goals, considerations, and motives while noting the project's limitation. In addition, it is very useful if the project outline can be underpinned by a clear government policy. The position of the project in the wider goals and in the regional/national strategy on energy transition. This, we must conclude, is seldom the case due to a lack of robust policy. Only a handful of countries can pride themselves on a real transition policy—including countries such as Denmark, Spain, Germany, and perhaps China. These countries have a clear long-term energy policy with clearly defined courses of action that are unwavering and stable for a longer period of time.

In most other countries and regions the context of a project can be characterized as a complex mix of arguments over combined with and a political scrum that, is far from settled and fails to be conclusive.* There is interference between political and ideological assessments of (the human role in) climate change,† the confusing arms race of information and disinformation (Oreskes and Conway, 2010), conflicting interests, the rate of depletion of

---

* The political stamina is sometimes politely being doubted. "A non-fossil world may be highly desirable, but getting there will demand great determination, cost and patience," (Vaclav Smil, OECD Observer, #258/259, December 2006).

† Doubts and fears are sometimes being fueled deliberately to keep the suggestion alive that there is a balanced "scientific controversy" that could legitimate nonaction.

fossil fuels* and the technical (in)possibilities of the transition,† uncertainty about the best production modalities (Pimentel, 2008), and geopolitical considerations. For the time being, in most countries, energy transitions will be bound to be the "science of muddling-through" (Lindblom, 1959).

The resulting problem with public participation trajectories on wind-turbine parks is that in most countries and regions, all the *why, where,* and *how* questions at all levels emerge simultaneously at the project level, thus creating confusion even when it comes to facts and figures. An authoritative voice and a clear policy vector are dearly missed. Add all this to the complex emotional landscape illustrated earlier, and one can imagine the babel of voices that are heard on public meetings on wind-turbine projects. There is no alternative than to address all these questions and to establish basic assumptions and figures. This will not always be easy. Even "facts" remain deeply contested as they are framed in terms of emotional significance.

In addition, public gut feelings must be taken very seriously. These are often treated as signs of insufficient knowledge or forms of lower rationality, but we will better treat them for what they are: the swiftest form of cognition. As stated earlier, the debate is layered with many triggers for anxiety, uncertainty, or otherwise to raise alarm. For specialists, these fears might seem exaggerated or even illegitimate in the face of "the facts"—but they are nonetheless very real, and all sensible voices should be included in the debate. But how does one determine which voices are sensible and which are not? We illustrated earlier how public inquiry tends to end in a complete stalemate.

Smits (2002) observed that bringing in ethics or philosophers does not provide a clear outcome either—in turn choosing similar positions to the layperson while providing more articulated arguments. Some scholars position themselves among those who think that "natural or landscape boundaries" are being overstepped by this new technology. In the other camp, the nature skeptics point to the fact that "overstepping natural boundaries" happens all the time—this school of thought is inclined to portray public fear as subjective. "Either public fear is rationalized and fascination is trivialized, or public fear is played down while fascination is being legitimized." Both antagonists seem to overlook the possibility that both fear and fascination could stem from the confusion that the new technology does not easily fit in existing cultural categories (or seems to also fit in more categories than were thought compatible) and that pragmatic assimilation or adaptation may be able to change or shift existing cultural prejudice. This context is formative for our cognition and that like tectonic plates is not rigid and stable but slowly changing and evolving.

---

* The rarity of authoritative sources and voices in this field has a lot to do with the fact that most big energy firms and fossil fuel exporting countries keep their data on the stocks to themselves because this knowledge can be harmful for their market or negotiation positions.
† Technical assessment of the transition ranges from "It can be done" (ECF, 2010) to "sharp warnings of the difficulties ahead" (Fridley, 2010).

In this chapter, we showed that active mediation combined with the domestication of new technologies on the societal level offers a third road. This appropriation of new technology though could be a slow and gradual process requiring much patience. At a government level, the challenges of social dialogue will be rewarded through supporting this kind of cultural analysis. It is a necessary step to a second, more vital opportunity for technology policy—analysis of cultural perceptions will unveil ways to act through pragmatic mediation.

A possible way to fast forward this process is to move from the traditional public inquiry to a pluralistic and hands-on participatory process. Real public participation in actual projects offers perspectives for overcoming fears and doubts. Interactive sessions, where local stakeholders and other interested parties actively enter the design process and are invited to sketch out their ideas, appear to offer a way out of the traditional stalemate (Venhuizen, 2009). State-of-the-art graphics programs make it possible to simulate alternative options in real time during the public session.

Most significantly, people must be offered the possibility to take at least some responsibility or collective ownership for their energy future—while taking pride in the project too. Referring back to the "homo economicus," sharing the advantages of decentralized sustainable production is of great importance. Outcomes may involve sharing responsibility for the project and/or reduction in the collective electricity bill. This really helps to overcome hesitation and mitigate the fears of new energy technology. A positive side of decentralized, community-initiated energy systems is a renewed confidence that individuals are no longer intimidated by a defeatist attitude toward the energy transition problem. Therefore, participation strategies are vital in the development of projects, while simultaneously acknowledging resistance toward change (Johansson and Laike, 2007).

3.4.2 Role of Design

The role of design in this process can be threefold. In the first place, designers must of course perform their core business: making good designs at every relevant scale. On a national level, this could amount to defining or preserving strategic difference between wind concentration areas and "voids" that are in a planning fashion kept free from turbines, perhaps as a sole consolation for people that are hindered by turbine parks (Rijksadviseur voor het Landschap, 2007). On a regional and local level, designers must focus their efforts to make a sensible connection to the existing landscape and, in some cases, develop larger master plans. Designers must find ways to mediate between the "monsters" and Arcadian notions of landscape. Good design, to some extent, can integrate technology into the landscape by making the landscape legible, accentuating phenomena like roads, canals, landscape gradients, etc. It is possible that good design may even help expose the monumentality of grids and lines in the landscape (Feddes, 2010).

Design might also be able to play a role in terms of "mediation," in "re-centering" people and "domestication" by giving a spatial expression to assimilation and or adaptation of wind turbines by giving more attention to the receiving landscape or even by redesign of the turbines (e.g., Sir Norman Foster is the designer of the Enercon turbines). In a research by design project in 2007, we experimented with challenging four bureaus and an artist to react to the four attitudes toward technological innovations as defined by Smits (2002). Expelling, assimilation, adaptation, and embracing of monsters were all researched on their spatial characteristics and their specific strong and weak points (Sijmons 2007).

The third and perhaps least obvious role is that in designs and in the design discourse, a consistent storyline can be constructed that can play a role in communication between policymakers and the public. An open-design discourse seems to be able to indirectly fill the niche that modern politics is leaving wide open. In Western countries, forms of political voluntarism have become increasingly rare and sometimes are even absent. Ideals and ideologies do not seem to play a role as the backbone of the debate. In some countries, like the Netherlands, this concurs with the tendency of politics becoming increasingly technocratic,* focused on legal aspects and saturated with cost-benefit thinking. This political attitude makes it hard to articulate visions and ideals. An open-design discourse sometimes, and under specific circumstances, can fill this niche (Hajer et al., 2006).

For the elaboration of these three roles for design and the application on both a national and regional scale, we refer to the upcoming book kWh/m² on the spatial effects of the complete transition to sustainable sources.†

References

Barrett, N. 2012. Getting the Wind Up, Exploring the concern about adverse health effects of wind power in Australia and Europe. Castlemaine, Australia.

Bartholomew, L. K., Parcel, G., Kok, G., and N. Gottlieb. (eds.) (2001). *Intervention Mapping: Designing Theory- and Evidence-Based Health Promotion Programs.* Boston, MA: McGraw-Hill.

van den Berg, F. E. Pedersen, J. Bouma, and R. Bakker. 2008. *Windfarm Perception—Visual and Acoustic Impact of Wind Turbine Farms on Residents.* Groningen, the Netherlands; Science Shop for Medicine and Public Health Applied Health Research.

Chuihua, J. C. J. Inaba, R. Koolhaas, and T.L. Sze. 2002. *The Harvard Design School Guide to Shopping.* New York: Taschen.

---

* One can indeed argue if there isn't a cause and effect relationship here.
† D. Sijmons and A. van den Dobbelsteen, kWh/m². On the spatial impact of energy transition, NAi Publishers (in preparation, 2013).

Churchill, W. S. 1943. Speech, 28 October, 1943 to the House of Commons (meeting in the House of Lords).

Douglas, M. 1973. *Natural Symbols: Explorations of Cosmology*. London, U.K.: Berry & Jenkins.

Douglas, M. 1978. *Cultural Bias*. London, U.K.: The Royal Anthopological Institute.

European Climate Foundation. 2010. *Energy Roadmap 2050*. Brussels, Belgium: ECF.

Feddes, Y. 2010. *Windturbines Hebben Een Landschappelijk Verhaal Nodig, [Windturbines need a Landscape discourse]*. Den Haag, the Netherlands: Atelier Rijksbouwmeester.

Fridley, D. 2010. Nine challenges for alternative energy in: Heinberg, R. and D. Lerch (eds.) *The Postcarbon Reader, Managing the 21st Century Sustainability Crises*. Los Angeles, CA: Watershed Media, The University of California Press.

Hajer, M., D. Sijmons, and F. Feddes. 2006. *Een Plan dat Werkt, Ontwerp En Politiek in De Regionale Planvorming [A Plan That Works, Design and Politics in Regional Planning]*. Rotterdam, the Netherlands: NAi Publishers.

de Heer, J. 2000. *Age without Architecture, the Towers of Hendrick de Keijser and the Horizon of Amsterdam*. Amsterdam, the Netherlands: Uitgeverij Duizend & Een.

Hoen, B., W. Ryan, C. Peter, T. Mark, and S. Gautam. 2009. *The Impact of Wind Power Projects on Residential Property Values in the United States: A Multi-Site Hedonic Analysis*. Berkeley National Laboratory, & San Diego State University.

Johansson, M. and T. Laike. 2007. Intention to respond to local wind turbines: The role of attitudes and visual perception. *Wind Energy*, 10: 435–451. doi: 10.1002/we.232.

Kaplan, R. and S. Kaplan. 2005. Preference, restoration, and meaningful action in the context of nearby nature in Barlett, P. (ed.) *Urban Place—Reconnecting with the Natural World*. Cambridge, U.K.: The MIT Press.

von Kardorff, E. and H. Ohlbrecht. 2008. Overweight, obesity and eating disorders in adolescents-a socio-somatic reaction to social change? *Journal of Public Health* 16(6): 429–438. doi: 10.1007/s10389-008-0192-y.

Knopper, L. D. and C. A. Ollson 2011. Health effects and wind turbines: A review of the Literature. *Environmental Health*, 10: 78.

Kockelkoren, P. 2003. *Technology, Art, Fairground and Theatre*. Rotterdam, the Netherlands: NAi Publishers.

Ladenburg, J. and A. Dubgaard. 2009. Preferences of coastal zone user groups regarding the siting of offshore wind farms. *Ocean and Coastal Management*, 52: 233–242.

Latour, B. 2005. *Reassembling the Social*. New York: Oxford University Press.

Lindblom, C. 1959. Planning, the science of muddling through. *Public Administration Review*, 19: 79–88.

Marsman, H. 1936. *Memory of Holland Poem Translated by Longley in 1939*. Belfast, U.K.

Matton, T. and R. Wamka. 2004. *De dans der turbines [The Dance of the Turbines]*. Roterdam/Wendorf, the Netherlands: Mattonoffice.

Merleau Ponty, M. 1962. *Phenomenology of Perception*. New York: Routledge.

MSU (Michigan State University). 2004. *Land Use and Zoning Issues Related to Site Development for Utility Scale Wind Turbine Generators*. East Lancing, MI: Michigan State University.

NRC—National Research Council. 2007. 4 impacts of wind-energy development on humans, in *Environmental Impacts of Wind-Energy Projects*. Washington, DC: The National Academies Press.

Oreskes, N. and E. Conway. 2010. *Merchants of Doubt*. New York: Bloomsbury Press.

Pedersen, E., LR.-M. Hallberg, and K.P. Waye. 2007. Living in the vicinity of wind turbines—A grounded theory study. *Qualitative Research in Psychology*, 4: 1–2, 49–63.

Pedersen, E. and P. Larsman. 2008. The impact of visual factors on noise annoyance among people living in the vicinity of wind turbines. *Journal of Environmental Psychology*, 28: 379–389.

Persky, J. 1995. Retrospectives: The ethology of homo economicus. *The Journal of Economic Perspectives*, 9(2): 221–231.

Pimentel, D. (ed.). 2008. *Biofuels, Solar and Wind as Renewable Energy Systems. Benefits and Risks*. New York: Springer.

Rijksadviseur voor het Landschap, *Windturbines in het Nederlandse Landschap* [State Advisor on Landscape, Windturbines in the Dutch Landscape. Three tomes: Backgrounds, vision and Advice to the Government) Atelier Rijksbouwmeester, Den Haag, the Netherlands, March 2007.

Schivelbuch, W. 1977. *Geschichte der Eisenbahnreaise. Zur Indusrialisierung von Raum und Zeit*, München [*The Railway Journey, Train and Travel in the 19th Century*, Chicago, IL, 1986].

Schöne, M.B. 2007. *Windturbines in Het Landschap*. Wageningen, the Netherlands: Alterra and WUR.

Schuler, A.J. 2003. Overcoming resistance to change: Top ten reasons for change resistance in What's Up, Doc? *Schuler Solutions Newsletter*, 3(9). Available: www.schulersolutions.com/resistance_to_change.html

Sijmons, D. 2007. *Windturbines in het Nederlandse Landschap*. Den haag, the Netherlands: Atelier Rijksbouwmeesters.

Sijmons, D. 2010. *Betwiste horizon: windmolenpark voor de kust van Urk [Contested Horizon: A Turbine Park off the Coast of Urk]* in Blauwe Kamer, April 18, 2010.

Sloterdijk, P. 2009. *Du Musst dein Leben ändern. Über Anthropotechnik*. Frankfurt, Germany: Suhrkamp Verlag.

Smits, M. 2002. *Monster bezwering. De culturele domesticatie van nieuwe technologie [with English summary: Monster Treatment: The Cultural Domestication of New Technology]*. Amsterdam, the Netherlands: Boom Publishers.

Thoreau, D.H. 1854. *Walden, or Life in the Woods*. Boston, MA: Houghton Mifflin.

Ulrich, R.S. 1983. Aesthetic and affective response to natural environment in I. Altman and J.F. Wohlwill (eds.) *Human Behavior and the Environment: Volume 6* (pp. 85–125). New York: Plenum Press.

Venhuizen, H. 2009. *The making of....* Letter to the municipality of the Wieringermeer on public participation in the Wind Weekend 9, 10 and 11 October.

Vrijlandt, P. and K. Kerkstra. 1984. *Infrastructuur en Landschap als teken van leven*. In Plan 10/1984, pp. 6–19.

Vrijlandt P. et al. 1980. *Elektriciteitswerken in het Landschap: Probleemverkenning en konceptvorming*. Wageningen, the Netherlands: Dorschkamp.

Vroon, P. 1990. *Psychologische aspecten van ziekmakende gebouwen*. Den Haag, the Netherlands: Ministerie van VROM.

Vuyk, K. 1992. *De esthetisering van het wereldbeeld: Essays over filosofie en kunst*. Kampen, Overijssel, the Netherlands: Uitg. Kok Agora.

Young, J.G. et al. 2012. Touch-screen tablet user configurations and case-supported tilt affect head and neck flexion angles. *Work*, 41(1): 81–91.

Zee, J.M.H. 1999. *Van vrijheid. Een studie naar motieven voor kust toerisme en vrijetijds ervaringen aan de kust*. Wageningen, the Netherlands: Afstudeerscriptie Werkgroep Recreatie en Toerisme.

Part II

Methods

4

Energy Potential Mapping and Heat Mapping: Prerequisite for Energy-Conscious Planning and Design

Andy van den Dobbelsteen, Siebe Broersma, and Michiel Fremouw

CONTENTS

4.1 Introduction

Although estimates of remaining resources fluctuate, it is apparent to both energy experts and oil companies that the economic end is near. According to the Industry Taskforce on Peak Oil and Energy Security (ITPOES 2010), we have passed peak oil: these days we consume more oil than can be produced. Meanwhile, few people from the West understand how dependent they have become on energy and that a collapse in the provision would have devastating effects to everyday life. That this is a literally dangerous situation was demonstrated by the two gulf wars and recent turmoil around gas from Russia (first: Ukraine disconnected and second: Belarus threatening to halt the throughput of Russian gas).

As Mackay (2009) demonstrated and Sijmons and Van Dorst pointed out in Chapter 3, it is very difficult to establish a society fully run on renewables. Nevertheless, Cullen and Allwood (2010) showed that most of the energy we use is lost as nonfunctional waste energy. So the initial demand for useful energy can be reduced by more effective usage, such as by low-exergy means (to be discussed further on).

The Western craving for energy limits the opportunities of developing and emerging regions to catch up in prosperity, one of the main goals of the World Commission on Environment and Development (Brundtland et al. 1987). As Figure 4.1 indicates, Western countries owe their prosperity to limited use of energy in other parts of the world.

The abundance—until now—of relatively cheap fossil energy has made the world lazy and inactive to search for local possibilities. We need to learn planning and designing again in such a way that local resources are optimally seized before any import from elsewhere.

Energy potential mapping (EPM) can support this. This method was developed to support spatial planning for more energy-efficient urban or rural environments. Locally available renewable energy sources are charted to steer spatial developments where they are most effective in terms of the energy system. Since its first introduction, EPM has been applied at various scales.

Heat mapping is the latest development part of the EPM methodology. In a project carried out for the Dutch energy research agency, Agentschap NL, the Netherlands was heat mapped (Broersma et al. 2011) in order to help develop an online heat atlas. During the project, three-dimensional (3D) maps were drawn for heat (and cold) demands and potentials. Not only the quantity of these was visualized but also the energy quality (exergetic value).

This chapter discusses the methodology for both EPM and heat mapping, exemplified by case studies in the Netherlands.

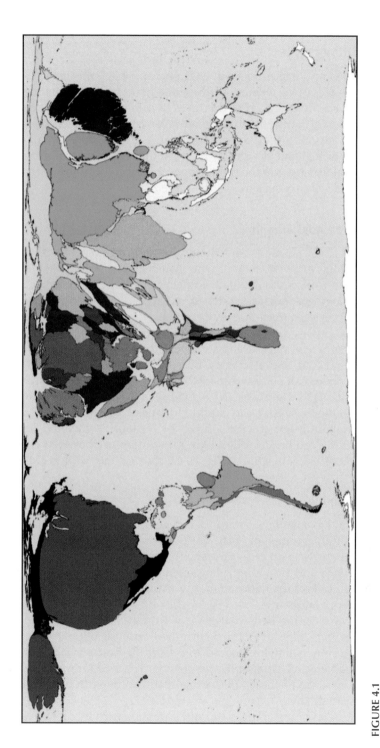

FIGURE 4.1

Countries and the area of land respective to the amount of fuel they consume (Dorling et al., 2009, downloadable from www.worldmapper.org); developed countries above the equator infest on other regions for energy.

4.2 Methodology

In order to develop a sustainable built environment, it is important to know locally available recourses when designing and making plans. The method of EPM was especially designed to do this. The aim of EPM is to chart and quantify all different local potentials for sources of renewable energy. This can be done on various scales, depending on the area of study. In an early stage of the planning and designing process, the maps can contribute to locating different functions at the right geographical positions in congruence with a sustainable energy system.

4.2.1 Energy Potential Mapping

With its origins in the Grounds for Change study (Roggema et al. 2006), the method of EPM has evolved from a cartoonish charting of climatic features with energy consequences to detailed 3D stacks of energy potential maps.

The methodology encompasses a structured approach. It commences with the analysis of basic sources and maps containing climatic, topographic, geophysical, and other possibly relevant information of local properties and characteristics. As a next step, these are converted into energy potential maps for fuels, heat and cold, and electricity, possibly divided by the level where these energy sources can be harvested (from high altitudes of wind to deep geothermal wells). All potentials can be quantified for each specific map.

All maps together visualize the energy "richness" or "poverty" of areas and identify the locality of specific sources, by means of which the process of spatial planning can be steered, making the difference between energy-positive and energy-negative plans. Figure 4.2 clarifies the method of EPM.

4.2.2 Heat Mapping

Heat mapping is a particular form of EPM. The purpose of a heat map is to provide a geographical imprint of the various thermal sources and sinks as well as infrastructures in an area, showing the net energetic—or even better, exergetic—balance and providing planners and other users a visual catalogue by which to design a thermal energy plan, which essentially forms a sustainable energy landscape.

In order to gain a comprehensive overview of an area's heat characteristics, first, all sources and sinks have to be determined. Second, following the EPM method, literature needs to be studied to obtain the best available data on quantities, qualities, and geographic dispersion of heat sources and sinks. In case of the absence of these, alternative methods need to be applied to calculate or estimate this.

The available data on heat sources and sinks are subsequently converted into equivalent units. The uniformity of data (encompassing one type of energy)

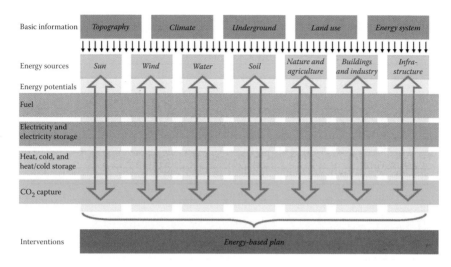

FIGURE 4.2
Schematic overview of the method of EPM.

offers the opportunity to present this by 3D-visualization techniques in which exergetic values can be incorporated.

We will zoom in on the heat mapping methodology a bit further.

4.2.3 Heat Sources and Sinks

Figure 4.3 lists the input used for the heat maps and indicates which of the variables of location, quantity, and temperature (quality) are defined for each type, although not all the data desired turned out to be available during the secondary research phase. For the various natural resources, the annually available amount of heat that can be harvested in a sustainable way was calculated using present-day techniques, avoiding single land use.

Locations can either be a point or a diffuse area, expressed by GJ or GJ/ha per year (or by suitability, in case of underground storage or heat exchange). Data are divided by the categories of demand, supply, and infrastructure. The right-hand side of the table shows which data are suitable for which scale of map.

With each of the regional and national maps, data were collected in large spreadsheets detailing each subregion, before translation into visual form.

4.2.4 Exergy of Heat

The exergy of heat is the theoretical maximum work that can be obtained by bringing the heat in thermal equilibrium with the environment using a reversible process (Jansen and Woudstra 2010). Originating from the second law of thermodynamics, exergy quantifies the quality of energy, which in case of heat relates to temperature. When assessing the heat potential of an area, it is beneficial to take this quality of heat into account.

Input: Desired Data and Variables					
	Location	Quantity (Units)	Temp.	NL	Region
Heat demand					
Bulk consumers					
Industries	Area	GJ/PJ/ha	V		•
Greenhouses	Area	GJ/ha	V	•	•
Hospitals	Point	GJ	V	•	•
Swimming pools	Point	GJ	V		•
Large-scale consumers					
Residential area	Area	GJ/ha	V	•	•
Commercial buildings	Area	GJ/ha	V		
Heat supply					
Residual heat sources					
Powerplant	Point	GJ/PJ	V	•	•
Incinerator plant	Point	GJ/PJ	V	•	•
Supermarkets	Point	GJ	V		•
Specific heavy industries	Point	GJ/PJ	V		
Biomass					
Sewage treatment plant	Point	GJ	V	•	•
Residual biomass	Area	GJ/ha	/		
Solar collectors					
Global radiation	Area	GJ/ha	/	•	
Global radiation on roofs	Area	GJ/ha	V		•
Global radiation on roads	Area	GJ/ha	V		•
Geothermal					
Underground (0–50 m depth)	Area	Suitability	/	•	•
Shallow (50–300 m depth)	Area	Suitability	/	•	
Deep (>2000 m)	Area	GJ/ha	V	•	•
Heat infrastructures					
District heating	Layout		/	•	•
CHP	Point		/		•
Biomass installation (CHP/incinerator)	Point		/	•	•

FIGURE 4.3
Example of input values for a regional heat calculation.

Local industries, for example, may require higher temperatures than dwellings, and similarly the heat generated in greenhouses may not have a temperature high enough to heat a living room. Upgrading the generally ubiquitous low-temperature renewable heat to a (less available) higher temperature by means of a heat pump requires additional energy, whereas industries using high-temperature heat may have lower-temperature residual heat available to start a "heat cascade." The resulting exergy landscape will thus make optimal use of the quality of valuable high-temperature heat (Stremke et al. 2011).

For the Heat Maps project, sources and sinks were divided into a number of common temperature ranges (30°C–50°C, 50°C–70°C, 70°C–110°C, and 110+°C) as well as a separate entry for renewable fuels (which, if used for

	Area size:	149 ha				OUD-CHARLOIS (1574)		
	Name	Description	Total		30–50	30–70	70–110	110–…	Fuel
Supply — Area sources	Geothermal source		467	GJ/ha			467		
	Biomass/gas source		6	GJ/ha					6
	Solar thermal roofs (housing)		1,218	GJ/ha	1,018	200			
	Solar thermal roofs (businesses)		150	GJ/ha	150				
	Area sources total:			GJ	0	0	0	0	
1D/0D sources	Roads		20,992	GJ	20,992				
	Supermarkets	2x	1,632	GJ	1,632				
	Sewage treatment facilities		50,000	GJ			50,000		
	Supply (total)			GJ	22,624	0	50,000	0	0
Demand — Area sources	Demand dwellings		2,314	GJ/ha		314	2,000		
	Demand utility/business		200	GJ/ha			200		
	Area sources total:			GJ	0	0	0	0	
1D/0D sources	Miscellaneous		0	GJ					
	Miscellaneous		0	GJ					
	Demand (total):			GJ	0	0	0	0	0

FIGURE 4.4
Example of area exergetic heat analysis: Oud-Charlois, a Rotterdam neighborhood.

heating purposes, should only provide the highest temperature range). The resulting area overview (Figure 4.4) shows whether or not it is necessary to import high-temperature heat or biofuels, or whether it is perhaps even possible to export these to surrounding areas.

4.2.5 Visualization

As part of the EPM method, the area analysis results in a stack of energy potential maps, each of which show the geographical characteristics of a specific type of energy potential, for example, geothermal heat availability or electricity potential for wind turbines based on the average local wind speed at nacelle height.

Heat mapping builds upon this, taking advantage of its focus on a single type of energy by introducing a cumulative third dimension in these maps (Figure 4.5). Heat demand is depicted as elevated contours; heat supply fills the resulting sink. When all these individual 3D maps are stacked, the resulting 3D landscape quickly indicates where demand may potentially outstrip supply and vice versa. Point sources, such as supermarkets, swimming pools, industry, and power plants, are depicted by cylinders, where the color denotes heat or cold and hollow cylinders represent the net demand for an installation. Larger roads can be equipped with integrated pipes to become giant solar collectors; these are depicted by yellow bands. When present, a district heating grid providing heat transport opportunities is hatched in red.

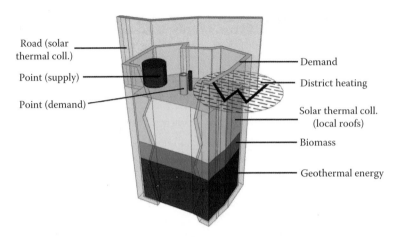

Road (solar thermal coll.)
Point (supply)
Point (demand)
Demand
District heating
Solar thermal coll. (local roofs)
Biomass
Geothermal energy

FIGURE 4.5
Example of a single section of a 3D heat map.

FIGURE 4.6
Evolution of the exergy component.

The next step, visualizing the exergetic gradient of available and required heat—without reducing overall legibility—proved to be a challenge. The initial stack, for example, would be less visible if neighboring areas have a higher total amount available, and the net balance for individual temperature ranges may be hard to read. To eliminate these issues, the exergy data were pictured as a 3D pie chart placed on top of each area (Figure 4.6, rightmost model). Divided into equal sections, the outer ring corresponds to the demand for a temperature range, with the slices on the inside showing available supply.

4.3 EPM Case Studies

4.3.1 Province of Groningen

The EPM study for the new provincial environmental plan (POP: Provinciaal OmgevingsPlan in Dutch) for Groningen was the first occasion where energy was taken into account in spatial planning. Energy potential maps for fuels, heat and cold, electricity, and carbon capture were drawn to determine the best locations for new (re)developments from the viewpoint of sustainable energy. The study produced interesting geological directions, for instance, to optimally use waste heat from industries or where heat and cold could be easily stored in the underground (Dobbelsteen et al. 2007).

Figure 4.7 presents some of the energy potential maps made for Groningen: wind, hydroelectric, power plants, and geothermal heat. All were meticulously charted by Geographical Information Systems (GIS). Not shown is the overlay map of all best energy potentials.

The energy potential maps of Groningen were eventually translated to a map of proposed spatial developments and interventions. Figure 4.8 depicts

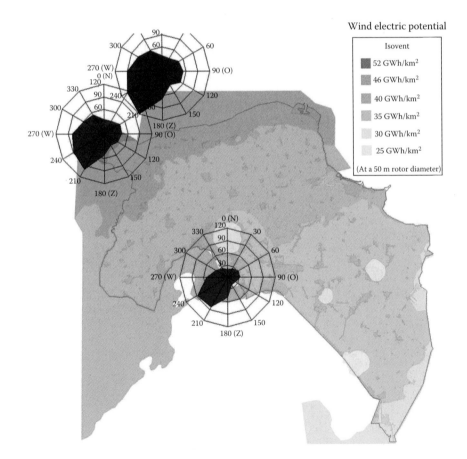

FIGURE 4.7
Some of the energy potential maps of the Dutch province of Groningen, from top to bottom: wind potential, hydroelectric potential, best places for power plants, and geothermal potential.

(continued)

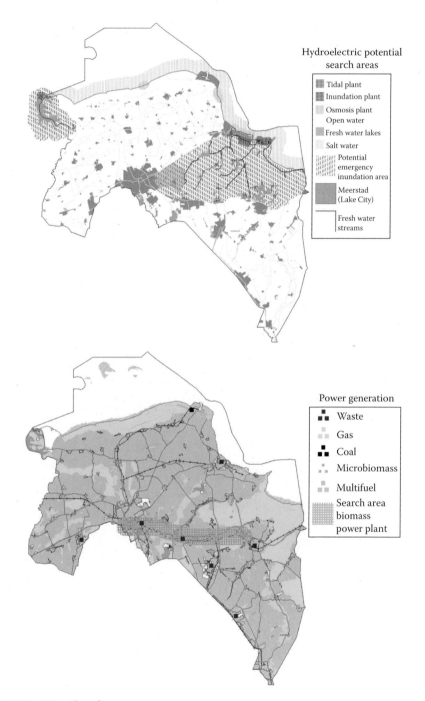

FIGURE 4.7 (continued)
Some of the energy potential maps of the Dutch province of Groningen, from top to bottom: wind potential, hydroelectric potential, best places for power plants, and geothermal potential.

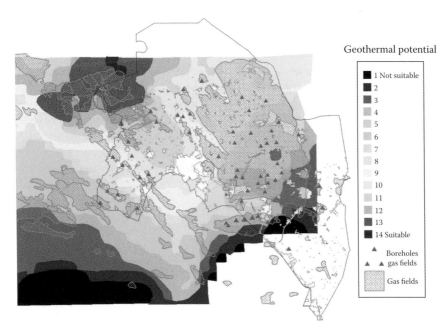

Geothermal potential

- ■ 1 Not suitable
- ■ 2
- ■ 3
- ▨ 4
- ▨ 5
- □ 6
- □ 7
- ▨ 8
- 9
- 10
- 11
- ▨ 12
- ■ 13
- ■ 14 Suitable
- ▲ Boreholes
- ▲ ▲ gas fields
- ▨ Gas fields

FIGURE 4.7 (continued)
Some of the energy potential maps of the Dutch province of Groningen, from top to bottom: wind potential, hydroelectric potential, best places for power plants, and geothermal potential.

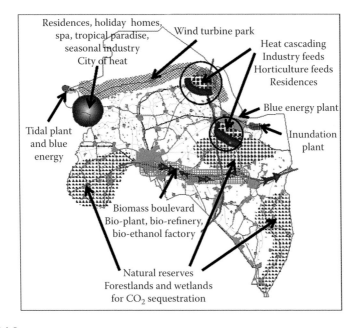

Residences, holiday homes, spa, tropical paradise, seasonal industry
City of heat

Wind turbine park

Heat cascading
Industry feeds
Horticulture feeds
Residences

Blue energy plant

Tidal plant and blue energy

Inundation plant

Biomass boulevard
Bio-plant, bio-refinery, bio-ethanol factory

Natural reserves
Forestlands and wetlands for CO_2 sequestration

FIGURE 4.8
Map of proposed spatial developments based on the energy potential study of Groningen.

this proposal, clearly related to the underlying energy potentials: a tidal plant where (for the entire Netherlands as a matter of fact) tide differences are greatest, a wind park along the coast, new urbanization on the spot of great geothermal heat and aquifer storage capacity, greenhouses and housing next to industries that cascade residual heat, a blue energy plant where fresh water runs into the salt sea, biomass processing facilities at a maximum distance of 50 km from all agricultural lands, and new natural reserves that apart from having a nature and recreation purpose serve to sequester carbon dioxide.

The influence on the landscape of the spatial developments proposed for the POP Groningen are evident, but the plan, as we presented it, carefully took into account existing landscape values, even enhancing these where new nature reserves were projected, as opposed to the magnified urbanization around industries.

4.3.2 Region of the Peat Colonies

The Veenkoloniën ("Peat Colonies") is an agricultural region of Groningen and the neighboring province of Drenthe, with a past as a large-scale extraction area for Peat. So the Peat Colonies boast a landscape vaguely reminding of that era, now being under less favorable economic conditions. This is why the provincial governments strived to come with new redevelopment plans for the future, including new energy perspectives. As a joint project with the University of Wageningen, we charted the energy potentials of the area and translated these into two visions, totally different but both sustainable in their own way: Alleenkoloniën ("Alone Colonies") and Veenkometro ("Peatcometro").

4.3.2.1 Alone Colonies

In this vision, small-scale energy provision from renewable resources is the goal, decentralized to the scale of separate villages and towns with their direct surroundings. Each of these conglomerates is self-sufficient in energy. The vision more or less coincides with the European trend to solve issues at the local scale, introverted to the local community. Energy is locally generated, by means of PV (photovoltaic) panels and small wind turbines but predominantly from bio-fermentation plants that can feed local heat grids. Some areas have the luxury of geothermal sources, to be won through former gas-drilling sites. Others may invest in sustainable greenhouses for food, materials, biogas, and thermal solar energy.

This vision includes mobility largely (75%) run on electricity. For heavy vehicles, algae are grown for the production of biodiesel. Local buses can drive on biogas.

For these localized self-sufficient communities, optimal interexchange (through local heat and power grids) and interseasonal storage of energy (in underground aquifers or salt caverns) are quintessential.

Figure 4.9 illustrates the Alone Colonies' energy vision.

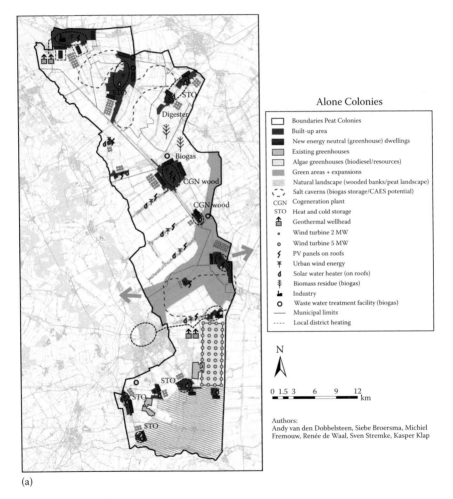

Alone Colonies

☐	Boundaries Peat Colonies
■	Built-up area
■	New energy neutral (greenhouse) dwellings
▦	Existing greenhouses
☐	Algae greenhouses (biodiesel/resources)
▦	Green areas + expansions
▨	Natural landscape (wooded banks/peat landscape)
⌇	Salt caverns (biogas storage/CAES potential)
CGN	Cogeneration plant
STO	Heat and cold storage
⌂	Geothermal wellhead
∘	Wind turbine 2 MW
⊙	Wind turbine 5 MW
ϟ	PV panels on roofs
⼂	Urban wind energy
◗	Solar water heater (on roofs)
⼂	Biomass residue (biogas)
⬤	Industry
○	Waste water treatment facility (biogas)
—	Municipal limits
----	Local district heating

N

0 1.5 3   6   9   12
km

Authors:
Andy van den Dobbelsteen, Siebe Broersma, Michiel Fremouw, Renée de Waal, Sven Stremke, Kasper Klap

(a)

FIGURE 4.9
Energy vision of the Alone Colonies (a).

(*continued*)

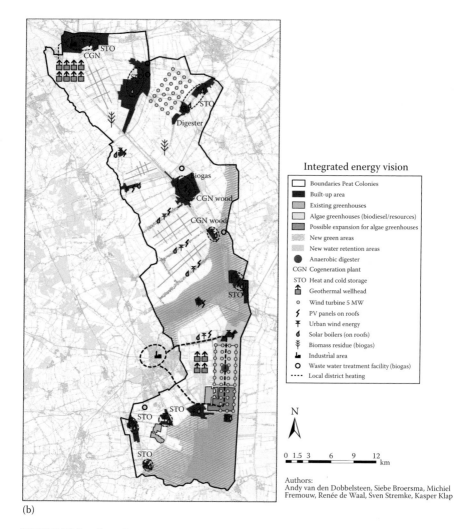

Integrated energy vision

☐ Boundaries Peat Colonies
■ Built-up area
▨ Existing greenhouses
☐ Algae greenhouses (biodiesel/resources)
▨ Possible expansion for algae greenhouses
▨ New green areas
▨ New water retention areas
● Anaerobic digester
CGN Cogeneration plant
STO Heat and cold storage
▥ Geothermal wellhead
○ Wind turbine 5 MW
⚡ PV panels on roofs
🌱 Urban wind energy
⬥ Solar boilers (on roofs)
⚘ Biomass residue (biogas)
⬛ Industrial area
◯ Waste water treatment facility (biogas)
---- Local district heating

N

0 1.5 3   6   9   12
                                km

Authors:
Andy van den Dobbelsteen, Siebe Broersma, Michiel
Fremouw, Renée de Waal, Sven Stremke, Kasper Klap

(b)

FIGURE 4.9 (continued)
Energy vision of the Peatcometro (b).

4.3.2.2 *Peatcometro*

This version of the future Peat Colonies envisions a more ambitious goal of becoming energy productive rather than just energy neutral (which was the case with the Alone Colonies vision). In the near future, energy equals space, and the Peat Colonies are one of the few regions in the overpopulated Netherlands still abundant in nonbuilt area. So if energy-exporting areas can be identified, this is one of them. The Peatcometro vision therefore entails a large-scale, centralized energy system, also based on renewables.

Most of the energy produced in Peatcometro comes from a concentrated area with modern greenhouses producing bio-based materials and food (of which nonusable waste material is fermented to biogas). Also, algae can be effectively grown inside the greenhouses, fed on nutrient-rich waste water and being the source for proteins and biodiesel. The greenhouses are equipped with spectrally selective PV foil on south-facing shed roofs and still leave space for wind turbines in between the premises. Thermal solar energy collected by the greenhouses is interseasonally stored in the underground, providing plenty of heat for residences surrounding the productive area.

The large-scale production of energy of all sorts enables a wider-spanning heat grid that connects the different villages and towns. This vision also includes the optimal usage of biomass from nonglazed areas. In order to store excess production of electricity, Peatcometro also presents a so-called fall lake on the higher ridge of the Hondsrug. In times of excess, water is pumped up, while turbines produce a constant quantity of power from water flowing back down.

Figure 4.9 presents the Peatcometro energy vision.

4.3.2.3 Energy Landscape

The consequence of both energy visions for the Peat Colonies on the landscape is evident: whereas Alone Colonies stays close to the traditional dispersed organization of built-up area in the agricultural landscape—only locally modified by smaller wind turbines, greenhouses, and bio-fermentation tanks—Peatcometro implies a radical change to the traditional land use and hence the landscape. The latter fills vast agricultural land with a modern agrohorticulture guided by large wind turbines.

It probably does not surprise the reader that at a seminar discussing the energy visions, the local community found no charm in the Peatcometro vision, whereas people from outside the region sometimes strongly advocated the plan, understanding the value to the rest of the Netherlands. Following this event, an integrated proposal was made, attempting to combine the best of both earlier visions.

4.3.3 District of the Green Campaign

De Groene Compagnie ("The Green Campaign") is a new planned district south of the town of Hoogezand in the province of Groningen. An EPM study was executed to contribute to the urban plan (Dobbelsteen et al. 2011). In a study prior to this one, stacked energy potential maps had appeared for the first time, showing all energy potentials in one overview. For the Green Campaign study, the stacked maps clarified the local potentials yet even clearer (see Figure 4.10). Not only the theoretical potentials were visualized, but also the realistically harvestable amounts of energy, taking into account technical efficiencies.

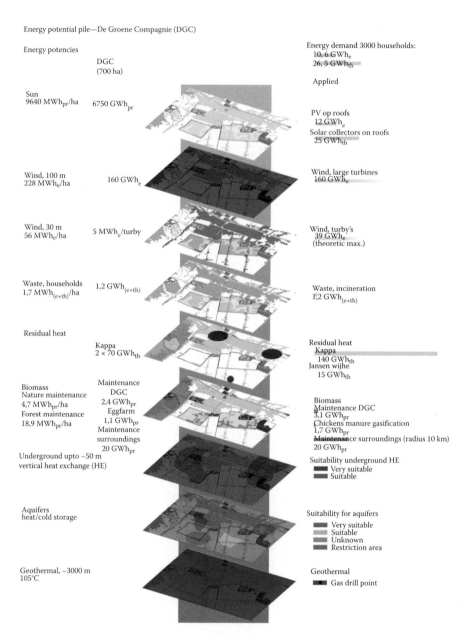

Energy potential pile—De Groene Compagnie (DGC)

Energy potencies

DGC
(700 ha)

Energy demand 3000 households:
10,6 GWh$_e$
26,5 GWh$_{th}$

Applied

Sun
9640 MWh$_{pr}$/ha   6750 GWh$_{pr}$

PV op roofs
12 GWh$_e$
Solar collectors on roofs
25 GWh$_{th}$

Wind, 100 m
228 MWh$_e$/ha   160 GWh$_e$

Wind, large turbines
160 GWh$_e$

Wind, 30 m
56 MWh$_e$/ha   5 MWh$_e$/turby

Wind, turby's
39 GWh$_e$
(theoretic max.)

Waste, households
1,7 MWh$_{(e+th)}$/ha   1,2 GWh$_{(e+th)}$

Waste, incineration
1,2 GWh$_{(e+th)}$

Residual heat

Kappa
2 × 70 GWh$_{th}$

Residual heat
Kappa
140 GWh$_{th}$
Jansen wijhe
15 GWh$_{th}$

Biomass
Nature maintenance
4,7 MWh$_{pr}$/ha
Forest maintenance
18,9 MWh$_{pr}$/ha

Maintenance
DGC
2,4 GWh$_{pr}$
Eggfarm
1,1 GWh$_{pr}$
Maintenance
surroundings
20 GWh$_{pr}$

Biomass
Maintenance DGC
3,1 GWh$_{pr}$
Chickens manure gasification
1,7 GWh$_{pr}$
Maintenance surroundings (radius 10 km)
20 GWh$_{pr}$

Underground upto –50 m
vertical heat exchange (HE)

Suitability underground HE
███ Very suitable
███ Suitable

Aquifers
heat/cold storage

Suitability for aquifers
███ Very suitable
███ Suitable
███ Unknown
▨▨▨ Restriction area

Geothermal, –3000 m
105°C

Geothermal
███ Gas drill point

FIGURE 4.10
Stacked energy potential pile of "De Groene Compagnie" in Hoogezand.

Designing an adjusted urban plan for the district, the Green Campaign turned out to be possibly energy producing. This could be achieved without large interventions such as solar power plants or converting agriculture from producing food into energy crops.

4.3.4 Town of Almere

An interesting different study we conducted concerned the Dutch town of Almere, not far from Amsterdam, which was selected by the national government as growth area that will have to expand from 75,000 to 100,000–135,000 households by 2030 (this was before the economic crisis). The Municipality of Almere had pointed out four areas where the city could expand or grow through redevelopment. TU Delft was commissioned with two EPM studies, one of which was for Almere-Oost, the new eastern district of Almere. The EPM study of Almere-Oost (Dobbelsteen et al. 2008) had to provide input for the new urban plan, involving local sustainable energy resources. Beforehand, four urban alternatives had already been proposed by the municipal urban planner.

We found that—because of the presence of relatively large-scale farms in this area, each with a considerable produce of organic or animal waste—energetically autonomous clusters could be formed of one farm with approximately 30 dwellings. Only the northern edge of the area turned out to be suited for the storage of heat and cold in an open aquifer, which allows a mix of functions (housing combined with offices, retail, and light industry) in high density. Because of this proposition, the required number of residences could be met while leaving the agricultural land open. This strongly contrasted with the urban plans made earlier. Figure 4.11 shows two of our alternatives based on the energy potential study.

4.4 Heat Map Case Studies

4.4.1 The Netherlands

For a heat map of the Netherlands, the most significant heat sources and sinks were charted: heat demand (from greenhouses and residential areas), solar collector potential (of residences and greenhouses), point sources (power plants, sewage treatment facilities, etc.), biomass (manure, domestic organic waste, and trimmings from public parks), geothermal potential, ground-source heat-exchanger potential, and underground heat storage potential. The 3D composite map (Figure 4.12) shows a large potential heat surplus in the Northern Netherlands (mainly to be provided by geothermal heat) as well as a large net demand in the Rotterdam—the Hague area (mostly caused by greenhouses). Also visible are the many point sources and heat-related utilities (such as power plants, anaerobic digesters, and sewage treatment facilities).

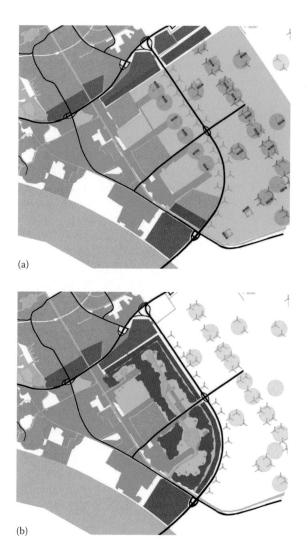

(a)

(b)

FIGURE 4.11
Two urban extension variants for Almere East, based on the local energy potentials. Plan (a) uses existing farms to create energy-neutral farm-court dwelling units and concentrate high-density housing, offices, and light industry in the north, where heat and cold can be stored in the underground. Plan (b) concentrates developments in the western part, alongside the motorway and leaves open the internal space for a park.

4.4.2 Emmen Region

Being of the size of 100,000 inhabitants, the regional center of Emmen provides many facilities for the surrounding areas, such as a hospital, swimming pools, a zoo, and a large sewage treatment facility. Several district heating systems can be found in the area, although some are reported to currently not be in use. Geothermal potential is moderate but adequate.

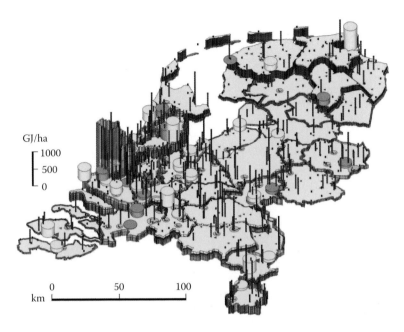

FIGURE 4.12
Composite heat demand and potential map for the Netherlands.

Comprehensive data on heat in industrial processes within the commercial zones were incomplete, but, where possible, individual factories and other installations have been included. In order to get a similar size area as the other city level case studies, the map contours do not always follow the city limits, but form a smaller, somewhat rectangular area. The composite 3D map (Figure 4.13) shows a net demand in the densely populated areas to the west, a significant demand and supply in the greenhouse complex to the east, and a net surplus in most of the sparsely populated areas in the middle.

4.4.3 Rotterdam

At around 600,000 inhabitants, Rotterdam, the second largest city in the Netherlands, has a combination of high-rise and low-rise buildings, industry, harbors, an extensive network of (main) roads, and various other facilities associated with a densely populated urban area, such as hospitals, a zoo, and a large sewage treatment facility. As high-rise areas have far less roof surface per dwelling and are overall somewhat less effective due to shading lower premises, a multiplier was used for some areas to factor this in. As dwellings have a relatively limited range, it was possible to estimate heat demand and supply per area. Comprehensive data on heat in industrial processes within the harbor areas along the Meuse river were, however, unavailable due to the highly diverse nature of commercial activity there. Nonetheless, where possible, individual factories and other installations were included. The 3D composite map

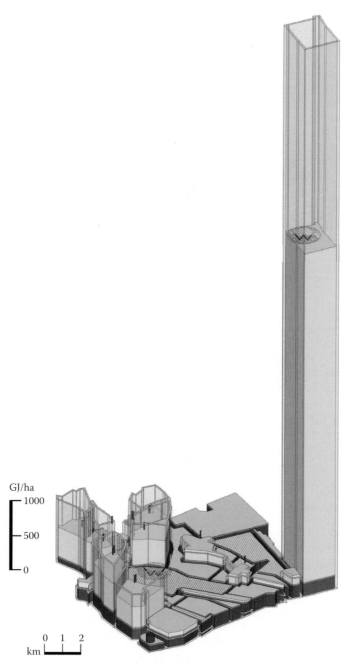

FIGURE 4.13
Composite heat potential map for the Emmen region.

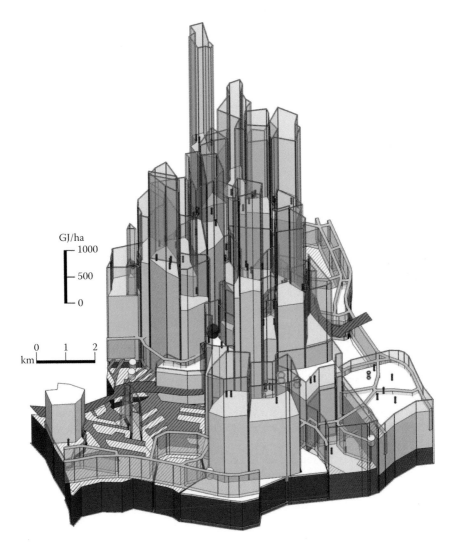

FIGURE 4.14
Composite heat potential map for Rotterdam.

(Figure 4.14) shows a heat surplus in the relatively sparsely populated southern areas as well as a large net demand in the densely populated city center.

4.5 Conclusions and Discussion

EPM has proven to be very useful in spatial planning at various scales: country, region, city, district, and neighborhood. It facilitates a built environment that more effectively seizes local energy opportunities before requiring import from elsewhere. Thus, these areas can become more self-sufficient and in some cases even energy producing, as De Groene Compagnie study indicated.

The methodology of heat mapping provides an accessible visualization of the balance between heat demand and supply in an area. The focus on low-temperature heat (which cannot be transported over long distances) implies that heat maps on a regional scale will be more useful than the national map. This is also the reason why the maps work visually well: the energy scale for low-temperature heat sources is of course relatively limited. Although the 3D maps developed during the project were intended as a proof of concept and are static in nature, they do give a clearer overall view than in the current situation where this information is hidden in many reports and drawings, and the information compounded in the map can easily be accessed and separated by clicking on objects or turning off layers. This will speed up transition to more sustainable sources of heat, and this will support planners making informed decisions on infrastructure investments and technology selections.

4.5.1 Limitations

The main limiting factor with both EPM and heat mapping was the availability of comprehensive data on energy production and demand. Although nationwide geothermal data are relatively easy to access, detailed information on, for example, age, type, and energy efficiency of dwellings or the internal heat cycles of various types of industrial plants is often unavailable. Having access to this data would greatly increase the accuracy of each map.

Even though the 3D landscapes can be explored in detail by panning, tilting, and turning layers on and off, they are still static models. More interactive parametric heat map models were briefly investigated with tools such as Rhino and Grasshopper. Given a base vector map, sufficient data, and an expanded script, it is certainly possible to use the heat mapping methodology to investigate various future scenarios and their geographical impact on heat distribution, by generating heat maps instantly.

Many (renewable) heat sources and sinks have outputs and requirements that fluctuate over time and will operate at different temperature ranges. Although studies so far have focused on lump sums of GJ and TJ per year for each category, investigating this temporal component of renewable heat and including an exergetic analysis would make it possible to quantify energy storage requirements, for example, the necessary amounts of biofuels or aquifer storage, thus increasing accuracy of projections.

The energy potential maps and heat maps only concern types of energy that are connected to their location, since thermal energy can only be exported over small distances via heat networks. Other forms of energy, such as electricity or biofuels, are easier to transport over a longer distance. They may be available or suitable to be produced on the spot but not necessary to be consumed nearby. Therefore, localized energy potentials not necessarily tell the whole story of future sustainable energy systems or energy landscapes, for that matter.

4.5.2 Future Directions

A logical next step for EPM and heat mapping could be digital 3D charts that can be updated real time by different stakeholders such as municipalities, provinces, and research centers. Thereby, new existing amounts of residual heat from industries would be mapped each time data are available. Amounts of energy that are already used and connected with heat grids would be mapped as well. A digital map in the form of Google Maps would offer possibilities to zoom in and out to different scales, where each time a new information could pop up. By clicking on it, more data could be found behind the different energetic interventions on the map.

Acknowledgments

The authors thank all commissioners involved with the development of EPM: the Province of Groningen, Municipality of Almere, Schiphol Area Development Corporation, Municipality of Hoogezand-Sappemeer, and the commissioner of the Heat Maps project, Agentschap NL. Furthermore, we express our gratitude to all researchers who have been involved in the whole process of EPM development, ever since 2005: Sabine Jansen, Arjan van Timmeren, Bram van der Grinten and Chris Hellinga (TU Delft), Sven Stremke and Renée de Waal (Wageningen UR), Rob Roggema (formerly Province of Groningen), Kees Stegenga (Stegenga urban design workshop), Steven Slabbers (Bosch Slabbers landscape architects), Jan Leemans (Adecs), Patrick Nan (Planmaat), Sebas Veldhuisen (BuildDesk), and Ronald Rovers (Zuyd University Heerlen).

References

Broersma S., Fremouw M., and Dobbelsteen A. van den 2011; Heat mapping the Netherlands—Laying the foundations for energy-based planning, in: *Proceedings SB11*; VTT, Helsinki, Finland.

Brundtland G.H. (ed.) et al. (World Commission on Environment and Development) 1987; *Our Common Future*; Oxford University Press, Oxford/New York.

Cullen J.M. and Allwood J.M. 2010; The efficient use of energy: Tracing the global flow of energy from fuel to service, *Energy Policy*, 38(1), 75–81.

Dobbelsteen A. van den, Broersma S., and Stremke S. 2011; Energy potential mapping for energy-producing neighborhoods, *International Journal of Sustainable Building Technology and Urban Development*, 2(2), 170–176.

Dobbelsteen A. van den, Grinten B. van der, Timmeren A. van, and Veldhuisen S. 2008; *Energiepotenties Almere—Energiepotentiestudie Almere-Oost*; TU Delft, Faculteit Bouwkunde.

Dobbelsteen A. van den, Jansen S., Timmeren A. van, and Roggema R. 2007; Energy potential mapping—A systematic approach to sustainable regional planning based on climate change, local potentials and exergy, in: *Proceedings of the CIB World Building Congress 2007*; CIB/CSIR, Cape Town, South Africa.

Dorling D., Newman M., and Barford A. 2009; *The Atlas of the Real World—Mapping the Way We Live*; Thomas & Hudson, London, U.K.

ITPOES (Industry Taskforce on Peak Oil and Energy Security) 2010; *The Oil Crunch—A Wake-Up Call for the UK Economy*, on: http://peakoiltaskforce.net

Jansen S. and Woudstra N. 2010; Understanding the exergy of cold: Theory and practical examples, *International Journal of Exergy*, 6, 695.

MacKay D.J.C. 2009; *Sustainable Energy—Without the Hot Air*; UIT Cambridge Ltd, Cambridge, U.K.

Roggema R., Dobbelsteen A. van den, and Stegenga K. (eds.) 2006; *Pallet of Possibilities—Bridging to the Future, Spatial Team, Grounds for Change*; Province of Groningen, Groningen, the Netherlands.

Stremke S., Dobbelsteen A. van den, and Koh J. 2011; Exergy landscapes: Exploration of second-law thinking towards sustainable landscape design, *International Journal of Exergy*, 8(2), 148–174.

5

Five-Step Approach to the Design of Sustainable Energy Landscapes

Sven Stremke

CONTENTS

5.1 Introduction

Climate change and depletion of fossil fuels necessitate an unprecedented transformation of today's physical environment into a more sustainable, robust, and resilient one. This transition presents many challenges to designers, planners, and managers of the larger physical environment, especially landscape architects and spatial planners. The adaptation to climate change and renewable energy sources, undoubtedly, will require decennia—a time frame that bears many uncertainties and that is clearly beyond that of conventional planning and design.

This chapter addresses critical uncertainties in long-term planning and design. Many scholars have stressed the importance of considering external trends and forces by means of long-term visions (e.g., Kunzmann 2000; Mintzberg 1994). Spatial planners as well as landscape architects compose long-term visions at the regional scale (see, e.g., Rodriguez and Martinez 2003; Weller 2008). Despite achievements in spatial planning and landscape architecture, the two disciplines have yet to explore the potentials of a joint approach to strategic regional design.

This chapter is motivated by the need to advance long-term thinking in regional planning and design. The objective is to discuss how to compose imaginative and yet realistic long-term visions. In doing so, we explore the so-called *five-step approach* that can be employed to compose long-term visions such as for the development of sustainable energy landscapes. The proposed methodological framework is constructed on the basis of existing planning and design methods and reflects upon the experiences from composing several long-term visions for research projects and commissioned design projects as well as from envisioning sustainable energy landscapes together with graduate students at Wageningen University.

5.2 Methods

The research and design reported in this chapter commenced with a study of existing approaches to strategic spatial planning, design-oriented planning, and landscape planning. Whereas the literature study provided the key building blocks for an advanced methodological framework, case studies and teaching* enabled us to test and refine the *five-step approach*. Over the past 7 years, the *five-step approach* has been employed to envision sustainable energy landscapes in various locations across the Netherlands (Figure 5.1). The members of the multidisciplinary project teams varied but consisted of architects, urban planners, spatial planners, and landscape architects. The close collaboration with commissioners, decision makers, stakeholders, and energy experts provided us with continuous feedback (Figure 5.2). In this chapter, a research and design project on the development of sustainable energy landscapes in South Limburg is used to illustrate the five steps of the methodological framework.

5.3 Literature Study

Any proposal for the long-term development of large territorial systems faces a great number of uncertainties due to the trends and forces that are beyond the control of the planner and designer. Despite the difficulties of

---

* The five-step approach has been applied in various graduate student design studios and MSc theses projects.

FIGURE 5.1
Overview of energy landscape projects between 2006 and 2012 (in blue). Projects where the five-step approach has been applied are indicated in dark blue. The approach has been applied in South Limburg in the very south of the Netherlands (2007–2008), in Southeast Drenthe in northern Netherlands (2009–2010), and in the Delta in the west of the country (2011). A number of graduate students have employed the five-step approach in their thesis projects (locations indicated with dark blue stars).

FIGURE 5.2
Impression from one of the workshops with stakeholders and energy experts in South Limburg, 2007.

working with such uncertainties, it is important to envision a desirable future (Rosenhead 2001) and identify actions that can help reaching that future (Albrechts 2004). Among the different approaches to spatial planning and landscape design, we now discuss a selection of three methods relevant for the composition of long-term visions (Table 5.1).

Design-oriented planners are among the professionals concerned with long-term development at the regional scale. They aim to influence the actions of those who shape the physical environment by, for example, discussing probable and desired futures (Carsjens 2009). Dammers et al. (2005) describe the so-called cyclic scenario approach to design-oriented planning. The cyclic scenario approach is relevant to the discussion on long-term visions because it helps to identify and focus on a number of key issues. Moreover, employing "external and policy scenarios avoid the problem to specifically predict the future, (an approach) which is appropriate in complex situations with high degree of uncertainty" (Carsjens 2009, p. 52). Three modes of change

TABLE 5.1

Comparison of the Cyclic Scenario Approach, the Four-Track Approach, and the Design Framework

	Cyclic Scenario Approach Dammers et al. (2005)	**Four-Track Approach** Albrechts (2004)	**Design Framework** Steinitz (1990, 2002)
Initial step	**Basic analysis** Analyze present situation, trends, and policies Identify focal issues	Analysis Analyze main processes that shape environment Agenda setting	Representation Analyze conditions Process Study relationships Evaluation Identify dysfunctions
First mode of change	*Analysis of current trends is part of analysis*	*No explicit reference to current projected trends*	Change caused by current projected trends Identify trends
Second mode of change	External scenarios Compose scenarios to identify possible futures	*No explicit reference to context scenarios and critical uncertainties*	*No explicit reference to context scenarios and critical uncertainties*
Third mode of change	Policy scenarios Explore alternative policy strategies	Long-term vision Represent values and meanings for the future	Change caused by implementable design Describe interventions
Final step	Recommendations and knowledge questions Support development of policy strategies Master plan with short-term actions Contingency plan with long-term actions	Short- and long-term actions Short-term actions to solve present problems Long-term actions to achieve desired future Budget and strategy for implementation Creation of commitment	Impact Estimate impact of alternative interventions Decision Support decision-making process

are addressed by the cyclic scenario approach: current projected trends, critical uncertainties, and intended change. One drawback of the *cyclic scenario approach*, from an operational perspective, is that the development of meaningful context scenarios requires substantial resources and special expertise. Employing existing socioeconomic scenarios may present an alternative that will be discussed later in this chapter.

The rationale of *strategic spatial planners* is to "frame the activities of stakeholders to help achieve shared concerns about spatial changes" (Albrechts 2004, p. 749). Sustainable development is among the shared concerns that require us to employ long-term visions (ibid.). Composing long-term visions may present an alternative to conventional planning because it is unlikely that a single (blueprint) plan can address the critical uncertainties and dynamics of large territorial systems. Several approaches have been described in order to structure the strategic spatial-planning process. The so-called *four-track approach* (Albrechts 2004) is a prominent example where much attention is devoted to the question on who should participate in the envisioning process. Very little information exists on how to actually compose the visions—that is, in other words, how to give shape to a desired future. A great value of strategic planning, in general, is the focus on critical uncertainties and implementable actions as two modes of change. Albrechts (2004), however, makes no explicit reference to context scenarios.

Landscape architecture is yet another discipline concerned with the conscious shaping of the human environment. Among the various approaches to landscape planning and design, the *design framework* has been chosen for discussion because it has been applied successfully in many research and design projects as well as teaching (see, e.g., Steinitz 2003). Without going into too much detail, we may assert that Steinitz (2002) refers to two modes of change—that is, change due to current trends and change due to implementable design. Steinitz makes no explicit reference to critical uncertainties that are beyond the control of the planner/designer but that cannot be disregarded in long-term planning and design.

All three approaches, to begin with a similarity, aim to support decision making. The first three steps of the *design framework* correspond to a great extent with the "analysis phase" of the planning frameworks. The estimation of the impact of alternative interventions, suggested by Steinitz (2002), is similar to the evaluation of policy strategies suggested by the planners. One substantial difference is that Steinitz (2002) and Albrechts (2004) make no explicit reference to the use of context scenarios in the planning and design process. Dammers et al. (2005), on the contrary, suggest the creation of context scenarios as part of the planning process.

Strategic spatial planners, generally speaking, recognize the significance of external trends and forces that influence the future of a study area (see, e.g., Friedmann 2004). Rosenhead stresses that "strategic planning cannot be firmly based on an attempt to predict what will happen

[...] identifying a range of versions what might happen, would be a modest and supportable basis for planning analysis" (2001, p. 185). We propose to adopt a similar approach to long-term regional design: critical uncertainties should be integrated in the design process. Whether globalization will continue in the future is a critical uncertainty; globalization influences land use patterns and consequently affects the design of sustainable energy landscapes. A further enlargement of farm plots, for example, would reduce the number of small-scale landscape elements and, subsequently, decrease the amount of biomass that could be harvested from these landscape elements.

5.4 Methodological Framework

In order to compose imaginative and yet realistic long-term visions (Healey 2009), we argue that *current projected trends*, *critical uncertainties*, and *intended change* must be integrated in the design process. Each of these three modes of change gives rise to one step in the proposed methodological framework, namely, near-future developments (step 2), possible far-futures (step 3), and integrated visions (step 4). Those three steps are preceded by the analysis of present conditions (step 1) and complemented by the identification of robust spatial interventions (step 5). The methodological framework of the *five-step approach* is depicted in Figure 5.3. Table 5.2 presents an overview of the different activities and means of representation.

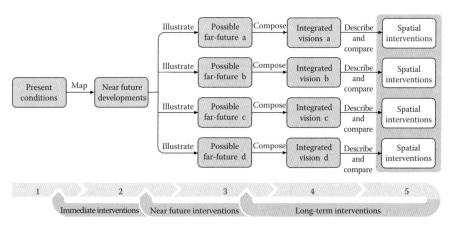

FIGURE 5.3
Methodological framework of the five-step approach. The study commences with the analysis of present conditions (step 1) and leads to the identification of robust spatial interventions (step 5) that can be further distinguished into long-term, near feature, and immediate interventions.

TABLE 5.2

Overview of the Five Steps, Respective Activities, and Means of Representation

Step	Activity		Representation
1	Analyzing present conditions	Landscape characteristics	Topographic map
			Land-use map
			Infrastructure map
		Present energy system	Energy provision map
			Transport, conversion, and storage
			Energy consumption map
		Renewable energy potentials	Solar energy map
			Wind power map
			Hydropower map
			Biomass map
			Heat–cold storage map
			Geothermal energy map
2	Mapping near-future developments		Near-future base-map
3	Illustrating possible far-futures	Four scenario base-maps	Global market base-map
			Secure region base-map
			Global solidarity base-map
			Caring region base-map
4	Composing integrated visions	Four energy visions	Global market energy vision
			Secure region energy vision
			Global solidarity energy vision
			Caring region energy vision
5	Identifying energy-conscious spatial interventions		Tables, text, and reference images

Note: This example is taken from a study on sustainable energy landscapes in South Limburg.

The methodological framework is organized around a set of five questions, each question is subject to one step of the design process. The sequence of five steps should be passed through (at least) twice. During the first cycle, the context and scope of the study are defined. Also, maps and data are gathered, and stakeholders and decision makers are invited to participate. During the second cycle, the visions are composed and robust spatial interventions identified. Although the framework consists of five consecutive steps, the process is iterative. It may be necessary to return to an earlier step in order to answer all questions. The five steps are now described in more detail.

5.4.1 Analyzing Present Conditions

The first step centers on the question "How does the present region function and how can it be evaluated in comparison with other regions?" The analysis

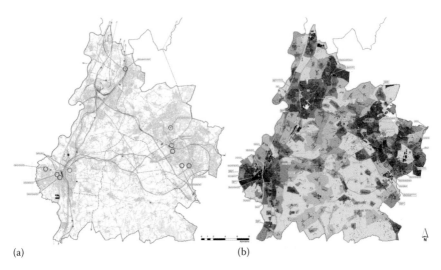

(a)   (b)

FIGURE 5.4

Two of three maps representing the present-day energy system of South Limburg: Energy storage and transport (a) and energy demand (b). The map on the left depicts the extensive road and natural gas network. The gray shades in the map on the right represent population densities as an indicator for energy use (the darker the area, the more energy is consumed per square kilometer). Business parks (blue) and industrial parks (brown) consume the most energy per square kilometer. Please note that the high-resolution maps at the scale 1:25,000 including legends, can be accessed at www.exergieplanning.nl.

includes both the study of present landscape conditions and historic developments. While envisioning sustainable energy landscapes, it is also necessary to analyze the present energy system and to map renewable energy potentials.* The first step of the study can best be conducted by a multidisciplinary team consisting of planners, designers, energy experts, and experts from the study region.

Figure 5.4 depicts two results of the regional analysis in South Limburg—a map of the present-day energy storage and transport systems (left) and the spatial distribution of energy demand (right). Figure 5.5 depicts two of the six energy potential maps that were composed for South Limburg and, in a similar manner, for all of our studies on sustainable energy landscapes. These GIS maps require substantial efforts in data mining and processing. They are, however, of paramount importance both for the envisioning process and the communication with the commissioners and stakeholders. Studies that are limited to spreadsheets and pie chart diagrams are insufficient for the initiation of a long-term transformation process, at least at the regional scale.

---

* For more information on how to map renewable energy potentials, see, for example, Dobbelsteen et al. (2011).

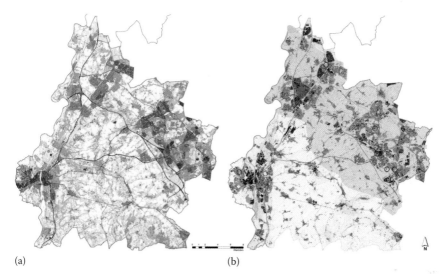

(a) (b)

FIGURE 5.5

Two of the six maps illustrating energy potentials in South Limburg: Biomass potentials (a) and underground/residual heat potentials (b). There are about 2100 km of linear landscape elements such as road verges and hedges with biomass potentials. The total land surface with biomass potential amounts to 480 km²—that is, more than two-third of the total land surface. Approximately 7% of the area has high potential for heat–cold storage, 33% good, and 60% moderate potential.

5.4.2 Mapping Near-Future Developments

The guiding question of the second step is "How will the region change in the near-future?" To answer that question, one must analyze current trends and policies, identify planned developments, and consult key decision makers in the study region. Near-future developments can be illustrated by means of a so-called *near-future base-map*. Many of the near-future developments shown on this map may not yet have left any marks in the physical environment but will influence long-term development of the region (see, e.g., Table 5.3).

5.4.3 Illustrating Possible Far-Futures

The guiding question of step three is "What kinds of possible long-term developments are expected in the study region, and at which locations?" A selection of possible far-futures can be studied by means of existing scenario studies, in the case of South Limburg a socioeconomic scenario study (Hanemaaijer et al. 2007).* Each scenario storyline needs to

---

* In many cases, context scenario studies explore four possible futures, a manageable number of scenarios that can serve as input for energy visions.

TABLE 5.3

Overview of Near-Future Land-Use Changes for the
Municipality of Margraten in South Limburg

Land-Use Type	Current Size (ha)	Relative Share (%)	Planned Change (ha)	Relative Change (%)
Forest	555	10	+142	+26
Residential area	241	4	+16	+7
Pasture	2106	36	−214	−10
Arable land	2433	42	−214	−9
Other land uses	465	8	+270	+58

Note: All numbers based on RPB (2008), Margraten (2008) and Kerkstra
et al. (2007).

be illustrated by means of a *scenario base-map*. Note that more explicit
scenario studies are easier to concretize and illustrate. The analysis of
existing context scenarios and mapping of possible far-futures can be conducted by experts but should be verified by representative stakeholders.
This is especially crucial if the resolution of the existing context scenario
study is coarse.

Figure 5.6 depicts four scenario base-maps for South Limburg. To avoid
unnecessary complexity, we decided that these scenario base-maps should
only feature long-term developments that may affect the design of sustainable energy landscapes.

5.4.4 Composing Integrated Visions

The objective in step four is to compose a set of integrated visions. Each
vision should reveal "How to turn a possible future into a desired future."
This question should be further specified to meet the objective of the
respective study, in our case, exploring possible pathways for the development of sustainable energy landscapes. Note that each vision should
explore possible pathways under the conditions established by the respective scenario. In other words, the global market energy vision should
depict interventions that are feasible, given the socioeconomic circumstances in a global market world. It is important to stress that the goal
of this "exercise" is not to render the ideal future but to reveal different pathways of reaching a desired future. In order to identify a wide
range of possible interventions, while maintaining a sense of realism, we
suggest conducting this normative step in a transdisciplinary manner.
Workshops and design charrettes can facilitate the collaboration between
experts, decision makers, and stakeholders.

Figure 5.7 depicts a set of four regional energy visions that were the result
of an intensive and iterative planning and design process. Over the course of

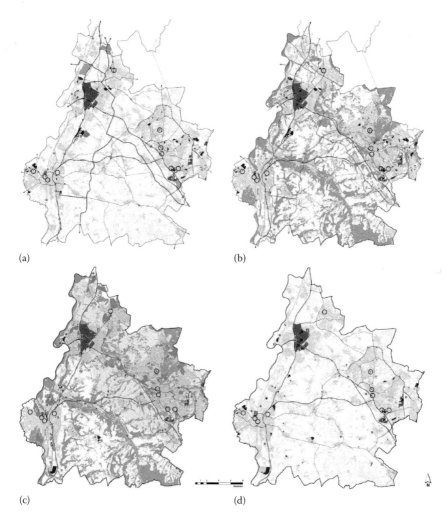

(a) (b)

(c) (d)

FIGURE 5.6
Set of four scenario base-maps illustrating possible long-term developments in South Limburg (starting left above, clockwise): global market, global solidarity, caring region, and secure region. In a "global world" (a and b), South Limburg remains connected to international energy networks. In a "regional world" (c and d), South Limburg aims to reduce energy imports. In the two "sustainability scenarios" (b and c), nature and landscape development will continue to expand whereas this is put to a hold in the two "market scenarios" (a and d).

the different projects, we realized that the possibility of alternative proposals helps to sustain a very constructive interaction between the various participants of the design charrettes. Rather than engaging in predetermined and often politically motivated discussions, participants contributed to the visions and therefore (perhaps unintentionally) to the incipience of a long-term and widely supported transition process.

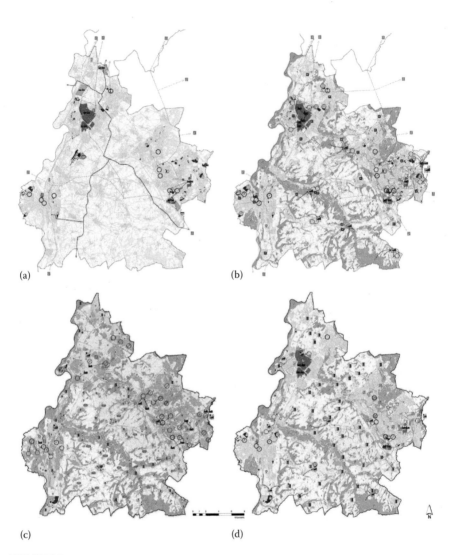

(a)   (b)

(c)   (d)

FIGURE 5.7
Set of four energy visions for the region of South Limburg (starting left above, clockwise): global market, global solidarity, caring region, and secure region. In the global market vision (a), energy-conscious interventions are limited to heat-cascading and energy-saving measure at the building scale. In the global solidarity vision (b), a part of the renewable electricity is generated making use of solid biomass; much of the remaining electricity is imported from outside the region. Extensive energy networks and the usage of all renewable energy potentials characterize the caring region vision (c). The secure region vision (d) depicts an energy system that draws on energy crops, production forest, wind parks, and hydropower in the Meuse valley.

5.4.5 Identifying Spatial Interventions

The final question that needs to be answered as part of the envisioning process is "Which possible intervention should be implemented?" First, possible energy-conscious interventions should be identified and illustrated in a comprehensive manner; maps, tables, and reference images are helpful in the discussion with decision makers.* Second, the robustness of possible interventions should be assessed, for instance, through comparative analysis of the four visions (see Table 5.4).

It is true that robust interventions can be implemented in the short term because they are less dependent on critical uncertainties (Carsjens 2009). In many situations, however, less robust interventions are necessary to perform a full transition to a desired future, for instance a 100% renewable energy supply. That is why both robust and less robust interventions need to be considered in the further decision-making and implementation process.

TABLE 5.4

List of Selected Energy-Conscious Interventions Proposed for South Limburg

Energy-Conscious Intervention	Global Market	Secure Region	Global Solidarity	Caring Region	Relative Robustness
Use of residual heat from industrial parks	7×	4×	6×	7×	4
Use of second generation biomass (maintenance)	—	50,000 ha	50,000 ha	50,000 ha	3
New district heat networks in the denser settlements	—	10×	10×	22×	3
Convert wastewater treatment plant (biogas)	—	8×	10×	8×	3
CHP[a] based on regional organic and solid waste	—	3×	1×	3×	3
CHP based on solid regional biomass	—	4×	4×	—	2
Search area for large wind park in South Limburg	—	4,400 ha	—	1,300 ha	2
Areas with heat–cold pumps (closed system)	—	>39×	—	22×	2
CHP powered by biogas from wastewater plant	—	2×	—	2×	2
Large-scale hydropower plant in the Meuse Valley	—	3×	—	3×	2

Note: The right column lists the relative robustness of interventions based on a comparative analysis of the four energy visions. The robustness of an intervention is considered high when it appears in multiple visions.

[a] CHP stands for combined heat and power plant.

---

* Please refer to Stremke et al. (2012b) for visualizations of sustainable energy landscapes.

5.5 Conclusions

The transition to renewable energy requires both strategic thinking and long-term visions. The far-future, most will agree, depends on critical uncertainties and therefore considered rather unpredictable. Yet, there is growing consensus that a range of possible futures can be explored by means of scenario studies. In this chapter, it is argued that socioeconomic scenario studies—readily available for many countries and regions—can and should be utilized in long-term spatial planning and landscape design.

Based upon literature research and design practice, the *five-step approach* has been put forward—a methodological framework that allows to integrate a selection of critical uncertainties into the design process.* The initial step is to analyze the present conditions in the study area. Today's reality is however not the only "starting point"; current projected trends and critical uncertainties are integrated as well. Consequently, a set of integrated visions can be composed. Each vision depicts how to turn a possible future (described by a context scenario) into a desired future, for example, a regional energy landscape that can be sustained on the basis of locally available renewable energy sources. The final step, for the designer, is to examine the array of possible interventions that can help reach a desired future.[†]

Employing the *five-step approach* does not necessarily lead to a single spatial plan or design in the conventional sense. Rather, it results in a set of integrated spatially explicit visions and a list of possible interventions that should be visualized.[‡] Empirical data such as the relative reduction of CO_2 emissions for each possible energy-conscious intervention can further facilitate decision making and implementation.[§]

The *five-step approach* presents a flexible framework that can be adapted to suit the objective of many studies between the municipal and regional scale. Each of the five steps of the framework can be operationalized through methods such as the 'multilayer approach' (Sijmons 2002), 'energy potential mapping' (Dobbelsteen et al. 2011) or other means presented in this book.

A final note concerns the degree of sustainability of possible interventions identified by employing the *five-step approach*. The fact that the one or the other possible intervention may be considered robust—that is, less dependent on critical uncertainties—has no significance for the actual degree of sustainability of that particular intervention. A robust intervention, in other words, may still compromise landscape quality, biodiversity, food production, or other ecosystem services. In order to identify robust and yet sustainable

---

* For a detailed elaboration of the *five-step approach* and another case study, see Stremke et al. (2012a,b).
[†] Methods to further assess possible interventions are presented, for example, in Chapter 6.
[‡] The visualization of sustainable energy landscapes is discussed in Chapter 7.
[§] The modeling of and empirical research on energy landscapes is discussed, for example, in Chapters 17 through 19.

energy-conscious interventions, it is vital that environmental designers (1) explore alternative futures together with the stakeholders, (2) make the design process transparent and comprehensible, and (3) carefully assess the impact of any proposal. A selection of innovative methods and tools to assess the sustainability of energy-conscious interventions is presented in the following chapters of this book.

Acknowledgments

This chapter presents results of the *Synergies between Regional Planning and Exergy* research project funded by NL Agency—the Dutch agency for innovation and sustainable development. I would like to thank my fellow researchers Wouter Leduc and Ronald Rovers (then Wageningen University); Andy van den Dobbelsteen, Leo Gommans, and Siebe Broersma (Delft University of Technology); Gert de Roo and Ferry Van Kann (University of Groningen); and Rob van der Krogt (TNO). I also thank my colleague Adrie van't Veer for helping with the figures.

References

Albrechts, L. 2004. Strategic (spatial) planning re-examined. *Environment and Planning B: Planning and Design* 31:743–758.

Carsjens, G. J. 2009. *Supporting Strategic Spatial Planning: Planning Support Systems for the Spatial Planning of Metropolitan Landscapes*. PhD dissertation. Wageningen, the Netherlands: Wageningen University.

Dammers, E., Evers, D., and De Vries, A. 2005. Spatial scenarios in relation to the ESDP and cohesion policy. In *Proceedings AESOP Congress 2005*. Vienna, Austria: AESOP.

Dobbelsteen, A. van den, Broersma, S., and Stremke, S. 2011. Energy potential mapping for energy-producing neighborhoods. *International Journal of Sustainable Building Technology and Urban Development* 2:170–176.

Friedmann, J. 2004. Strategic spatial planning and the longer range. *Planning Theory and Practice* 5:49–67.

Hanemaaijer, A., De Ridder, W., Aalbers, T. et al. 2007. *Nederland en een duurzame wereld: Armoede, klimaat en biodiversiteit*. Bilthoven, the Netherlands: MNP.

Healey, P. 2009. In search of the "strategic" in spatial strategy making. *Planning Theory & Practice* 10:439–457.

Kerkstra, K., Vrijland, P., De Jong, H., and Houwen, J. 2007. *Landschapsvisie Zuid-Limburg [Landscape Vision South Limburg]*. Maastricht, the Netherlands: Province Limburg and Wageningen University.

Kunzmann, K. 2000. Strategic spatial development through information and communication. In *The Revival of Strategic Spatial Planning*, eds. W. Salet and A. Faludi, pp. 53–65. Amsterdam, the Netherlands: Royal Netherlands Academy of Arts and Sciences.

Margraten. 2008. *Land-Use Plan of Margraten Municipality.* http://www.margraten.nl (July 2008).

Mintzberg, H. 1994. *The Rise and Fall of Strategic Planning: Re-Conceiving Roles for Planning, Plans, Planners.* New York: Free Press.

Rodriguez, A. and Martinez, E. 2003. Restructuring cities: Miracles and mirages in urban revitalization in Bilbao. In *The Globalized City: Economic Restructuring and Social Polarization in European Cities,* eds. F. Moulaert, A. Rodriguez, and E. Swyngedouw, pp. 181–207. Oxford, U.K.: Oxford University Press.

Rosenhead, J. 2001 Robustness analysis: Keeping your options open. In *Rational Analysis for a Problematic World Revisited,* eds. J. Rosenhead and J. Mingers, 2nd edn., pp. 181–207. Chichester, U.K.: John Wiley.

RPB. 2008. *Nieuwe Kaart van Nederland [New map of the Netherlands].* Den Haag, the Netherlands: Ruimtelijk Planbureau.

Sijmons, D. 2002. The programme guide. In *Landscape,* eds. D. Sijmons, H. Venema, and N. van Dooren, pp. 11–22. Amsterdam, the Netherlands: Architecture + Nature.

Steinitz, C. 1990. A framework for theory applicable to the education of landscape architects (and other design professionals). *Landscape Journal* 9:136–143.

Steinitz, C. 2002. On teaching ecological principles to designers. In *Ecology and Design: Frameworks for Learning,* eds. B. Johnson and K. Hill, pp. 231–244. Washington, DC: Island Press.

Steinitz, C. 2003. *Alternative Futures for Changing Landscapes: The Upper San Pedro River Basin in Arizona and Sonora.* Washington, DC: Island Press.

Stremke, S., Neven, K., Boekel, A. et al. 2012b. Integrated visions (part II): Envisioning sustainable energy landscapes. *European Planning Studies* 20:609–626.

Stremke, S., Van Kann, F., and Koh, J. 2012a. Integrated visions (part I): Methodological framework. *European Planning Studies* 20:305–319.

Weller, R. 2008. Planning by design: Landscape architectural scenarios for a rapidly growing city. *Journal of Landscape Architecture* 2:18–29.

6

Multicriteria Decision Analysis for the Planning and Design of Sustainable Energy Landscapes

Adrienne Grêt-Regamey and Ulrike Wissen Hayek

CONTENTS

6.1 Introduction

The pressure to use more renewable energies to mitigate climate change and to cope with still growing energy demands while fossil energy resources deplete brings new challenges for landscape planning (Dobbelsteen et al. 2011; Stremke and Koh 2011). A massive expansion of new infrastructures for renewable energy exploitation will modify the landscape and can affect the production of ecosystem services—the benefits people derive from ecosystems (TEEB 2010). For example, biomass production may cause negative impacts on habitats and particularly on native species (Thornley 2006). Solar energy technologies and wind turbines can affect the visual landscape aesthetic (Tsoutsos et al. 2005; Wolsink 2007), the latter potentially also reducing the recreational landscape quality by noise made by rotating turbine blades (Pedersen and Larsman 2008). In turn, the fear of these impacts obstructs the fulfillment of possible renewable energy targets (MacKay 2009).

Stremke (2010) has stressed the paramount necessity for sustainable (energy landscape) design rather than just implementing renewable technologies. A formulation of effective goals for future landscape development with the use of renewable energies is required in order to ensure sustainable management of a multifunctional landscape that supports the well-being of people (Kienast et al. 2009). Due to the absence of concepts and methods to evaluate the impacts of these infrastructures on the landscape, it is not known yet how the new energy systems can be integrated into the landscape in order to secure the provision of required ecosystem services and, at the same time, address the technical-economic requirements related to energy generation. There is a lack of approaches that allow for spatially explicit identification of suitable areas for these new elements in the landscape, recognizing the value of ecosystem services (Peters and Graumann 2006). Not only one future spatial landscape pattern may be sustainable but alternative solutions are possible (Opdam et al. 2006).

Furthermore, a tendency to top-down, technocratic planning approaches in the planning of renewable energy technology can be noticed. For example, driven by energy policy targets the Welsh Assembly Government chose an expert-dominated approach of zoning spatial energy development (Cowell 2010). Top-down approaches have been determined as a major obstacle to successful implementation, causing very slow development of renewable generation capacity in many countries. In the United Kingdom, for example, central planning problems were the most important factors interfering biomass implementation (van der Horst 2005; Wolsink 2007). Rather open, democratic decision making is necessary, which takes into account multiple views and thus allows for learning and creating perceived fairness (Szarka 2006; Wolsink 2007; Higgs et al. 2008).

Summing up, approaches for the analysis of spatial potentials for renewable energy technologies and infrastructure (RET) are required, which (1) consider the spatial potential for a mix of the different RET, (2) show different possible alternatives of exploiting the maximum capacity with an explicit formulation of the trade-offs between the systems' technical-economic requirements and the values of ecosystem services (ES), (3) integrate the relevant stakeholder knowledge and values into the evaluation, and (4) provide methods for utilizing the broad range of spatial indicators on different spatial scales in participatory spatial-planning processes (Wissen and Grêt-Regamey 2009; Wissen Hayek 2011; for further reading about participatory spatial planning see, e.g., Arnstein 1969 or Selman 2004).

In this chapter, we present a methodological framework for optimizing the integration of a mix of RET into the landscape by considering technical-economic aspects of RET and ES potentially affected by the renewable energy infrastructures. The application of the methodological framework is illustrated in a case study in Switzerland—the region of the Entlebuch, a UNESCO Biosphere Reserve. Finally, the results are discussed with regard to their contribution to developing more sustainable energy landscapes. These are landscapes adapted to locally available renewable energy exploitation while the adequate provision of other required ecosystem services is also secured (cf. Stremke 2010: 1).

6.2 Methods

Landscape development with RET needs collaborative approaches to planning that are well informed about the impacts of these facilities on ES (Kienast et al. 2009) and offer possibilities to integrate stakeholder valuations into the decision-making process (Nassauer and Corry 2004). Figure 6.1 presents the decision-support system for evaluating regional potentials for a sustainable mix of RET.

First, the potential locations for different RET, such as thermal solar and photovoltaic systems, hydropower, wood and moist biomass, and wind turbines, are mapped by overlaying the results from an analysis of the general technical-economic spatial potential for the individual RET. This step is followed by the spatially explicit quantification of selected ES quantified using

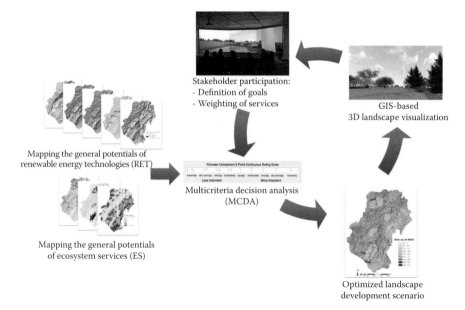

FIGURE 6.1
Methodological framework for an optimized landscape impact assessment of renewable energy.

established methods (e.g., Grêt-Regamey et al. 2008). Not all services supplied by the ecosystems have to be assessed but a wide range of values with a focus on criteria particularly important for the local stakeholders (Meyer et al. 2008). Ecosystem services that might be impacted by the new infrastructures include timber, food, erosion prevention, habitat, enjoyment of scenery, nonrecreational appreciation of landscape features, travel to natural ecosystems for ecotourism, heritage value, and of course raw materials for energy generation (for an overview on ecosystem services, see, for example, TEEB 2010).

The mapped regional potentials of energy generation and provision of ecosystem services are then input to a multicriteria decision analysis. A trade-off key is developed that integrates goals with regard to the level of energy and service provision into the simulation and thus allows for modeling optimal location potentials for RET depending on the set goals. The results of this analysis are spatially explicit maps of land-use change combinations that differ by the amount and allocation of RET according to the balancing of values.

The valuation of the ecosystem services—the "weighting" of the criteria that influence the evaluation of the potentials—is however hard to define. It is often arbitrary and does not necessarily reflect wider public opinion or preferences (Higgs et al. 2008). To overcome this problem, stakeholder feedback has to be integrated (Park et al., 2004).

We propose to use the abbreviated pairwise comparison method (Stranger and Rosenberger 2006) to assign the importance of each criterion relative to other criteria. Paired comparison methods are well-developed methods of ordering attributes or characteristics of a given set of items (Saaty 1987a) and applied in many studies for spatial planning (Kline and Wilchems 1996, 1998; Alho and Kangas, 1997; Duke and Aull-Hyde 2002). In the abbreviated format, not all possible pairings of the criteria are presented to the participant, allowing comparison of all criteria. Pairs are sequentially assigned as A–B, B–C, C–D, etc., where the initial criterion and the second criterion in each subsequent pair are randomly assigned. A complete ranking of criteria is based on the actual choices made and assuming transitive preferences. Peterson and Brown (1998) show that people are transitive in their preferences revealed through a method of paired comparison. Consistency ratios, as measures of consistent (transitive) preferences, are redundant when transitivity is assumed as in the case of the abbreviated pairwise comparison format.

Ideally, a stakeholder group of nonacademic stakeholders (society) and scientific stakeholders (experts) should be brought in a transdisciplinary dialogue in this mutual learning and knowledge integration process (Scholz 2011). In order to efficiently communicate the resulting quantitative spatially explicit information to the stakeholders, we implement (Geographic Information System) GIS-based three-dimensional (3D) visualizations of the possible energy landscapes (Wissen Hayek 2011). These are linked to corresponding indicators of the energy generation and the provision of ecosystem services to facilitate a stakeholder-based definition of optimization goals and the weighting of the different services. Overall, this approach allows for an iterative optimization of the possible future landscape patterns based on the analysis and adaptation of alternative goal sets, which finally can result in exploitation potentials of a region's renewable energy resources broadly accepted by the stakeholders.

In the following sections, it is demonstrated for a case study region, the Entlebuch UNESCO Biosphere Reserve, how to (1) map possible locations of RET based on technical-economic criteria with GIS analysis, (2) quantify the value of relevant ES, and (3) use resulting maps of spatial potentials as input for a multicriteria decision analysis. Furthermore, it is explained how to determine possible trade-offs between the different economic and ES services implementing GIS-based 3D visualizations of alternative energy landscape development scenarios. Please note that the focus in this chapter is on the quantitative modeling approach, while the qualitative, participatory approach has not been implemented.

6.2.1 Case Study Area

The Entlebuch UNESCO Biosphere Reserve (http://www.biosphaere.ch) in the Canton of Lucerne comprises the main valley between Lucerne and Berne in central Switzerland with an area of 395 sq km and an altitude that ranges from 590 to 2350 m above sea level (Figure 6.2). Its cultural landscape is of international significance because it contains important habitats for plants and animals, for example, karst areas, forests, and unique mire landscapes (Figure 6.3).

FIGURE 6.2
Location of the Entlebuch UNESCO Biosphere Reserve in central Switzerland (approximate scale 1:2.500.000).

(a)

(b)

(c)

FIGURE 6.3
Impressions of the Entlebuch's cultural landscape: (a) Village of Escholzmatt, (b) Mire landscape
and characteristic karst mountain, and (c) Skiing area in the winter. (Photos by J. Egli, 2004.)

Approximately 17,000 inhabitants live in this prealpine region, which is shaped
mainly by agriculture and forestry. Agriculture and tourism each employ about
one-third of the working population.

As a UNESCO Biosphere Reserve, Entlebuch serves as a model region where
sustainable development concepts are elaborated in participatory planning
processes. Particularly notable are Entlebuch's established and sophisticated
participation structures, including collaborative planning forums on topics such as business and industry, tourism, agriculture, energy, wood, and
education. Participation in these forums is characterized by shared power,
group learning, transparency, and a consensus-oriented style of communication (Schroth et al. 2011).

As other regions, too, the Entlebuch is challenged by balancing the often conflicting requirements of sustainable socioeconomic development while developing new markets, such as the generation of renewable energy, strengthening the multifunctionality of agriculture, and sustaining the landscape services for the public, including preservation of landscape beauty and conservation of biodiversity. Thus, it offers an ideal case for implementing the suggested approach.

6.2.2 Mapping the General RET Potential

The first step was to map the maximum spatial capacity for renewable energy exploitation in the case study area. The conditions of the region Entlebuch offer generally major potentials for the exploitation of wind, solar, wood, and moist biomass (manure for biogas generation), hydropower, and geothermal power (heat pumps). The latter two were not included in this study, and we focused on the other terrestrial RET to demonstrate the approach exemplarily.

Rather than calculating and assigning general potentials to a political unit (see, for example, Ramechandra and Shruthi 2007), we calculated the spatially explicit potential of these RET using ESRIS's ArcGIS ModelBuilder generating for each type of RET a GIS-based model. These models were applied on a GIS data base including a digital terrain model (DTM) with a resolution of 25 m, the land cover/use data of the status quo on a scale of 1:25,000, current land-use regulation maps (national and local inventories of nature protected areas), and specific additional data according to the respective RET's technical-economic basic requirements, for example, data on the wind speed 100 m above ground, suitability maps regarding soil and climate, or spatially explicit harvesting costs of wood. Figure 6.4 gives an

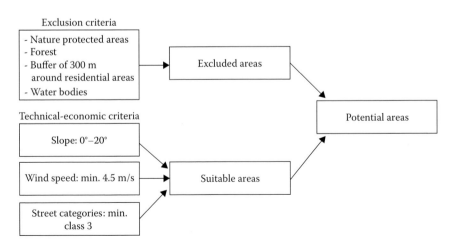

FIGURE 6.4
Model for calculating the spatially explicit potential areas for wind turbines in the case study area.

TABLE 6.1

Power that can be produced per square meter by the individual
renewable energy sources at the identified locations in the
Entlebuch region

Renewable Energy Source	Rated Power (kWh/m² per Year)
Wind energy[a]	41,700
Solar energy[b]	400
Wood as energy source[c]	0.26–1.30
Moist biomass[d]	0.07

[a] With a wind turbine requiring an area of 10 m² and producing 4.17 GWh
   per year. This calculation only considers the physical footprint of the wind
   turbine and excludes the minimum distances between turbines and
   between them and other land uses.
[b] BE Netz AG: http://www.benetz.ch/docs/31/273_1098_de.pdf (accessed
   November 24, 2011).
[c] Jungbluth, N., Frischknecht, R., and Faist, M. 2002. Ökobilanz für die
   Stromerzeugung aus Holzbrennstoffen und Altholz. Forschungs- und
   P + D Programm Biomasse des Bundesamtes für Energie, Bern.
[d] Ernst Basler + Partner AG 2006. Evaluation von Standortregionen für
   Biomasseanlagen im CKW-Gebiet.

example of the model used for calculating the areas for potential wind turbines from a technical-economic perspective (suitable areas) and according to current land-use regulations (excluded areas). Based on the mapped
potentials for each renewable energy source, the potential energy that can
be produced per square meter was calculated (Table 6.1).

6.2.3 Quantification and Valuation of Ecosystem Services

Relevant ES in the Entlebuch are the capacities for agricultural production in yield per hectare (CHF/ha per year) and the attractiveness of the
landscape for ecotourism and recreation (sensitivity of the close up and
distant view with regard to the enjoyment of scenery) as these are the
dominant sources of income in the region. Furthermore, the Entlebuch has
the responsibility to maintain environmental conditions supporting biodiversity, in particular, preserving species of mire landscapes. The latter
service was described by the provision of habitats with sufficient quality
for target species.

The quantification of these ES was carried out using GIS-based process models (Grêt-Regamey et al. 2008). For the valuation, agricultural
production was directly assessed in monetary units according to the
market price for these goods. In this study, we used the mean income of
the farms in the Canton of Lucerne provided by AGRO-Treuhand Sursee
(http://www.atsursee.ch) and distributed the amount on the average
agricultural area of a farm in the region Entlebuch (16 ha). Thereby, it was

distinguished between the incomes of farms in two mountain zones. The operating costs in the hilly zone are smaller (*Bergzone I*) than in the mountainous zone (*Bergzone II–IV*). The delineation of the zones was given as GIS data.

The quantification of services such as landscape aesthetics and habitat provision required different approaches. In order to identify areas that the public values as "sensitive" with regard to landscape aesthetics, the participatory instrument of a "cognimap" was used. This cognimap was already available for the region Entlebuch and was established during an exhibition in the set-up phase of the UNESCO Biosphere Reserve. All visitors were asked to delineate those areas that particularly contribute to their identification with the region.* According to the frequency of an area's selection, the sensitivity was assigned; this means, the more often it was indicated by different people, the higher the sensitivity. This map was spatially joined with buffers of 150 m along hiking paths for assessing the sensitivity of the close-up view. The visible areas were assigned the sensitivity values of the cognimap, and the remaining areas were assigned the value zero. Furthermore, since wind turbines are visible over far distances, a viewshed analysis was carried out mapping all areas in which wind turbines on the mapped possible locations (cf. Section 6.2.3) are visible, and then spatially joined with the cognimap. The resulting map provides the degree of impact on landscape aesthetics when implementing the potential wind turbines (Figure 6.5).

The quality of the different habitat types available in the region Entlebuch was ranked in an interview with a local expert (ideally, this step should be done with a group of experts). These values were then assigned to the attribute tables of the available GIS-data of the habitat inventories and land uses.

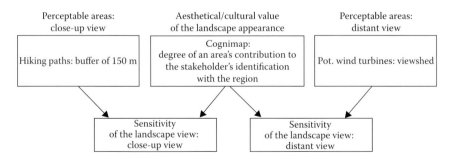

FIGURE 6.5
Model for calculating the spatially explicit degree of the landscape view's sensitivity in the case study area.

---

* Communication with F. Knaus, scientific coordinator of the Entlebuch UNESCO Biosphere Reserve.

6.2.4 Multicriteria Decision Analysis

The maps for potential RET and the impacts on ES were integrated into a multicriteria decision analysis (MCDA). Different methods are available for weighting the values in MCDAs, such as ranking and rating methods, simple additive weighting, trade-off analysis method, or the analytical hierarch process (AHP) (Malczewski 1999; Joerin et al. 2001). We chose the AHP method since it offers a transparent and comprehensive approach. Each criterion is weighted and compared pair wise with the others to compute a best-fit set of weights (Saaty 1987b). We used the software IDRISI Taiga (http://www.clarklabs.org) for conducting the MCDA.

The overall goal of the MCDA was to identify a land-use pattern that allows for highest possible energy output with lowest impact on the landscape (Figure 6.6). According to this goal, two submodels were defined. The first model quantifies the potential for ES provision. The second model computes the potential energy output per unit considering the spatially explicit potentials for the alternative renewable energy sources. The results of both models are raster data containing for each cell information on the suitability for the alternative land uses. In combining the results of these two models, locations with the highest energy output and the lowest impact on the landscape are assessed.

The suitability for the resulting land-use patterns can be modified by changing the weight given to the different services. In this study, we assigned equal weight to all services in models one and two and in the combination of these two models' results. Changing the weighting of the different factors would result in alternative spatially explicit scenarios of land-use patterns.

6.2.5 GIS-Based 3D Visualization of Alternative Scenarios

Effective and structured communication of the outcomes of a multicriteria analysis to stakeholders is crucial for collaborative planning (de Groot 2006). In the context of participatory landscape planning, GIS-based virtual landscapes

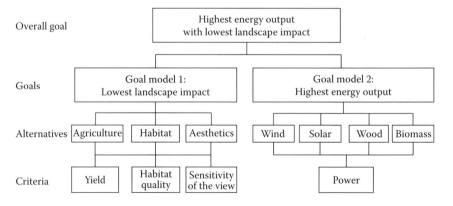

FIGURE 6.6
Hierarchy of the multicriteria decision-making model applied in the case study region.

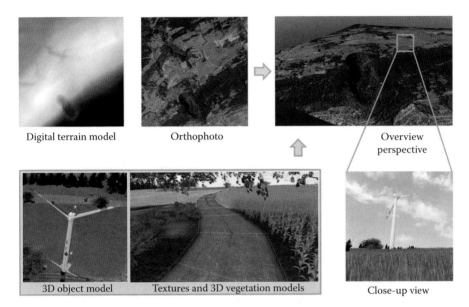

Digital terrain model   Orthophoto   Overview perspective

3D object model   Textures and 3D vegetation models   Close-up view

FIGURE 6.7
General workflow of GIS-based 3D landscape visualization. This example shows the development of computer-generated representations of perspective landscape views of the wind park at Mont Crosin (CH). First, a DTM providing a 3D representation of the topography and GIS vector data with the information of the land cover/use are imported. Second, 2D corresponding aerial imagery are draped over the DTM. Third, specific textures and geometries of natural and man-made structures are added, raising the level of detail of the landscape representation. The digital 3D model allows changing the perspective from overview to close-up views.

have proved to be the media that provide a common communication basis and support mutual concept development (Hehl-Lange and Lange 2005; Wissen Hayek 2011). Therefore, the resulting alternative suitability maps of exploiting the regional renewable energy resources were prepared as detailed 3D visualizations using the software CryEngine (http://mycryengine.com) (Figure 6.7).

6.3 Results

In the following, the results of the Entlebuch's computed RET potential and ES provision maps as well as the scenarios resulting from the MCDA are presented. Furthermore, we demonstrate a possible setting for participatory workshops with 3D landscape visualizations of the scenarios.

6.3.1 Mapped Renewable Energy Potentials

The maps in Figure 6.8 depict the spatial potentials for the RET considering technical-economic criteria: The region Entlebuch provides about 16 locations suitable for wind turbines; an overall wind energy potential of 60 GWh

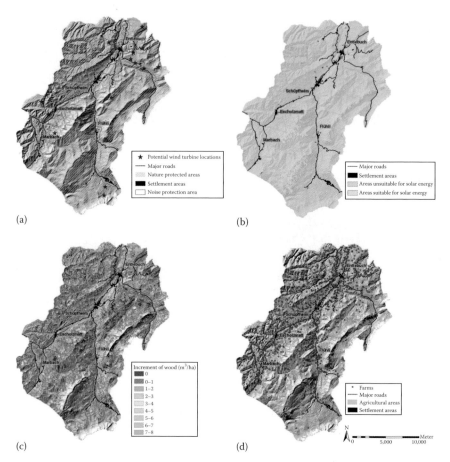

FIGURE 6.8
Potential areas for (a) wind energy, (b) solar energy, (c) wood energy, and (d) moist biomass energy.

per year. A total area of 1087 square meter of building roofs* can be exploited for solar energy, resulting in a total of 435 MWh of thermal energy per year. The wood utilization potential of the predominant spruce forests was determined by the increment of wood suitable for energy provision (m³/ha), resulting in a potential of about 24 MWh per year.[†] The potential of energy from moist biomass (manure) adds up to 210 kton or 12 MWh per year. Since

---

* In Switzerland, it is not yet allowed to build solar farms on open spaces (except test farms for research and development purposes) but on roofs only. The amount of suitable roof area in potential areas for solar energy approximate 0.1% of the generally utilizable total roof area that was assumed to be 50% of the total roof area in the case study region.

[†] Note that the total growth of wood is based on 112%, since official growth rates do not take into account the percentage of branches, which are relevant for energy wood exploitation. The branches approximate 12% of the total growth of wood (Hofer and Altwegg 2008).

for the production of moist biomass the agricultural areas for dairy farming are required, the potential was evenly distributed across these areas.*

6.3.2 Mapped Potentials of Ecosystem Services' Provision

Figure 6.9 shows the provision of the three selected ES in the case study region, namely, agriculture production, habitat, and landscape aesthetics. With regard to agricultural production, the results show that the region comprises areas in three zones providing different yields: the hilly zone has a yield potential of about CHF 3757 per hectare per year, the mountainous zone of CHF 3346 per hectare per year, and the zone

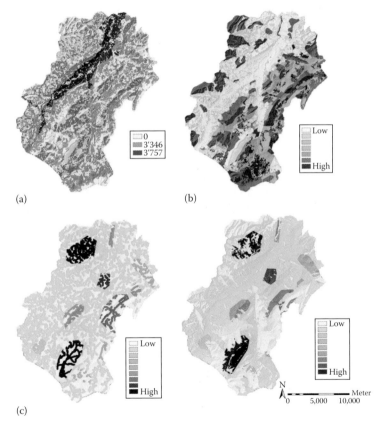

FIGURE 6.9
Potential of ecosystem services provision: (a) yield per hectare, (b) habitat quality, and (c) sensitivity of the landscape view (left: close-up view; right: impact on the distant view by potential wind turbines).

---

* Note that for calculating the amount of moist biomass, a simplified model was implemented, which does not take into account times when cattle are grazing on pastures and no moist biomass can be collected.

of nonagricultural areas has no yield potential. The habitat quality map indicates in dark colors those areas with high quality of living spaces and reproduction habitats for target species. The sensitivity of the aesthetical landscape appearance is given in the lower two maps. Regarding the close-up perspective, well-developed areas with hiking paths and highly valued by the public are assigned higher quality than less-developed areas. The results of the overlay of the viewshed analysis of the 16 potential locations for wind turbines with the landscape sensitivity map give the potential impact on the distant view. The more sensitive a visible area is, the higher is the potential impact of RET on landscape aesthetics.

6.3.3 Scenarios Resulting from Multicriteria Decision Analysis

Figure 6.10 presents the results of the MCDA model 1 rendering the lowest landscape impact. Raster cells with high values are more suitable

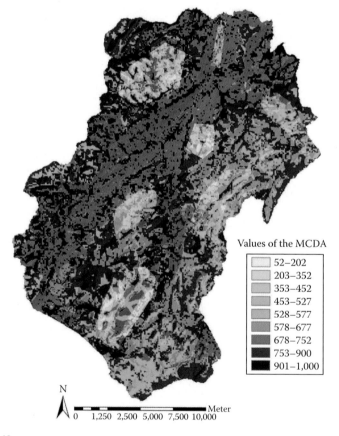

FIGURE 6.10
Result MCDA model 1 with goal lowest landscape impact. The higher the value, the smaller the impact on the selected ecosystem services.

for the implementation of RET. Areas with lower values provide higher potentials for ES provision and would cause higher land-use conflicts than the other cells.

The resulting map of model 2, with the goal of highest energy output, shows highest values on potential areas for wind and solar energy since these provide highest energy output per unit area.* Combining the results of these two models in the MCDA with the objective of highest energy output with lowest impact on the landscape resulted in the map given in Figure 6.11. The higher the computed value, the better the target is met. Most of the potential areas for wind energy got highest values, as, for example, the red cell in

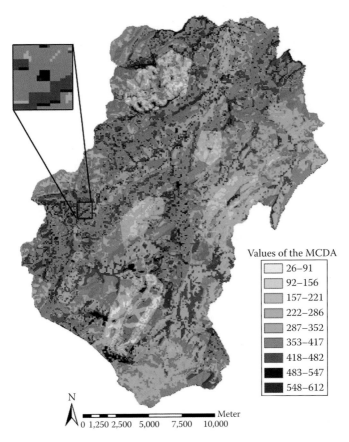

Values of the MCDA

☐ 26–91
☐ 92–156
☐ 157–221
☐ 222–286
☐ 287–352
☐ 353–417
■ 418–482
■ 483–547
■ 548–612

N

Meter
0 1,250 2,500   5,000   7,500   10,000

FIGURE 6.11
Result of the MCDA aiming at highest renewable energy output for a mix of wind, solar, wood, and moist biomass energy and lowest landscape impact considering yield, landscape aesthetics, and habitat quality. The higher the values, the better the goals are met. The 25 m × 25 m raster cell fulfilling the criteria best is a wind turbine site.

---

* This map, similar to some other results, is not presented here due to lack of space.

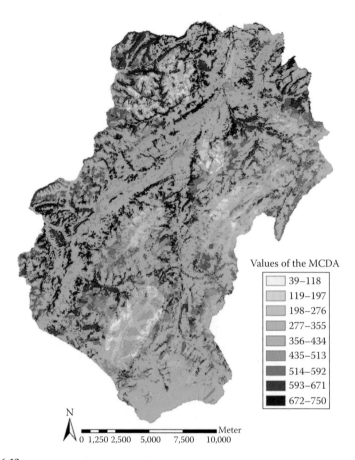

Values of the MCDA

☐	39–118
☐	119–197
☐	198–276
☐	277–355
☐	356–434
☐	435–513
☐	514–592
☐	593–671
■	672–750

N

0 1,250 2,500   5,000   7,500   10,000 Meter

FIGURE 6.12

Result of the MCDA aiming at highest energy output only taking into account wood and moist biomass as well as lowest landscape impact considering yield, landscape aesthetics, and habitat quality. The higher the values, the better the targets are met.

the enlarged map section. However, some potential areas for wind energy achieved only medium values.

Including only the potentials for wood and moist biomass in the MCDA resulted in an alternative scenario (Figure 6.12). Again, the highest energy output with lowest impact on the landscape is fulfilled by the raster cells with highest values, which now are predominantly forests in low altitudes.

6.3.4 GIS-Based 3D Landscape Visualization for Participatory Decision Making

The rather abstract results of the MCDA have to be interpreted collaboratively with diverse stakeholders in order to support the development of sustainable energy landscapes. In Figure 6.13, we demonstrate how GIS-based 3D landscape visualizations of the RET scenarios could be used as suitable stimulus

FIGURE 6.13
Possible participatory workshop setting implementing GIS-based 3D visualization of alternative landscape developments with renewable energy technologies.

for an evaluation by individuals or in groups of stakeholders. Linking the 3D visualizations with the quantitative indicators describing the provision of energy as well as the provision of ES can facilitate weighting the different services and assessing the societal acceptance of the scenarios (Bateman et al. 2009; Meyerhoff et al. 2010).

6.4 Discussion and Conclusions

The presented approach contributes to enhancing methods and instruments for participative processes on sustainable landscape development with a special focus on integrating RET into the landscape. In particular, it fosters the consideration of ES demand and supply in the negotiation process on land-use change in order to provide a better, more informed, and comprehensible decision basis. The modeling approach is principally applicable also to other spatial planning and landscape design tasks, for example, integration of pylons in the landscape. Further, from a methodological perspective, the method provides a framework for inverse modeling approaches enhancing the spatial explicit analysis of sustainable land-use potentials by planning from a future vision (Grêt-Regamey and Crespo 2011; Crespo and Grêt-Regamey 2012).

The exemplary implementation of the presented here approach for the Entlebuch region proves its applicability in heterogeneous landscapes. The GIS models for quantifying spatial potentials for both the provision of ES as well as for the RET are simplified and could be further developed. Further parameters determining the utilization limitations of RET, for instance, "maximum system size" depending on energy quantity and quality (Stremke and Koh 2011), should be integrated. Moreover, possible changes of regional conditions relevant for locating RET should be considered by using alternative regional development scenarios and expanding them with the aspects of renewable energy exploitation. Thus, the analysis of possible landscape patterns takes place in a richer context, taking into account the uncertainties (Kok et al. 2011).

The limits of ecosystems to provide certain services are included in the optimization model as restrictions and define the boundaries of the optimization. In our study, all services were weighted equally. Stakeholder knowledge and experiences should be integrated into both the selection of relevant services and the weighting of their importance in order to reflect societal values and preferences for landscape development. Using the capacity of GIS-based 3D landscape, visualizations linked to quantitative indicators and spatially explicit site optimization models provide new powerful means for such integrated and collaborative landscape impact assessments. Furthermore, in order to enhance the suitability of possible scenarios, a fuzzy logic system can be implemented in the MCDA, providing human-like reasoning capabilities in determining optimal locations (see, e.g., Bittermann 2009).

Core contention of critics against a massive implementation of RET into the landscape is that it entails economic, social, and environmental costs (Szarka 2006). The integration of ES in the decision-making process for determining regional potentials for RET using an MCDA approach takes these costs partially into account. Furthermore, the proposed framework provides a systematic approach that strongly incorporates the societal dimension into the quantitative simulation using GIS-based 3D visualizations.

According to the definition proposed in this book, sustainable energy landscapes shall be sustained on the basis of locally available renewable energy sources without compromising landscape quality, biodiversity, food production, and other ecosystem services. Our approach can support strategy development on regional level answering the question if and how particular landscapes should be transformed into energy landscapes. To this end, it provides a spatial explicit overview on the general potentials for provision of different ecosystem services. If the option of developing an energy landscape is supported, the iterative optimization modeling and visualization of alternative landscape development scenarios are a sound planning and design basis. It enables stakeholders to make informed trade-off decisions on the agreed maximum exploitation of their landscape's renewable energy sources and desired quality of the other ecosystem services. To select a sound compromise is the objective of a participative negotiation process. The outlined proactive approach can significantly reduce the environmental impact of RET and enhance their public acceptability, supporting the development of an optimized, more sustainable energy landscape.

Acknowledgments

We thank the Canton of Lucerne, Switzerland, for providing digital data from their geographical information system and key stakeholders of the Entlebuch UNESCO Biosphere Reserve for providing valuable information required for calculating the regional spatial potentials. Special thanks go to Korintha Bärtsch, who applied the presented approach in her master's thesis.

References

Alho, J. M. and Kangas, J. 1997. Analyzing uncertainties in experts' opinions of forest plan performance. *Forest Science* 43/4: 521–527.

Arnstein, S. R. 1969. A ladder of citizen participation. *Journal of the American Planning Association* 35/4: 216–224.

Bateman, I. J., Day, B. H., Jones, A. P., and Jude, S. 2009. Reducing gain-loss asymmetry: A virtual reality choice experiment valuing land use change. *Journal of Environmental Economics and Management* 58: 106–118.

Bittermann, M. S. 2009. *Intelligent Design Objects (IDO)—A Cognitive Approach for Performance-Based Design*. PhD thesis, Department of Building Technology, Delft University of Technology, Delft, the Netherlands.

Cowell, R. 2010. Wind power, landscape and strategic, spatial planning—The construction of 'acceptable locations' in Wales. *Land Use Policy* 27: 222–232.

Crespo, R. and Grêt-Regamey, A. 2012. Spatially explicit inverse modeling for urban planning. *Applied Geography* 34: 47–56.

Dobbelsteen, A., van den Broersma, S., and Stremke, S. 2011. Energy potential mapping for energy-producing neighborhoods. *SUSB Journal* 2/2: 170–176.

Duke, J. and Aull-Hyde, R. 2002. Identifying public preferences for land preservation using the analytic hierarchy process. *Ecological Economics* 42: 131–145.

Grêt-Regamey, A., Bebi, P., Bishop, I. D., and Schmid, W. 2008. Linking GIS-based models to value ecosystem services in an Alpine region. *Journal of Environmental Management* 89: 197–208.

Grêt-Regamey, A. and Crespo, R. 2011. Planning from a future vision: Inverse modeling in spatial planning, *Environment and Planning B* 38: 979–994.

de Groot, R. S. 2006. Function-analysis and valuation as a tool to assess land use conflicts in planning for sustainable, multi-functional landscapes. *Landscape and Urban Planning* 75: 175–186.

Hehl-Lange, S. and Lange, E. 2005. Ein partizipativer Planungsansatz für ein Windenergieprojekt mit Hilfe eines virtuellen Landschaftsmodells. *Natur und Landschaft* 80/4: 148–153.

Higgs, G., Berry, R., Kidner, D., and Langford, M. 2008. Using IT approaches to promote public participation in renewable energy planning: Prospects and challenges. *Land Use Policy* 25: 596–607.

Hofer, P. and Altwegg, J. 2008. *Holz Nutzungspotentiale im Schweizer Wald auf Basis LFI3*. Zürich, Switzerland: Bundesamt für Umwelt (BAFU). http://www.news.admin.ch/NSBSubscriber/message/attachments/10954.pdf (accessed May 21, 2012).

van der Horst, D. 2005. UK biomass energy since 1990: The mismatch between project types and policy objectives. *Energy Policy* 33: 705–716.

Joerin, F., Thériault, M., and Musy, A. 2001. Using GIS and outranking multicriteria analysis of land-use suitability assessment. *International Journal of Geographical Information Science* 15: 153–174.

Kienast, F., Bolliger, J., Potschin, M., de Groot, R. S., Verburg, P., Heller, I., Wascher, D., and Haines-Young, R. 2009. Assessing landscape functions with broad-scale environmental data: Insights gained from a prototype development for Europe. *Environmental Management* 44: 1099–1120.

Kline, J. and Wilchems, D. 1996. Public preferences regarding the goals of farmland preservation programs. *Land Economics* 72: 538–549.

Kline, J. and Wilchems, D. 1998. Measuring heterogeneous preferences for preserving farmland and open space. *Ecological Economics* 26: 211–224.

Kok, K., van Vliet, M., Bärlund, I., Dubel, A., and Sendzimir, J. 2011. Combining participative backcasting and exploratory scenario development: Experiences from the SCENES project. *Technological Forecasting and Social Change* 78/5: 835–851.

MacKay, D. J. C. 2009. *Sustainable Energy—Without the Hot Air*. Cambridge, U.K.: UIT Cambridge.

Malczewski, J. 1999. *GIS and Multicriteria Decision Analysis*. New York: John Wiley & Sons.

Meyer, B. C., Phillips, A., and Annett, S. 2008. Optimising rural land health: From landscape policy to community land use decision-making. *Landscape Research* 33/2: 181–196.

Meyerhoff, J., Ohl, C., and Hartje, V. 2010. Landscape externalities from onshore wind power. *Energy Policy* 38: 82–92.

Nassauer, J. I. and Corry, R. C. 2004. Using normative scenarios in landscape ecology. *Landscape Ecology* 19: 343–356.

Opdam, P., Steingröver, E., Rooij, and van, S. 2006. Ecological networks: A spatial concept for multi-actor planning of sustainable landscapes. *Landscape and Urban Planning* 75: 322–332.

Park, J. R., Stabler, M. J., Mortimer, S. R., Jones, P. J., Ansell, D. J., and Parker, G. P. D. 2004. The use of a multiple criteria decision analysis to evaluate the effectiveness of landscape and habitat enhancement mechanisms: An example from the South Downs. *Journal of Environmental Planning and Management* 47/5: 773–793.

Pedersen, E. and Larsman, P. 2008. The impact of visual factors on noise annoyance among people living in the vicinity of wind turbines. *Journal of Environmental Psychology* 28: 379–389.

Peters, J. and Graumann, U. 2006. Regenerative Energien und Kulturlandschaft. Chancen für Schutz und Entwicklung von Kulturlandschaften durch den Ausbau erneuerbarer Energien. *Stadt + Grün* 12: 48–53.

Peterson, G. L. and Brown, T. C. 1998. Economic valuation by the method of paired comparison, with emphasis on evaluation of the transitivity axiom. *Land Economics* 74/2: 240–261.

Ramachandra, T. V. and Shrithi, B. V. 2007. Spatial mapping of renewable energy potential. *Renewable and Sustainable Energy Reviews* 11: 1460–1480.

Saaty, R. W. 1987b. The analytic hierarchy process—what it is and how it is used. *Mathematical Modelling* 9/3–5: 161–176.

Saaty, T. L. 1987a. Concepts, theory, and techniques rank generation, preservation, and reversal in the analytic hierarchy decision process. *Decision Sciences* 18: 157–177.

Scholz, R. W. 2011. *Environmental Literacy in Science and Society: From Knowledge to Decisions.* Cambridge, U.K.: Cambridge University Press.

Schroth, O., Wissen Hayek, U., Lange, E., Sheppard, S., and Schmid, W. A. 2011. A multiple-case study of landscape visualizations as a tool in transdisciplinary planning workshops. *Landscape Journal* 30/1: 53–71.

Selman, P. 2004. Community participation in the planning and management of cultural landscapes. *Journal of Environmental Planning and Management* 47/3: 365–392.

Stranger, M. P. and Rosenberger R. S. 2006. Incorporating stakeholder preferences for land conservation: Weights and measures in spatial MCA. *Ecological Economics* 57/4: 627–639.

Stremke, S. 2010. *Designing Sustainable Energy Landscapes: Concepts, Principles and Procedures.* PhD thesis, Wageningen (NL), the Netherlands: Wageningen University.

Stremke, S. and Koh, J. 2011. Integration of ecological and thermodynamic concepts in the design of sustainable energy landscapes. *Landscape Journal* 30: 2–11.

Szarka, J. 2006. Wind power, policy learning and paradigm change. *Energy Policy* 24: 3041–3048.

TEEB 2010. The economics of ecosystems and biodiversity: Integrating the ecological and economic dimensions in biodiversity and ecosystem service valuation. www.teebweb.org (accessed November 24, 2011).

Thornley, P. 2006. Increasing biomass based power generation in the UK. *Energy Policy* 34: 2087–2099.

Tsoutsos, T., Frantzeskaki, N., and Gekas, V. 2005. Environmental impacts from the solar energy technologies. *Energy Policy* 33: 289–229.

Wissen, U. and Grêt-Regamey, A. 2009. Advanced analysis of spatial multi-functionality to determine regional potentials for renewable energies. In Schrenk, M., Popovich, V. V., Engelke, D., and Elisei, P. *Real Corp 2009* 15–21. http://www.corp.at/archive/CORP2009_37.pdf (accessed May 21, 2012).

Wissen Hayek, U. 2011. Which is the appropriate 3D visualization type for participatory landscape planning workshops? A portfolio of their effectiveness. *Environment and Planning B* 38: 921–939.

Wolsink, M. 2007. Planning of renewables schemes: Deliberative and fair decision-making on landscape issues instead of reproachful accusations of non-cooperation. *Energy Policy* 35: 2692–2704.

7

Energy Landscape Visualization: Scientific Quality and Social Responsibility of a Powerful Tool

Sören Schöbel, Andreas R. Dittrich, and Daniel Czechowski

CONTENTS

7.1 Introduction

Landscape architects generally produce effective landscape images of views or plans—visualizations—that today are almost exclusively generated with the help of computer programs. Software tools have been produced in the last few years that have opened up entirely new dimensions of objectivity and performance in the understanding, development, and communication of space—both in design practice and in research.

Notwithstanding the enthusiasm for "all the things machines can now do," it is fair to say that visualizations, despite greatly advanced technology, are still comparatively simple graphics that rely on a few limited aspects of landscape representation. *Renderings* in particular do not really add to the imagination of an experienced landscape architect. But it are precisely these visualizations that prove so much more effective in communication, compared to other media, presumably because laypeople tend to associate suggestive pictorial effects with the reputed superior objectivity of machine-produced information. By contrast, maps generated through a process of research by design—*mappings*—are less suited for public communication because they are more abstract, although they are currently extremely fashionable in professional circles.

Two groups are distinguishable in research and in "research-led practice": "modelers" or "designers." While the former use computing power to identify mechanisms of action within complex spatial structures, the latter seek to expose and consolidate complexity and to open it up for different interpretations.

When viewed as a whole, it becomes clear that landscape visualizations may be least suited to do what the available software promises, which is to rationalize design processes. *Renderings* are more likely to serve communication, public debate, and political decision making, whereas *mappings* open up new experimental fields in research and practice. The importance of a flexible communication was also recognized by the software industry, whose visualization tools in recent years have focused on intuitive, even experimental, input (e.g., arc*sketch*), on the one hand, and on renderings that are particularly realistic (e.g., arc*scene*), on the other. Experimental mapping, otherwise, is supported by a range of data interfaces that allow including pixel or qualitative data. Chiefly, these tools prioritize the opening of human–machine interfaces rather than the "objectification" of a design.

In the following, we argue that visualization tools do not receive design legitimization by means of their technology, but by the mode of their application. For this reason, we will adopt a doubly critical stance in our consideration of designs by landscape architects of sustainable energy landscapes as we ask the question, what distinguishes different types of computer-aided visualization and what potentials do they offer both in design practice and in academic work on sustainable landscapes?

7.1.1 Renewable Energies, Sustainable Landscape, and Landscape Architecture

Renewable energies represent a predominantly new form of land use (see also Chapter 2). Developed industrial nations, in the past, tended to relocate their power generation to often far-flung regions where there is a high concentration of fossil fuels, thereby "exempting" their own "native" landscapes from the necessity of providing energy. Oil in particular has at the same time facilitated complete industrialization of the native landscapes by means of very productive, though extremely energy-intensive, agricultural methods and a high degree of mobility and increased spatial access (cf. Sieferle 1997). This changed the flows of materials and energy that had been present in landscapes for centuries; cities and settlements became disconnected from their surroundings in terms of energy supplies. This has led to the homogenization and depletion of landscape structures in both the remote regions where energy is extracted (i.e., fossil fuels) and in the "native" landscapes.

A closer examination of the ongoing energy transition reveals a "return" of energy generation to quotidian landscapes, to the immediate living environments of many Western societies. This is where landscape development has reached a crossroad; as the renewable energy sector expands, it can either continue along the current industrial logic whereby this "new" land use has a homogenizing effect or it can create unique landscapes with great diversity and relations to existing landscape patterns—the latter is what we call sustainable energy landscapes. We therefore maintain that landscape architects are needed for this—the development of sustainable energy landscapes—and not merely for the visualization of renewable energy landscapes.

Hence, we have the first criterion for a legitimate application of computer-generated visualizations in design and planning practice; visualizations should contribute to depicting landscape diversity and interrelationships both as quality and design objectives. In order to achieve this, they need to be sufficiently complex and, based on the complex interrelationships, develop a new quality—in other words, emergence. In this respect, visualizations in general are counterproductive in that they lead to a reduction, homogenization, or curtailment of structural diversity in the landscape.

$$\text{Reduction} \rightarrow \text{Emergence} \leftarrow \text{Complexity}$$

7.1.2 Spatial Turn: Space and Landscape

The discussion on visualization in landscape architecture needs to take into consideration another aspect that affects our fundamental understanding of space. Since the "spatial turn"* in the cultural sciences,† the investigation of space has ceased to be the exclusive domain of physical geography, environmental planning, and, in particular, landscape architecture. Philosophy, sociology,‡ psychology, and anthropology too provide a broad field of study within which space is considered the prerequisite for cultural production and social conditions.

---

* Also referred to as "topographical turn" or "topological turn."
† Introduced by Henri Lefèbvre, in his "Production de l'éspace" [The Production of Space] in the early 1970s.
‡ For example, in the German-speaking area: Martina Löw: Raumsoziologie (English review: http://raumsoz.ifs.tu-darmstadt.de/pdf-dokumente/rezension.pdf), and Stefan Kaufmann: Soziologie der Landschaft (2005).

Since it has also been possible to refer to scientific findings on the relativity of space (Leibniz, Einstein), space is no longer understood as an absolute three-dimensional (3D) container but as a complex system of social relationships, spatial relations, and distances.

The reinterpretation of the notion of space has also had an impact on the discipline of landscape architecture. However, the discipline can neither adopt a purely cultural, relativistic meaning of space, nor can it insist on the old, absolute conception of space. Landscape architects who desire not only to understand space but also to design and change it, in other words, wish to build space *and* maintain it (cf. Heidegger 1971), must apply their own combined "relational" understanding of space (Löw 2001), which lies between the absolute and relativistic views of space.[*]

Absolute → Relational ← Relativistic

We can also distinguish between these different notions of space in the two types of programs used by landscape architects. While computer software from the field of computer-aided design treats space as if it were a body or a structure and therefore can only show space as a container or object (or an organism[†]) (Henges and Mensel 2008), geographic information system (GIS) programs can link specific spatial locations to information that is not directly bound to properties of objects. Programs that are GIS based (or include GIS) therefore open opportunities for creating and processing "relational" spatial properties. We will concentrate on the latter type of programs.

The second criterion for the application of computer-aided visualizations in design is therefore that the opportunities for a relational consideration of space are exploited. In particular, this means that a limited range of data is used only because it is easily accessible. There is a risk of over-reliance on quantitative, vectoral, and public data, simply because they are (1) easy to arrange in systems and hierarchies (standardized layers, cardinal scales), (2) easy to save, file, and compute (preprocessed), and (3) easy to obtain (widely available, open access, up-to-date). Often this leads to the loss of other, potentially more important, landscape-related information, such as data on the sociocultural significance of landscape and historic data, most of which is only available in maps that are analogue and considered "obsolete" by many.

Entire new worlds of data have opened up in recent years. In addition to standardized topographical maps,[‡] which are relatively readily available, there is a myriad of comprehensive, specialist databases (cadastral maps, land-use information)—although access to these is restricted for the legal protection of data and property—as well as commercial geodata (navigation systems).

---

[*] Martina Löw called the dual meaning of space in sociology the "relational model."
[†] cf. parametric designing using CAD in Hensel and Menges (2008).
[‡] For example, the German Amtliches Topographisch-Kartographische Informationssystem (ATKIS).

In addition, many historic maps are now available on the Internet.* Parallel to the world of functionally and hierarchically organized geodata, a whole new world of free spatial information is becoming available, ranging from free map servers of large internet companies[†] to open-source software[‡]. These are linked to a infinitive number of images and factual data by open user communities and so facilitate a wholly new approach to cultural as well as personal and community research (e.g., neography—see the examples in Section 7.3).

7.2 Methodology of Visualization

Planning tools and procedures, in which computer-aided visualizations have an increasingly important role to play, are readily available and make this wealth of spatial data accessible for the furtherance of landscape architectural insight and concepts. Table 7.1 illustrates the different levels and respective methods for processing and generating spatial information, as distinguished by landscape architectural research and research-led planning and design practice.

As a general rule, visualizations can be applied at all these levels. However, if we take the previously described "quality criteria" as a benchmark for landscape architectural research on and design of sustainable energy landscapes,

TABLE 7.1

Levels and Methods for Processing and Generating Spatial Information in Landscape Research and Practice

Level	Method
Concept	Communication of concepts (*communication*) Development of ideas (*conceptualizing, designing*)
Experiment	Challenge to the known (questioning knowledge) (*doubting*) Discovery of the unfamiliar (discover serendipities) (*abducting*)
Model	Generation of models and variants (*modeling*)
Assessment	Creation of order (*systematizing*) >Quantitative: structure according to criteria (*assigning/scaling*) >Qualitative: find higher contexts of meaning (*clustering*)
Analysis	Establish layers of information (*layering*) >Quantitative: find differences (*separating*) >Qualitative: find interrelationships (*approaching*)
Information	Data transfer (*tracing*)

---

* In Germany, mostly historic ordnance survey maps and nineteenth-century post-secularization land surveys.
† Above all, Google Maps and Microsoft Virtual Earth.
‡ For example, open street map.

that is to say an emergent and relational understanding of landscape, we must critically assess the potentials and limits of each tool. This will be demonstrated by means of discussing two types of visualization: rendering and mapping.

7.2.1 Rendering

Computer-aided visualization of landscape as images entails the virtual conversion of spatial data by rendering them into a *scene*, a 3D pictorial model. To achieve this, it is necessary to determine the amount of information and the way information is ordered (scene graph) as follows:

1. The selected *area* (an extract as in part of a landscape or place)
2. The selected *perspective* (extract from location, line of vision, height of viewpoint, horizontal alignment, and proportion of sky), the 3D *projection* (in perspective or orthogonal), and the *frustum* (depth and angle of vision and thus the visible area in the shape of a truncated cone or pyramid, the sectional planes at the beginning and end, as well as the setting for the depth of focus)
3. The *elements* shown (objects and the properties of object surfaces)
4. The *situation* shown (time of day, season, weather, and moods as represented by the selected lighting and colors, as well as consideration of reflections, shadows, etc.)
5. The *virtuality* (image quality, determined by factors that include resolution, simulation speed, quality of shading through interpolation, shading, etc.)
6. The *background information* (insertion of two- or four-dimensional images, e.g., diagrams and process schematics)
7. The additional *decorations* (birds and active or motionless people are often shown to illustrate scale but also to convey mood)

These parameters are either directly determined or can be influenced within certain limits by the observer. The list shows that great differences may occur in visualizations. These are caused in part by the leeway the planner/designer has when setting the parameters,* but also by the technical limitations of the software. Even specialists have problems in distinguishing the difference between what is caused by "objective," "subjective," or "technical" factors. This is, in any case, difficult for laypeople in particular. Therefore, design applications and, more importantly, research applications must ensure that these parameters are made *transparent*. This could be achieved by complementing particularly basic images with site photographs taken at different times of the day or by

---

* Albeit limited by his programming and pictorial design skills.

showing variations of different situations when images are photorealistic. The virtuality of an image can be demonstrated more effectively by "relativizing" it with the help of supplementary images rather than with written annotations.

Yet, with all the perfection of a rendering, it can only ever communicate some of the qualities of a space. Since the communication of design relies solely on perfect images and no longer necessitates explanation or elucidation—as Latz (2008) pointed out—acoustic, olfactory, flavor, and haptic qualities are neglected and so are individual variations in the experience of a space. This is because these qualities cannot be communicated through the semantics of images but only through the structure of space. "It is not the images, but the abstractions, schemata of information layers or single systems that are required for understanding *structure*. The images of perfect examples that aim at the semantic level no longer show how it should be done" (Latz 2008, 8, emphases added by the authors).

7.2.2 Mapping

Mapping is a term that describes the production of maps that, as well as containing topographical data, also spatially communicates nonphysical, for example, psychological or cultural, information, in map-like images like *mental maps*.* "The interrelationships between things in space, as well as the effects produced through such dynamic interactions, are becoming of greater significance for interventions in urban landscapes than the solely compositional arrangements of objects and surfaces. The experience of space cannot be separated from the events that happen in it; space is situated, contingent and differentiated" (Corner 2002, 227). The potentially infinite amount of such data requires that a selection be made. Objects or situations are selected, sorted, and grouped and usually shown within categories or types. It is virtually impossible to survey all the elements of a landscape, but a category, once it has been selected for illustration, will be covered area-wide thereafter, creating the impression of completeness. Such maps convey a high degree of factuality and reliability because they represent actual places and are abstract at the same time, as they show a selection of facts that are reduced in scale, standardized, and coded (cf. Corner 2002). Corner describes the process of mapping in three steps or operations:

1. Conception of the field, as the analogical equivalent to actual ground, with scope in the graphic system, the frame, and the scale
2. Selection of extracts by isolating observed things from a milieu, which initially means deterritorialization
3. Reterritorialization by "drawing out" connections between extracts

---

* Cf. Lynch, Kevin. The Image of the City.

Hence, mappings are not surveys or compilations, but the result of a *creative* process that both finds and produces things in the world. "The distinction here is between mapping as equal to what it is ('tracing') and mapping as equal to what it is and what it is not yet. In other words, the unfolding agency of mapping is most effective when its capacity for description also sets conditions for new eidetic and physical worlds to emerge.* This is why mapping is never neutral, passive or without consequence; on the contrary, mapping is perhaps the most formative and creative act of any design process, first disclosing and then staging the conditions for the *emergence* of new realties. Hence mapping, as an open and inclusive process of disclosure and enablement, comes to replace the *reduction* of planning" (Corner 2002, 214; emphases added by the authors).

7.2.3 Quality Criteria for Visualizations

Both methods—rendering and mapping—fundamentally rely on processes of selection and abductive logic. They are not provable, yet they should be designed to be *transparent* and comprehendible and thereby provide design legitimization and scientific value. Renderings should not illustrate "finished" semantically unambiguous situations, but should use abstraction to communicate the *structure* of a space in a way that is open to interpretation by the viewer and the prospective user. In this way, the quality of the background information becomes more important than the technical image quality (virtuality) and decorations. It is also essential that maps are structured to inspire many different interpretations, rather than lead to a reduction.

If we combine Corner's relational and creative position on mapping with Latz's critical structuralist position on rendering, we can speak of creative structuralistic requirements of visualizations on the basis of the following characteristics. To obtain both scientific and political legitimization, visualizations in research and design practice should not appear in a hermetic process but be "falsifiable," that is, all data sources, selections, and computer settings should be transparent. They should not pretend to represent reality but be the result of a creative process, and they should use the advantages of creativity. Visualizations in landscape architecture research and practice should be situated in—or applicable to—a concrete space instead of staying abstract. They should not distinguish landscape into distinct eternities but emphasize its contingence and capacity. Landscape qualities should not be homogenized or reduced but, on the contrary, visualizations should help to understand their variety, difference, and emergence. This basically means visualization in landscape architecture research and planning should not be pictorial on the surface but structural—getting to the bottom of the landscape (Table 7.2).

---

* Corner quotes the Rhizome theory: "What distinguishes the map from the tracing is that it is entirely oriented toward an experimentation in contact with the real. The map does not reproduce an unconscious closed in upon itself; it constructs the unconscious" (Deleuze, Guattari).

TABLE 7.2

Quality Criteria for Visualization
in Landscape Architecture
Research and Practice

–		+
Hermetic	→	*Transparent*
Scanned	→	*Created*
Abstract	→	*Situated*
Distinct	→	*Contingent*
Homogenic	→	*Differentiated*
Pictorial	→	*Structural*
Reduced	→	*Emerged*

7.3 Examples of and Discussion on Visualizing Space Models

It is not by coincidence that we have considered in detail the apparently widely disparate tools of rendering and mapping. Both of these tools are frequently described in both international research and design publications, although most authors omit a sufficiently critical view of their potentials and constraints. This, we will briefly discuss by the example of four projects and then return in the fourth section to the question of how visualization tools may be applied in research for and the design of sustainable energy landscapes.

7.3.1 Datascapes (MVRDV)

The Dutch architects MVRDV* use *datascapes* as a method for recording and visualizing all legal, technical, social, political, and natural influences relevant for the work of architects (cf. Lootsma 1999). All facts and parameters, including initially invisible data traces, are visualized in stacked layers of abstract spatial containers and in numerous diagrams. Overlaying the different *datascapes* creates a framework of new perspectives and opportunities in which the compiled information is translated into shapes. *Datascapes* are not visualizations of visions but simply applied research. They are only one possibility among many others for studying the complexity of our world by mapping the limitations of certain parameters (MVRDV 1999). In 2009, MVRDV included *datascapes* in their report on President Sarkozy's "Grand Paris 2030." Their project "Pari(s) Plus petit" is one of 10 proposals by international architecture and urbanism teams that envisioned the future of the French capital and its vast agglomeration. MVRDV's proposal is represented in four parts: (1) the *Synthesis*, which defines the spatial agenda for Paris; (2) the *City Calculator©*, a software and possible web tool, which quantifies the behavior and performance of a city and

---

* Founded by Winy Maas, Jacob van Rijs, and Nathalie de Vries in Rotterdam in 1991.

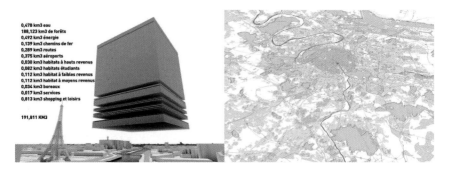

FIGURE 7.1
Diagram and plan of MVRDV's project Pari(s) Plus petit. (From MDRDV, 2009 Le Grand Pari de Grand Paris. Pari(s) Plus petit, http://www.mvrdv.nl/#/projects/grandparis. With permission.)

TABLE 7.3

Methodological Classification of *Datascapes* Regarding the Quality Criteria for Visualization in Landscape Architecture Research and Practice

–	Grading of Used Methods		+
Hermetic	Datascapes depict objective data	→	Transparent
Scanned	← Quantitative or quantifiable data are used exclusively		Created
Abstract	← Space factors are analyzed but space properties are not		Situated
Distinct	← Single factors are depicted to highlight their most extreme effect		Contingent
Homogenic	There are always several systems in development	→	Differentiated
Pictorial	Renderings, spaces, and diagrams are explanatory rather than for illustration	→	Structural
Reduced	The aim is to discover new and unfamiliar interrelationships	→	Emerged

connects qualitative to quantitative parameters; (3) the *Data*, which gives the overview of the detailed research; and (4) the *Observations* as a series of articles on the backgrounds, the history, the potentials, and the problems of the French metropolis (MVRDV 2009) (Figure 7.1) (Table 7.3).

7.3.2 Urban Portrait (ETH Studio Basel)

Switzerland: An Urban Portrait was published by the Studio Basel Institute (cf. ETH Studio Basel 2006). The publication is based on a theoretical examination of Henri Lefèbvre's 1970s theory of space. It also builds on numerous student studies of urban space phenomena in Switzerland (referred to as "boreholes") that are compiled in a comprehensive map and interpreted (Figure 7.2). Switzerland is no longer described as a rural idyll but, instead, is perceived as an urban space.

The central thesis is that Switzerland's settlement topography can be developed *differentially* rather than *generic* or featureless, despite the country's complete urbanization.

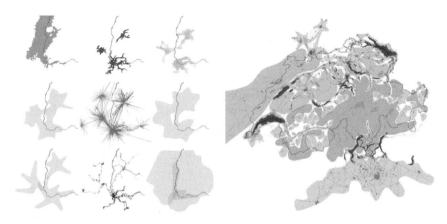

FIGURE 7.2
Layers of the metropolitan region Basel–Mulhouse–Freiburg and map of Switzerland's urban potential (From [ETH Studio Basel] Diener, R. et al., *Switzerland: An Urban Portrait*, Birkhäuser Verlag, Basel, 2006. With permission.)

The authors of *Switzerland: An Urban Portrait* presume that globalization does not lead to indistinguishable, generic cities but reinforces varying urban patterns of behavior. These patterns are inevitably found in the physical reality of cities and landscapes as specific forms of urbanity (cf. ETH Studio Basel 2006). In this context, they refer to the notion of "difference." According to Lefèbvre, the city is a differential space in which differences come to light. "Difference is informing and informed. It produces form" (Lefèbvre 2003, 133). Unlike features that depend on local conditions and are isolated one from the other, differences are active points of reference. Only the confrontation of typical features gives rise to understanding and thus difference (Table 7.4).

TABLE 7.4

Methodological Classification of *Urban Portrait* Regarding the Quality Criteria for Visualization in Landscape Architecture Research and Practice

–		Grading of Used Methods		+
Hermetic	←	"Borehole" and design development data are not documented		*Transparent*
Scanned		Data are interpreted by means of "archaic" sketches in maps	→	*Created*
Abstract		Spatial factors and spatial features are always closely linked	→	*Situated*
Distinct		Spatial shapes are always and repeatedly overlaid, "interpretable"	→	*Contingent*
Homogenic		Illustrations should typify the space and retain difference	→	*Differentiated*
Pictorial		Factual aerial photographs and maps that convince through "imagination"	→	*Structural*
Reduced		Several perspectives to highlight new urban qualities	→	*Emerged*

7.3.3 Systemic Design and Drosscape (Alan Berger et al.)

Landscape architect Alan Berger and his transdisciplinary team at MIT have applied the *systemic design approach* in their "Project for Reclamation Excellence" (P-REX), which assumes a substantial, complex, and active understanding of systems. This approach is based on the understanding of large-scale forces in the built environment and the merging of existing influences on the landscape. It requires innovative techniques of visualization in general and mapping in particular. So far, in order to reach conclusions, regional planning has largely relied on the analysis of regional systems by means of reduction. Berger, however, advocates new analysis tools, such as GIS in combination with the Internet, in order that small-scale spatial themes be adapted to and fed back into their regional context during the design process.

This process starts with the open probing of tangible knowledge, "combing" through literature and data in the sense of a search engine rather than a specialized branch library (cf. Berger 2009). On the basis of his own aerial photographs, diagrams, and maps, and by examples of 10 urban regions in the United States, Berger in his book *Drosscape* describes urban landscapes as waste landscapes that have emerged as a result of economic and industrial processes. This waste—or *dross*—can be redesigned (i.e., *scaped*), resurfaced, or reprogrammed for adaptive use.

"Entropic indicator maps" depict the spatial growth of urbanization between 1990 and 2000 as well as other specific large-scale land activities that have changed over this period (golf courses, military installations, airports, landfills, business parks, etc.). The entropic indicator maps are functionally associated to two other kinds of analytical diagrams: dispersal graphs show the peaks and dips of population densities in relation to distance from city centers, and spindle charts describe the growth and decline of commercial development. "If one correlates population density and distance from the city centre found on the dispersal graphs with areas of manufacturing activity found on the spindle charts and the land-use clusters found on the entropic indicator maps, the result is new, previously unforeseen spatial concentrations of a variety of waste landscapes from which future regional landscapes may be reclaimed for cultural or ecological benefits" (Berger 2006, 78).

Berger's mapping techniques permit a new understanding of urbanized regions. The visualizations combine multiple views and scales, resulting in new associations between disparate facts of urbanization over time. Aerial photographs help to explain and legitimize the abstract maps and diagrams (Shannon 2006) (Figure 7.3) (Table 7.5).

7.3.4 Shapes of Urban Landscapes: Between Typology and Idiography

Urban landscapes are one of the key areas of the research team Landscape Architecture and Regional Open Space (LAREG) at the Technical University

FIGURE 7.3
Entropic indicator map and dispersal graph, Boston-Lowell/Providence region (From Berger, A., *Drosscape: Wasting Land in Urban America*. Princeton Architectual Press, New York, 2006. With permission.)

TABLE 7.5

Methodological Classification of *Drosscape* Regarding the Quality Criteria
for Visualization in Landscape Architecture Research and Practice

–		Grading of Used Methods		+
Hermetic		The relevant systems depict objective data	→	*Transparent*
Scanned	←	All applied data are quantitative and current		*Created*
Abstract		Objects and interrelationships are localized	→	*Situated*
Distinct	←	Limited consideration of *waste landscapes* (stigmatization)		*Contingent*
Homogenic	←	Spatial types are described but not spatial characters	→	*Differentiated*
Pictorial		Objective aerial photography and expressive maps are shown	→	*Structural*
Reduced		New perspectives of *waste landscapes* are developed	→	*Emerged*

of Munich. On the one hand, the notion *urban landscapes* analytically describes the complete urbanization of space, that is, the wide-ranging sprawl
of buildings, infrastructures, and urban lifestyles, and, on the other, the
programmatic attempts to discern and design new interrelationships in the
fragmented spaces that are neither city nor countryside.*

In recent years, in many countries and particularly in Germany, several research projects have searched for generic—nomothetic—principles
and theories of the emergence and development of urban landscapes.
Diversity and interrelationships in urban landscapes, however, cannot be
explained by quantitative methods alone; they must also be recorded in
qualitative terms, in relation to actual places and forms—an idiographic
approach. The ETH Studio Basel's study (2006) has demonstrated urban
diversity by "revealing" differences. In order to advance the understanding of interrelationships of urban landscapes, Berger (2009) has looked at
both the "outer" logic of large-scale spatial interrelationships and regional
forces. Löw (2008) illustrated the factors that develop an "inner" or specific logic in each place. She coined the phrase "intrinsic logic" of cities,
which relates to something general—a logic such as urbanization, densification, and heterogenization—and develops intractable connections and
compositions.

The research project on urban landscapes at LAREG focuses on the study
of the components and interrelationships, structures, and layers of urban
landscapes using different mapping and data collection methods in order to
identify lasting generators of form for future developments and to reveal a
design repertoire that can be used to qualify urban landscapes. The research,

---

* This is the wording of our first German Wikipedia entry on "urban landscapes."

thus, not only aims to describe urban landscapes in general terms, but to design them typologically and specifically.

First, the structures of urban landscapes must be revealed: their specific characteristics, dependencies, and rules of formation as well as the interaction between *natural morphologie*s, the *textures of the cultural landscape*, and *(sub)urban elements*—assumed or observed components of urban settlement patterns, infrastructures, and uses. Phases of observation (i.e., induction) alternate methodically with targeted challenges and a rethinking of existing knowledge (i.e., abduction). Incidental observations (i.e., serendipity) can lead to new assumptions. Based on such assumptions, previously unknown rules can be developed, which in turn can be comprehended by means of new observations. Documenting all the sources and steps involved in reaching decisions during the selection process is of key importance. Selections and combinations made in the course of mapping must be communicated in a transparent way so that, although not verifiable, they are comprehensible. By combining new data with existing standardized topographical data, early conclusions about the relevant elements and interrelationships may be reformulated. Overlaying the data, either in a targeted way on the basis of substantiated assumptions or by testing, consolidates certain constellations of spatial characteristics and uses into typical situations composed of recurring interrelationships and structures. These situations give rise to urban "landscape types" and "landscape characters," which, in turn, help to describe and interpret urban landscapes. They are combined to form a map of Germany's urban landscapes (Figure 7.4) (Table 7.6).

A large part of the structures found in urban landscapes can be traced back to land uses that relate to fossil-based energy sources. The transition to renewable energy sources too is affecting and shaping urban landscapes across the world. Therefore, energy-conscious interventions (especially in the regional scale) can only be implemented effectively and sustainably when considering complex conditions in urban landscapes.

The aforementioned methodological overview on research projects reveals two problems faced by current studies that make use of visualizations by renderings or mappings. They either concentrate wholly on quantitative or quantifiable data—for example, *Datascape* and *Drosscape*—which omit or oversimplify essential spatial characteristics such as historic layers and cultural contexts, or they include the qualitative data but neglect to make transparent and comprehensible the process of reaching conclusions or design decisions on types and shapes—for example, *the Urban Portrait*. However, this is precisely the problem that needs to be overcome if visualizations are to become a legitimate part of academic work, research, and practice. Besides focusing on their core topics, the *Shapes of Urban Landscape* project aims to address this challenge similar to the following research projects on sustainable energy landscapes.

FIGURE 7.4
Layers of urban landscapes in Germany (ground-figure plan, rivers and valleys, topography, highways, and suburban train network) stacked to reveal a new "landscape type": *Big boxes in lowlands* (LAREG).

TABLE 7.6

Targeted Methodological Classification of *Shapes of Urban Landscape* Regarding the Quality Criteria for Visualization in Landscape Architecture Research and Practice

+		**Grading of Targeted Methods**		–
Hermetic		Comprehensible documentation of data sources, selections, and combinations	→	*Transparent*
Scanned	←	Applied data are both quantitative (standard) and qualitative (experimental)	→	*Created*
Abstract		Components and interrelationships are localized (georeferenced and "landscape-referenced")	→	*Situated*
Distinct		Spatial shapes are always and repeatedly overlaid, "interpretable"	→	*Contingent*
Homogenic		No classifications but *landscape types* and *landscape characters* are described	→	*Differentiated*
Pictorial		Maps and communicates actual spatial qualities including meanings and hidden structures	→	*Structural*
Reduced		The varieties of *urban landscapes* survive, coherences emerge	→	*Emerged*

7.4 Visualization of Energy Landscapes

The Earth will have to provide for the population's growing energy demand while drastically reducing the emission of greenhouse gases (GHG). The pursuit's significance for our future is one facet of what we try to define as sustainability. But if we plan this transition only with economy in mind, that is, in quantitative terms with homogenous and monofunctional land uses, mitigation measures will at best prevent us from completely losing sight of the 2° target defined by international climate policies.

In order to create truly sustainable energy landscapes instead, it is essential that renewable energies will become an everyday part of the landscape once again (see Section 7.1.1). This reintroduction into the landscape has to ensure that renewable energies are not only a commonplace part of the landscape, but also take on a mediating spatial function by creating new interrelationships between these alleged antipodes of what space is and what space means. Due to their high visibility (apart from perhaps geothermal energy) and their cultural significance for sustainability and thus for our future, renewable energies gain this potential to mediate between absolute (rational) and relativistic (relative) conceptions of space (see Section 7.1.2). Only in this way, in an intermediating position and responding to its surroundings, space can be created in an ecological, economic, and social context and thus with the meaning of sustainability for renewable energy landscapes.

To honor this promise, disciplines that intend to create space and in particular landscape architecture should merge these rational and relative inputs to achieve new, combined, and ultimately relational design that we have to demand from energy landscapes as being sustainable.

In the following, two preliminary design studies show how these rational data can be shifted to design tools for landscape elements and how relation as a response can be considered. Concluding this section, the current research project *Slowup Landscapes* shows how facts and meaning of space can be brought together as relational design of sustainable energy landscapes.

7.4.1 Rational Techniques as Landscape Elements

*Brandscape** is the title of a diploma thesis that examined the issue of new landscapes between energy provision and branding with a focus on biomass as an energy source for synthetic biofuels. The study site is situated in the region of Western Pomerania, near Greifswald. The area is one of the conceivable sites in Germany for a large-scale biomass power plant. Comprising an area of 1400 km², the anticipated output was 250 million liters

---

* Full title: Brandscape—new landscapes between energy provision and branding. Department of Landscape Architecture and Regional Open Space (LAREG) at the Technical University of Munich (LAREG 2007).

of fuel per year derived from 1.5 million ton of biomass. The study was based on the biomass to liquids process* that allows the processing of all types of carbonic biomass. The economic advantage of this complex technique results in explicit higher energy yields per unit of cultivated land. This process of energy provision "provides" the landscape architect with a complete palette of woody and stalk-containing biomass as design elements. Grassland, one and two crop harvests on arable land, short rotation crops, agro forestry, and permanent crops such as *Miscanthus* were studied in combination with different field crops and tree species. Different crop rotations and harvest cycles concentrated these design elements into processes. As a supplement, recycling residues from existing timber production and agriculture was considered (Figure 7.5).

Each chain in this (biomass-based) energy system has distinct visible effects on the landscape while factor L (factor landscape) finally lead to design decisions such as weather woody and stalk-containing biomass with their different kinds of cultivation that can be created as landmarks, barriers, and underlines (i.e., figure) or plain structures and textures in different acre sizes (i.e., ground).

The collection of information and search for spatial relations was achieved through GIS. ATKIS-data, Digital Terrain Model, Web Map Services, and geological maps were also integrated. Models of the distribution of use and time cycles were computed, crop yields and logistics estimated, wind and water erosion calculated, and, along with other data on, for example, water balance and transport concepts, depicted in analysis maps (Figure 7.6).

The current tourism infrastructure in the natural environment, which is focused mainly on the peninsula Usedom in the east of the project area, is supplemented by the development of centers of the new energy landscape more in the western part of the project (Figure 7.7). These centers are, on the one hand, pre- and interproducers within the process, dealing with questions like harvesting, biomass storage, or products of the manufacturing process (e.g., wood chips and liquid biomass). On the other hand, the centers are located like showcases and as figureheads along touristic routes to Usedom. Others are scattered around the energy landscape and increase touristic potentials by providing accommodation and serving as nodes between public and private transport systems.

Project visualization comprised analysis plans, graphs, 3D landscape models, and a master plan. Each of the approximately 40 types of biomass cultivation shown in the master plan was assigned a particular graphic design texture (Figure 7.8). This technique had three advantages over conventional plan keys: (1) Similar cultivation systems received similar textures on the plan, in order to

---

* In contrast to E10, which is currently debated in Germany, this second generation fuel has chemical properties and an energy density that presents an alternative to petrol and first generation biofuels.

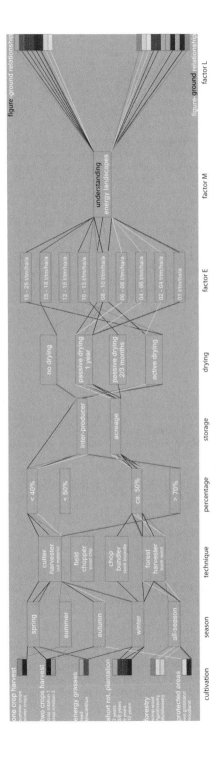

FIGURE 7.5
Scheme for an year-round, 24 × 7 energy provision including system of cultivation, season, and technique of harvest, anticipated percentage of dry biomass, location of storage within the process of production, the kind of biomass drying, resulting provision of energy (factor E-nergy), a channelling through perception of the observer targeting to afford an understanding for these new landscapes (factor M-en), and the basic definition within figure–ground relationship of being a design element of space (factor L-andscape).

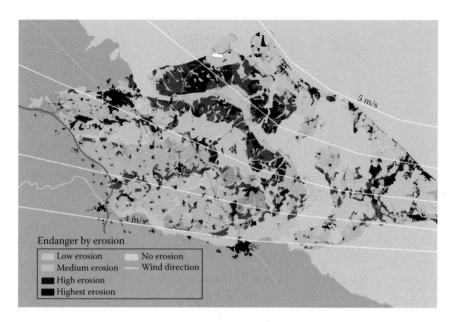

FIGURE 7.6
Analysis map of endanger by erosion and wind direction (scale approximately 1:500.000).

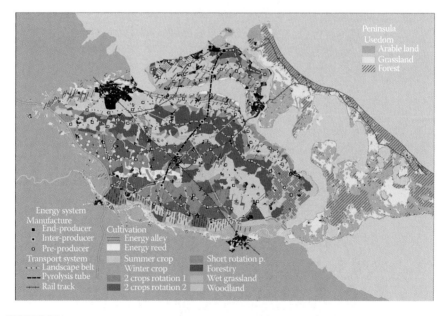

FIGURE 7.7
Analysis map for allocation of biomass, kinds of cultivation, and location of new centers of production and tourism (scale approximately 1:500.000).

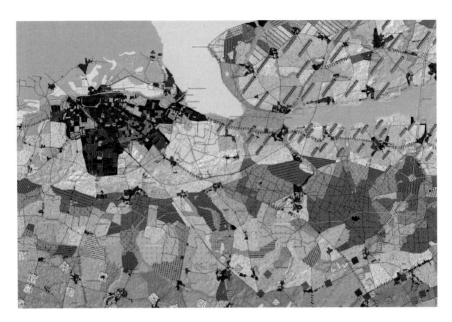

FIGURE 7.8
Zoom in of brandscape master plan (original scale 1:25,000) with design textures for different biomass cultivations.

emphasize those relations in the landscape for the viewer. (2) The graphics differentiating between different cultivation methods made it relatively easy to communicate reasons behind design decision and analysis results and made them comprehensible and retraceable. (3) The selected textures were more than mere symbols; they were directly derived from the graininess, textures, and line patterns associated with cultivation methods and plant species.

Ultimately, the *brandscape* project studied how rational data of renewable (biomass-based) energy provision can be applied to design elements. This first step toward the development of a sustainable energy landscape was linked with geomorphologic conditions and the enlargement of the existing touristic infrastructure.

7.4.2 Responding Techniques as Landscape Relations

Within the scope of the 2011 summer school themed *The Mediterranean challenge,** a first step was taken to study how renewable energies can be employed to produce synergies between environmental, economic, and social factors.† Across an area of 250,000 km², reaching from Andalucía across the Strait of Gibraltar to Morocco, the study sought to link renewable energies

---

* energylandscapes3.0 at the International summer school at the Bauhaus in Dessau/Germany; facilitator Studio 4: Stefan Tischer.
† Based on the three pillar model of sustainability.

with environmental questions, historic traces in the landscape while adding value for the population. Much of the work was based on scholars like Le Corbusier (Plan Obus) and utopians like Herman Sörgel (Atlantropa). The conceptual approach taken by the *Desertec** project served as a basis for the design scenarios. By means of analysis maps like solar radiation, water balance, desertification risk, and staking them, three approximately 180 km long corridors in Spain and Morocco came into the focus of further consideration.

Each corridor also had its specific design approach—how the technique for renewable energy-provision RET should respond to its surroundings. In the corridor between Cordoba and Seville (yellow in Figure 7.9), the relation to landscape was based on the implementation of decentralized energy provision by incorporating biomass, solar energy, and wind energy into the local networks of infrastructure (power lines and roads) and settlements (urban and rural areas). From the perception of a superimposed system, the many different and disperse energy cells provide as much energy as a centralized system.

Embedded into existing landscape structures and with a "democratic" tenure relation, the energy of the cells could also be used for "landscape" demands like water cleaning and pump storages when, for example, the prizes for electricity in the European power grid are low.

The corridor around Fes in Morocco was designed according to ancient Roman courses of roads and their former pulse of rest stops (blue in Figure 7.9). During the studio work, we found out that this "old" distance of around 18 Roman miles perfectly fits for the predicted interspaces of electromobility. Thus, rest stops were located along the existing street system and each conceptually designed with techniques for renewable energy provision. Some of them also had a direct link to archeological excavation. During their stops at those rest areas, consumers can directly see what they store in the batteries of their *E-cars*. At the same time, the renewable energy provision responds to the landscape as a relation to historic traces of this ancient cultural landscape (Figure 7.10).

The corridor south of Melilla (magenta in Figure 7.9) combined new solar thermal plants and concentrated solar power (CSP) on former woodland sites in eastern Morocco with the large-scale cultivation of a green wall using native tree species. This tool for counteracting desertification also functioned as a natural protective barrier. Solar panels are easily damaged by sandstorms and are therefore usually protected by walls. Furthermore, reforestation of this site with excellent insolation facilitated the stabilization of valuable groundwater resources that are required for energy generation and the cleaning of the solar panels (Figure 7.11).

Ultimately, the studio work in the summer school at Bauhaus Dessau looked at the question of how these synergies, as external relations of renewable energy provision to their surrounding landscape, could be designed

---

* The concept of *Desertec* is to produce renewable energies by wind and solar power in deserts and transport it to afar consumers, e.g., in Europe.

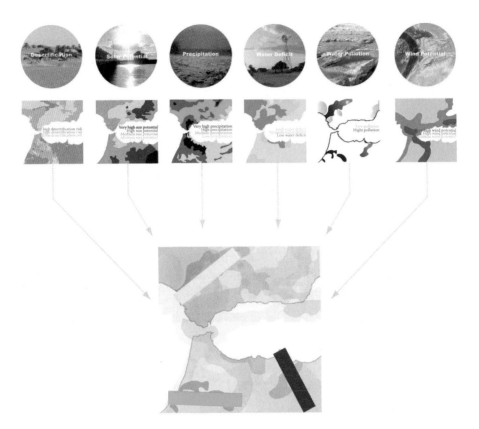

FIGURE 7.9
Analysis maps of desertification, solar potential, precipitation, water deficit, water pollution, and wind potential as well as the location of the three project corridors. (From Bauhaus Dessau Foundation, energylandscapes3.0., Report of the Summer School, 2011, Unpublished.)

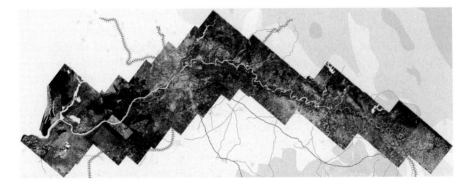

FIGURE 7.10
Conceptual design of the corridor (original scale 1:100,000) between Cordoba and Seville with decentralized energy cells as a superimposed system. (From Bauhaus Dessau Foundation, energylandscapes3.0., Report of the Summer School, 2011, Unpublished.)

FIGURE 7.11
Illustration of CSP in building phase and finally with a green wall against desertification and stabilized groundwater level. (From Bauhaus Dessau Foundation, energylandscapes3.0., Report of the Summer School, 2011, Unpublished.)

using the model of the three pillars of sustainability. The concept for their response was developed on the basis of existing networks and landscape structures as well as on cultural history at a regional scale.

7.4.3 Relational Design as Creating Sustainable Energy Landscapes

The *Slowup Landscapes* project* is based on the findings of *Brandscape* and *The Mediterranean challenge*, among others. Here, principles that establish how sustainability of European landscapes can be conceptualized in stages by adaption, by mitigation, and beyond climate change are studied. The main research question is how these two, landscape and sustainability, can be consolidated if we simultaneously speed up and slow down different natural processes, ecosystem services, modes of cultivation, and land use. In this project, renewable energy landscapes are conceptualized as *stage 2: retardation landscapes*. From mitigating GHG to achieve a closed CO_2 cycle, sustainable energy landscapes will hopefully contribute to the slowing down of climate change. This will give us time to adapt our environment in order

---

* *Slowup Landscapes*: Aspects of speedup and slowdown for sustainable landscapes is the PhD research project of Andreas René Dittrich in the Department of Landscape Architecture and Regional Open Space (LAREG) at the Technical University of Munich.

to reduce the effects of climate change (i.e., *stage 1: adaptation landscapes*) and develop systems that will act as carbon sinks (i.e., *stage 3: decubation landscapes*) and permanently bind GHG out of the atmosphere.

The methodology applied within the research topic and in these three concept stages is based on the assumption that sustainability and landscape can only be combined relationally, by overlaying rational (absolute) and relative spatial concepts (see Section 7.1.2). At the end of the research trajectory, exemplary design proposals will be given for selected project areas in Europe and organized according to the following four phases:

1. Rational quantitative phase—collect, sort, and explain as factual levels.
2. Relative qualitative phase—integrate, comprehend, and understand as interpretation levels.
3. Abductive critical phase—review, scrutinize, and transfer as European landscapes.
4. Relational design phase—connect, overlay, and consolidate as structural landscape levels.

The conceptual consolidation of the various internal and external relations at a regional scale is to lead to synergies between renewable energies and sustainability and manifest itself as additional value in the landscape. Visualizations as mappings or renderings are made legible and comprehensible by adhering to the methodical phases and worked out with planning steps for a research by design. The findings of such research on consistent energy landscapes that form a part of our everyday life should result in a succeeding relation of culture and nature (Schöbel 2012). It is especially the methodical organization of factual and interpretative levels and their relational concentration and hierarchical interpretation into new levels of decision making that promises its transferability to real planning processes in the context of renewable energies and further research work of *Slowup Landscapes*.*

7.5 Conclusions

Complex visualization tools can be legitimately used in academic research as well as planning and design practice provided (1) they enable the designer/planner to breach new horizons, that is, if they provide methods by which qualitative, experimental, and new knowledge are generated or (2) they serve the purpose of creating extra scope and alternatives in the process

---

* A transfer of methodologies and spatial derivatives to other thesis questions, conceptualized stages of sustainability, and respective landscapes in the project areas is intended.

of communicative planning, so that the public and/or the political stake-holders are able to make better decisions. These are the criteria for making visualizations—a powerful academic and democratic tool.

Both areas of applications require a high degree of transparency and comprehensibility of the selective and technical steps, leading to the visu-alization, but also of their basic perception of space. Particularly when look-ing at the larger scale of landscapes, GIS offers useful tools for the analysis and integration of data. The linkup of spatial with attribute data is attrac-tive, as is the availability of high-standard baseline data (from official and increasingly from open sources). Attempts to simply illustrate landscape elements from standardized topographical data or data available in Web 2.0 fall short. Only research for further data with qualitative methods will broaden the perspective. It facilitates the finding and integration of new and thus far unknown information. The aforementioned case studies from the Department of Landscape Architecture and Regional Open Space at the Technical University of Munich have illustrated the set objectives.

A methodological overview on current research projects in the field of large-scale landscape architecture has revealed problems with two kinds of visualizations: renderings and mappings. They either concen-trate wholly on quantitative or quantifiable data that omit or oversimplify essential spatial characteristics such as historic layers and cultural con-texts or they include the qualitative data but neglect to make transparent and comprehensible the process of reaching conclusions or design deci-sions on types and shapes.

The *Shapes of Urban Landscapes* project tries to overcome these weaknesses by merging standardized empirical GIS data on landscape morphology with qualitative and hermeneutical data on the significance of landscape in a transparent, comprehensible design process.

The *Slowup Landscapes* project superimposes contemporary European cultural landscapes, adaptations to climate change, and renewable ener-gies with their historic traces. With the help of qualitative methods and quantitative GIS data, we will explore how renewable energies can not only become an integral part of the landscape, but should even take on a mediat-ing function, by creating new interrelationships between what space is and what space means.

This book discusses the shift from *renewable* energy landscapes to *sus-tainable* energy landscapes, which describes and claims "landscape quality, biodiversity, food production and other life-supporting ecosystem services" (see definition in Chapter 1). It should be added that sustainable energy landscapes have to include a wide range of qualities and cultural benefits in the fields of civilizing progress, democratic inclusion, social welfare, and coherence. These qualities are not always quantifiable, yet proposals for sus-tainable energy landscapes have to address these complex issues and, never-theless, be creative, coherent, and holistic.

Cognition and action—including communication—of landscape architecture are bound to their methodological opportunities and attitudes. The *power of images* and the *trust in technology* provide the discipline with both influence and responsibility, for example, by using mappings and renderings as visualization tools in research, planning, and design.

References

Bauhaus Dessau Foundation. 2011. Energy landscapes 3.0. Report of the Summer School. Unpublished.

Berger, A. 2006. *Drosscape: Wasting Land in Urban America*. New York: Princeton Architectural Press.

Berger, A. 2009. *Systemic Design Can Change the World*. Amsterdam, the Netherlands: Sun Publishers.

Corner, J. 2002. The agency of mapping: Speculation, critique and intervention. In *Mappings*, ed. D. Cosgrove, pp. 213–252. London, U.K.: Reaktion Books Ltd.

[ETH Studio Basel] Diener, R., Herzog J., Meili, M., de Meuron, P., and Schmid, C. 2006. *Switzerland: An Urban Portrait*. Basel, Switzerland: Birkhäuser Verlag.

Heidegger, M. 1971. Building dwelling thinking. In *Poetry, Language, Thought*, ed. M. Heidegger, pp. 141–160. New York: Harper Colophon Books.

Hensel, M. and Menges, A. 2008. Form- und Materialwerdung. *ARCH+ 188*: 18–23.

Latz, P. 2008. Vision und Aktion. *Garten + Landschaft 3/2008*: 8–9.

Lefèbvre, H. 2003. *The Urban Revolution*. Minneapolis, MN: University of Minnesota Press.

Lootsma, B. 1999. Reality Bites. Forschung in der Zweiten Moderne. *Daidalos 69/70*: 136–141.

Löw, M. 2001. *Raumsoziologie*. Frankfurt (Main), Gemany: Suhrkamp Taschenbuch Wissenschaft.

Löw, M. 2008. *Die Soziologie der Städte*. Frankfurt (Main), Germany: Suhrkamp Taschenbuch Wissenschaft.

MVRDV, 1999. Irgendwann muss man sich gegen den Sprawl entscheiden. MVRDV im Gespräch mit ARCH+. *ARCH+ 147*: 56–66.

MVRDV, 2009. Le Grand Pari de Grand Paris. Pari(s) Plus petit. http://www.mvrdv.nl/#/projects/grandparis (accessed January 30, 2012).

Schöbel, S. 2012. *Windenergie und Landschaftsästhetik*. Berlin, Germany: Jovis Verlag.

Shannon, K. 2006. Drosscape: The darkside of man's cultural landscapes. *Topos 56*: 63–71.

Sieferle, R. P. 1997. *Rückblick auf die Natur*. München, Germany: Luchterhand Verlag.

8

Developing a Planning Theory for Wicked Problems: Swarm Planning

Rob Roggema

CONTENTS

8.1 Introduction

The provision of sustainable energy can be seen as a wicked problem. A wicked problem is accurately defined in the seminal paper of Rittel and Webber (1973):

- They have no definite formulation.
- They have no stopping rules.
- Their solutions are not true or false, however, better or worse.
- There is no immediate and ultimate test of a solution.
- Every solution is a "one-shot operation"; since there is no opportunity to learn by trial and error, every attempt counts significantly.

- They do not have an enumerable (or an exhaustively describable) set of potential solutions, nor is there a well-described set of permissible operations that may be incorporated into the plan.
- Every wicked problem is essentially unique.
- Every wicked problem can be considered as a symptom of another (wicked) problem.
- The causes of wicked problems can be explained in numerous ways; the choice of explanation determines the nature of the problem's resolution.
- With wicked problems, the planner has no right to be wrong.

Spatial planning, another scope of this chapter, is defined in many different ways. Dror (1973), for example, describes planning as a process: "Planning is the process of preparing a set of decisions for action in the future, directed at achieving goals by preferable means." In the course of this chapter, spatial planning is defined as the "co-ordination, making and mediation of space" (Gunder and Hillier, 2009: 4).

Current (and historic) discourses in spatial planning, such as on incrementalism (referring to Lindblom, 1959), postpositivism (as described in Allmendinger, 2002a), communicative planning (among others: Habermas, 1987, 1993; Healey, 1997; Innes, 2004), agonism (see Mouffe, 1993, 2005; Hillier, 2003; Pløger, 2004), reflexive planning (Beck et al., 2003; Lissandrello and Grin, 2011), or even the actor network approach (Boelens, 2010) do have considerable difficulties to deal with wicked problems or solutions or fail to accept wicked problems as the subject of planning. Hence, the need for an alternative theory emerges. In this chapter, one possible theory, swarm planning, is explored and developed.

8.2 Problem Statement

Our world becomes increasingly complex and turbulent (see, for instance, Ramirez et al., 2008), as reflected in the fields of energy (peak oil and consequences of oil prices: Campbell and Laherrère, 1998; Campbell, 1999, 2002a,b; Rifkin, 2002; Belin, 2008; Sergeev et al., 2009) or accelerated climate change (Tin, 2008; PBL et al., 2009; Richardson et al., 2009; Sommerkorn and Hassel, 2009). Meanwhile, climate adaptation, defined as a wicked problem itself, and energy systems planning (e.g., Van Dam and Noorman, 2005) are only marginally connected with the spatial planning domain. This means that, inevitably, energy planning and adaptation take place in separate worlds, where they actually are not "planned" as spatial systems. Meanwhile, regular planning (e.g., [urban] developments) continues to take place.

These problems are wicked and spatial planning lacks the processes, decision making, and tools to uptake them. Thus, the problem can be stated as follows:

1. Spatial planning is not used as a platform or framework for "solving" these problems.
2. Current spatial planning discourses themselves predominantly focus on decision making within the government for a well-described (planning) problem.
3. Within spatial planning theory, there is a lack of methods and planning approaches for wicked problems.

While changes increasingly appear in a nonlinear fashion, spatial planning increasingly lacks answers.

8.3 Approach

The research presented in this chapter distinguishes five steps (Figure 8.1). In the first step, a literature review about current planning discourses has been conducted in two different ways. In the first place, historical and current planning discourses have been identified and analyzed on their usefulness for wicked problems. Second, articles published in 2010 and 2011 in four international planning journals (*Planning Theory, Planning Theory and Practice, Australian Planner,* and *European Planning Studies*) were analyzed on the merits of containing theories useful to complex problems. This illuminates the common typology of current subjects in planning journals.

In the second step, complexity and planning for cities are explored, because cities are, on the one hand, seen as complex systems (Allen, 1996; Portugali, 2000; Batty, 2005; Dos Santos and Partidáro, 2011) and, on the other hand, because insights from complexity theory could be of use to develop a planning approach capable of dealing with wicked problems.

The central question after the first two steps has been if current planning discourses and/or in combination with scholarly writing on complexity and planning are sufficient of being able to make plans to solve wicked problems? The answer to this question led to the development of swarm planning theory, subject of the third step. A proposition for a planning theory, swarm planning, has been developed. The theory of swarms (Fisher, 2009; Miller, 2010) and tipping points (Gladwell, 2000), lessons from complexity (among others: Schwank, 1965), existing examples of creating plans using the understanding of swarms (Oosterhuis, 2006, 2011), and the use of

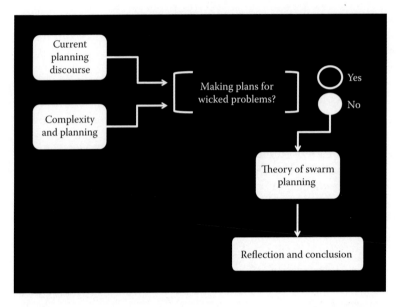

FIGURE 8.1
Schematic representation of the approach to the research presented in this chapter.

complex, adaptive system properties (Roggema et al., 2012) all were used to develop swarm planning theory, which enables planning to incorporate wicked problems.

Finally, the developed theory has been critically reflected upon, and conclusions are drawn.

8.4 Current Planning Paradigms

In this section, a brief overview of spatial planning paradigms is presented. In spite of the fact that it is hardly possible to do justice to existing planning theories and paradigms in one paragraph, this section attempts to capture the main characteristics in order to come to a judgment-*light* of the applicability of each to deal with wicked problems. In-depth study and elaboration are required to provide a more thorough basis for the judgments. For this study, the current planning paradigms are analyzed in two ways. In Section 8.4.1, a selection of well-known paradigms will be briefly described, and their eventual shortcomings in the face of dealing with wicked problems will be examined. Section 8.4.2 will examine all articles published in four international planning journals over the years 2010 and 2011.

8.4.1 Selection of Prevailing Planning Paradigms

In recent planning literature, sparse, but strong, signals can be found illuminating a change in planning paradigm. Scholars such as Newman, Boelens, Miraftab, Davy, and Gunder all, from different angles, point at (the need for) planning "moving away" from its traditional base: the government. In his pledge for postanarchistic or autonomous planning, Newman (2011) emphasizes the power of self-organizing groups and organizations, planning for their own environments outside the governmental and political arena and creating herewith a *disordered order* of spaces that are "becoming." In a debate-provoking paper, Boelens (2010) advocates planning to come from "outside inward," led by actors out of the normal governmental planning arena. Miraftab (2009) describes the informal, insurgent, planning taking place in slums in South Africa, and Davy (2008) promotes an "unsafe planning" in order to establish planning without tightening and dictating regulations. Gunder (2011) pledges to step away from the widespread code of what is unconsciously accepted "good planning," positioning the planner as the one "who knows," meanwhile creating, following Davy, a "non-innovative state of monorationality."

Therefore, based on these publications, the question may be asked: Are we moving toward a new planning paradigm? Before engaging in this discussion, however, a selection of well-known planning paradigms are summarized in Table 8.1.

Reflexive planning offers the framework of thought within which emerging debates take place, concerning the fluidity of relations and interactions in planning processes, and the ways in which these processes influence future developments (e.g., Hillier, 2007; Healey, 2009). In this context, Gunder (2011) calls to challenge all positions that seek the security provided by the planner "who knows." This is a call to challenge what is often, at best, a monorational practice of orthodox and repetitive universal. Already debated planning practices will stand out in different light and only can be opened up for wicked problems when the fundamental properties of Western planning monorationality (Davy, 2008) are left behind, being

1. *"Playing by the rules,"* which in the case of wicked problems no longer rule
2. *"Repeat habitual prior experiences,"* which in regard to wicked problem is useless, because every time the problem appears to be unique
3. Creating a *"non-innovative status quo,"* which is contraproductive if the wicked problem is "already changing again"

According to Davy (ibid.), monorationality must be replaced by an "unsafe" planning practice of polyrationality, where liquid, turbulent, or even wild boundaries of both planning thought and spatial territory can occur—literally, to do "it" without the safety of a condom! This is a planning practice that

TABLE 8.1

Summary of Current Planning Paradigms

Planning Paradigm	Key Characteristics	References	Comments
Positivism	Data inform planning One truth General laws, science based Top down. Prediction is possible once all characteristics are known Requires much data	Allmendinger (2002b) McLoughlin (1969)	Collection of data is hardly possible and requires large investments (Banfield, 1973). Not possible when problems are complex (De Roo, 2006). Simplification of problem is necessary (Lindblom, 1959). Suitable for unpredictable wicked problems?
Incrementalism	Step by step Muddling through Continually building out by small degrees Present policy is satisfying	Lindblom (1959), Dror (1964)	Status quo and compromises (Cates, 1979) Does not fit with a development with significant changes (Jones, 2010)
Post positivist	Subjective knowledge and endless possibilities. Rejection of "master narratives" Spaces and places are always in the process of being made and are unpredictable Structuring processes, stakeholder involvement, and aiming for consensus	Allmendinger (2002a,b), Farmer (1993), cited in Hillier and Cao (2011) Healey (1997), Innes (2004), Innes and Booher (1999, 2004): Collaborative planning Habermas (1987, 1993): communicative rationality Beauregard (1996), Jencks (1987), Allmendinger (2001): post-modernism Forester (1989): communicative planning	Formal neutrality and equality delegitimizes antagonism and dissent as irrational, violent, and undemocratic (Newman, 2011)

(continued)

TABLE 8.1 (continued)

Summary of Current Planning Paradigms

Planning Paradigm	Key Characteristics	References	Comments
Agonism	Acknowledges and respects permanent conflicts in political communication Compromises and consensus Respectful way of disagreement	Mouffe (1993, 1999, 2000, 2005), Hillier (2003), Pløger (2004)	Agonism always takes place within the unacknowledged framework of the state (Newman, 2011)
Reflexive planning	Considers the capacity to change to be the basis Learning by monitoring Actors become continuously engaged in an "actor consulting model" Participatory process of describing, evaluating, and reflecting on ongoing activities Projectivity, creativity, and change, always bearing in mind that the future is uncertain	Beck (1992, 1994), Beck et al. (1994, 2003), Amin (2004), Lombardi (1994), Sabel (1994), De Roo and Porter (2007), Grin and Weterings (2005), Lissandrello and Grin (2011)	Government in the lead The results depend largely on the actors involved

takes risks, accommodates differences, and encourages the new and creative. Several unsafe planning discourses are distinguished. The *actor-relational approach* (Boelens, 2010) allows for a planning approach from "outside inward." Leading actors outside the traditional governmental arena plan their environments more or less autonomously, keeping strong links with the government. *Insurgent planning* (Miraftab, 2009) is a way of informal planning providing structure in the slums of South Africa. Finally, *postanarchism* (Newman, 2011) is "an alternative theoretical model based on the politics of autonomy. The existence of autonomous movements, organizations and political spaces forces us to re-situate the political dimension away from the centricity of the state and toward alternative practices and forms of decision-making. A planning model of this kind would acknowledge and, indeed, construct itself around autonomous planning practices engaged in everyday by people and movements of resistance to statism and capitalism." As Seymour (2011) blogged, the planned manner of the happenings on and around Tahrir Square, Egypt, where networks of tents, toilets, guarded entrances, electricity supply, even garbage

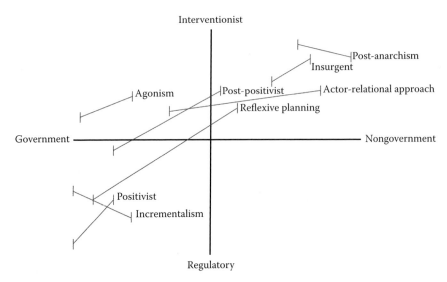

FIGURE 8.2
Framing planning paradigms.

collection and medical care had been arranged (planned) for can be seen as an extreme form of autonomous planning.

The aforementioned planning discourses are placed in a scheme (Figure 8.2), which is defined by two axes: the governmental (inside) or nongovernmental (outside) axis and the regulatory (dominated by rules, rationalism, and belief in data) or interventionist (aiming to influence and direct, make a difference) axis. Both axes were chosen on the merits where planning performs in fundamental different ways. Spatial planning within governments, as it is based on procedures and political consensus, has a different impact compared with spatial planning taking place outside the governmental arena. Also, a type of spatial planning which focuses on regulations and ticking the regulatory boxes is less aiming for spatial interventions and change, compared with spatial planning which intervenes in the spatial layout of particular physical environments.

This scheme shows, on the left hand and lower side, the central role for governments is maintained (among others postpositivism, reflexive planning), attempting to involve external stakeholders in their planning processes, meanwhile setting the rules, if not about the content, then at least in the description of the process or (even worse) procedures. On the other side of the scheme is a spontaneous, autonomous planning model that is capable of instantly reacting and planning immediately for necessities.

8.4.2 Review of Two Years of Planning Journals

The next step in the research is to examine whether unsafe, autonomous, and polyrational theories, concepts, and strategies are discussed in the planning community, and if, in relation with the former, wicked problems are addressed. In order to gain insight as to what extent this specific part exists within the spatial planning debate, two volumes of four spatial planning journals have been analyzed. In total, 275 articles, published in 2010–2011 in the journals of *Planning Theory* (43), *Planning Theory and Practice* (34), *The Australian Planner* (45), and *European Planning Studies* (153), being the leading theoretical and practice-oriented academic planning journals originating from two different continents, have been analyzed. The articles were judged on criteria informing whether in the articles theories, concepts, and strategies are discussed that potentially can deal with wicked problems. The following criteria have been distinguished:

1. *Integration (vs. thematic, specific, single subject)*: A wicked problem cannot be dealt with from a single narrow thematic perspective, because a singular solution for a problem that is wicked enables the problem to evolve into new forms the moment the thematic solution is executed. An integrated approach, in which themes and land-use functions are mutually connected and in which an area is approached as a whole, can deal much easier with unique, new, and suddenly changing problems. Does the article approach problems in an integrative way or is it focusing on a specific theme or subject?

2. *Dynamic (vs. static)*: A division can be made in the aim of planning to stabilize the future or to emphasize dynamic environments, which need to be planned for and/or even need to be created. When wicked problems are taken into account, spatial planning needs to recognize the existence of dynamic, continuous changing spatial settings and configurations. Does the article assume that planning tries to continue the current state or focuses it on dealing with changing environments or subjects?

3. *Intervention (vs. regulatory)*: Planning can be orientated on arranging general and objective regulations that prohibit or allow certain land use or it may aim for a deliberate physical change. A spatial intervention can be realized through design. In general, if problems are wicked, they normally are not dealt with by putting regulations in place, as these problems are essentially unique. Does the article discuss a design approach or a spatial intervention or does it focus on describing regulations and institutions?

4. *Paradigm shift (vs. status quo)*: When problems are new, especially if they are wicked, a new planning paradigm may emerge. The identification of these types of problems at an early stage illuminates their existence in the first place. If so, the early stages of a paradigm shift are announced, even if there are only small rudiments of it visible. In most of the cases, however, planning in its current state, a status quo, is described, which is less suitable in dealing with changing circumstances and wicked problems. Does the article describe planning as it is currently and/or was in the past or does the article focus on identifying a paradigm shift?

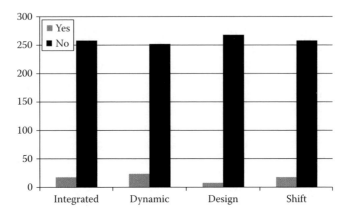

FIGURE 8.3
Number of articles in *Planning Theory, Planning Theory and Practice, Australian Planning,* and *European Planning Studies* (2010–2011) that do and do not discuss integration, dynamics, design, and a paradigm shift.

Having analyzed 275 articles, addressing the question if they contain integrated or thematic, dynamic or stable, interventionist or regulatory, and shift or status quo issues, the results are striking and shocking at the same time (Figure 8.3). The fundamental properties of Western planning monorationality are still around. Even stronger, articles that address dynamic, integrated, intervention topics, and paradigm shifts are hardly found.

The conclusion may be drawn that a very small portion of current planning discourse acknowledges fundamental changes in society, the changes in the environment, and the need to plan for wicked problems. However, the current debate is predominantly in the process of raising awareness and describing what is going on. It addresses the necessity to replace old rules with new ones, which can respond to more complex issues and are based on networks, interrelations, and connections. There are only a few scholars (e.g., Newman's postanarchism and Davy's unsafe planning) who discuss the necessity to start planning in a more "nonlinear" way. In this chapter, the exploration of a planning theory dealing with wicked problems draws upon these scholars and will search where wicked problems are closest related to complexity theory.

8.5 Exploring Complexity

In order to start planning for solving wicked problems, and more specifically planning sustainable energy landscapes, we need to take into account that it is likely that climate change, as well as resource depletion, will force (step) changes (Jones, 2010). That sustainable energy landscapes, as do climate adaptive landscape, have locality-specific characteristics and require to bridge impacts occurring over a wide time range. Therefore, it is useful to explore the potential of complexity theory in three ways. First, we need to understand complex (adaptive) systems, their nonlinearity and the idea that small changes might have big impacts, and the existence of bifurcation points and tipping points. Second, we need to understand cities as self-organizing systems, and third, we need to build upon the former to make knowledge on self-organizing systems available for planning.

8.5.1 Complexity Theory

Many scholars studied the complexity and self-organization of nonlinear dynamic (or adaptive) systems. Among these are the works of Prigogine and Stengers (1984), Gleick (1987), Mitchell Waldrop (1992), Lewin (1993), Cohen and Stewart (1994), and Kauffman (1995), which are further elaborated and explained by authors such as Johnson (2001), Miller and Page (2007), Johnson (2007), and Northrop (2011). Key concepts from complexity theory, which are seen as useful in a planning context, are the *self-organization* of complex systems, the surge for an actor to attractors, depicting a *fitness landscape*, the change and transformation of a complex system in times of crisis, and the existence of *bifurcation*, "the point in time where for identical external conditions various possible structures can exist" (Allen, 1996), and *tipping points*, "the point at which the system 'flips' from one state to another" (Gladwell, 2000).

Adaptation of (or within) systems is an internal process of *self-organization*, which is the tendency in complex systems to evolve toward order instead of disorder (Kauffman, 1993). The state of equilibrium is called attractor. Complex adaptive systems self-organize and adapt in order to remain within their current attractor. The system only shifts to other attractors (alternative states) after a shock that drives the system out of its current state (e.g., sky-rocketing energy prices). Major adjustments are needed, and after the shock, the system will self-organize to achieve those.

The process this system goes through can be represented in the form of a *fitness landscape* (Figure 8.4) (Langton et al., 1992; Mitchell Waldrop, 1992). This fitness landscape includes favorable (the mountaintops) and less-favorable (the valleys) positions. A complex system tends to move, while crossing less-favorable valleys, to the highest possible position in the landscape, the attractor.

At the mountaintop, the adaptive capacity is highest, which allows the system to adapt more easily to changes in its environment. The pathway of the system is represented in Figure 8.5. When a system self-organizes, it strives to reach a higher adaptive capacity by increasing order. When it reaches the mountaintop (B), it will continue to self-organize and increase order. However, by increasing order at this stage, adaptive capacity decreases, causing a less-stable system (the state of fixed and unchangeable regulations and standards), which starts to move toward a new attractor. At this stage, the

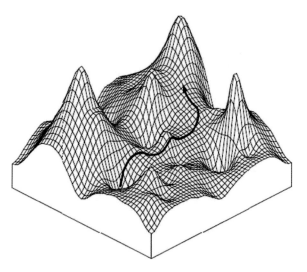

FIGURE 8.4
Fitness landscape showing a complex system moving from a less-favorable to a favorable position or attractor. (From Cohen, J. and Stewart, I., *The Collapse of Chaos: Discovering Simplicity in a Complex World*, Penguin Group Ltd., London, U.K., 1994. With permission.)

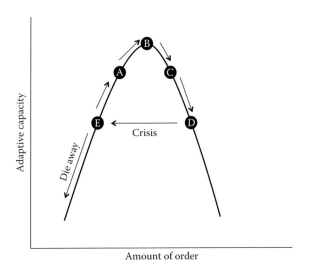

FIGURE 8.5
Typical pathway of a complex system toward the mountaintop (B), evolving toward instability (D), and dying or self-organizing again (E). (After Lietaer, B. and Belgin, S., Big change, In *Of Human Wealth: New Money for a New World*, Citerra Press, Boulder, CO, 2008.)

system is crossing the valley (from D to E) and searching for a new attractor, which can provide the system with renewed adaptive capacity. After reaching point E (a more chaotic state), two things can happen: the system dies (it did not reach/find the other attractor) or it self-organizes in a new way and starts to build up a transformed system by increasing its order again until it reaches its highest adaptive capacity (the mountaintop) again (B). Point E is defined as the bifurcation point or the point where the system fundamentally separates the pathway toward a new equilibrium from the one ending its existence ("die away"), also known as the tipping point, at which the system "flips" from one state to another (Gladwell, 2000).

8.5.2 Cities as Complex Systems

These theoretical concepts have been applied to cities. However, the majority of scholars (Allen, 1996; Portugali, 2000, 2006, 2008; Batty, 2005) use complexity theory mainly to understand self-organizing processes in cities through modeling of reality. Modeling remains a central activity at the intersection of complexity and spatial science (O'Sullivan, 2004) but there is a growing concern about the implicit limitations of this "orientation on modeling" as the relevance of the links between spatial and complexity theories becomes much wider (O'Sullivan et al., 2006). Still, the main attention in recent academic writings is focusing on different kinds of computational representations of spatial analyses, as represented in the thematic issue of *Environment and Planning A* (O'Sullivan et al., 2006) and the representation in models through agent-based modeling or cellular automata, as represented in the articles in the thematic issue of *Environment and Planning B* (Crawford et al., 2005). The question is whether this *mathematicalization* of the city offers more than only an understanding of self-organization in cities, but merely supports cities in dealing with wicked problems, as it lacks the tools to influence the performance of the city. Spaces (and places) are, as described in Portugali (2006), mainly seen as an object to study, analyze, explain, understand, describe, and model. But this understanding is, to my knowledge, hardly used to inform planning and design processes on how to improve the quality of the city or to better respond to and prepare for wicked problems.

8.5.3 Use of Complexity in Planning

As a bridge between the understanding of complexity in cities and planning for it, a key set of interrelated concepts that define a complex system (Manson, 2001) can be helpful.

At the core, the *relationships* between its components and its environment, forming an ever-changing internal structure, determine the whole of the system. Due to the number and complexities of these relationships, it is hardly possible to understand or predict the character of the whole system. Because of the wide array of complex internal relationships, the system is in most cases able to respond to novel, external relationships. In the case that there is no internal component capable of responding to novel, external circumstances, which are for instance induced through climate change, this may end in a catastrophe for the system.

A complex system exhibits *emergence*, for example, the system-wide characteristics stem from interactions among components (Lansing and Kremer, 1993) and are thus much more than a simple addition of component qualities. It is difficult to anticipate change beyond the short term, because other components of the system adjust to the intervention in addition to other changes in the environment (Youssefmir and Huberman, 1997). Any single change can have far-reaching large-scale effects due to not understanding emergence from complexity (Lansing and Kremer, 1993).

A complex system performs *change and evolution* through three different capabilities: (1) self-organization, for example, the capability to adjust its internal structures to better interact with a changing environment; (2) development of a dissipative structure, allowing the system to suddenly cross to a more organized state after being a certain period in a highly unorganized state (Schieve and Allen, 1982); and (3) self-organized criticality, allowing the system to keep the balance between nearly collapsing and not doing so, caused by an internal restructuring, almost too rapid to accommodate, but necessary for survival (Scheinkman and Woodford, 1994).

Finally, *path dependency* defines the development of a system as "a trajectory as function of past states" (O'Sullivan, 2004). This may be true for most systems. Portugali (2008) demonstrated that, in regard to planning, the fact a plan has been released causes a *reverse* form of path dependency in the sense that the trajectory is a function of (not yet realized) future states. All of the former properties are found through the study of ecological and, to a lesser extent, economic systems.

The question, however, is whether we can use the knowledge derived from, mainly, ecosystems for artificial systems, such as cities. As demonstrated by Simon (1999) (described in Portugali, 2006), we can use the findings of natural science to apply in artificial systems, but only to a limited extent. Humans are self-organizing their systems in the same way as ants do theirs, but only to modify the system for their own benefit. They both follow the principles of the *homo economicus* with only "maximizing profits" on their mind. As long as the individuals in the system have this singular relationship with the system (e.g., the interaction between the individuals shape the system), ants and humans are identical. However, as demonstrated by Portugali (2000, 2008), socioeconomic systems, such as cities and landscapes, exhibit a dual complexity: the city as a whole is a complex adaptive system as is each of its parts.* This means that the whole can no longer be explained by the singular behavior of individual components. As it is already difficult to encourage a single complex system to move toward increased adaptive capacity, the dual

---

* Portugali defines them as agents, e.g., human or organizational entities (Portugali, 2008).

complexity of interactions and feedbacks within landscapes makes it nearly impossible to predict the results from any intervention (Portugali, 2008).

Learning from nature again, most systems performing *swarm behavior* represent high resiliency, lessening the impact of uncertainties, complexity, and change through the development of emerging patterns and structures (Van Ginneken, 2009). Swarms (see Fisher, 2009; Miller, 2010) are self-organizing systems in preparing and responding to changing circumstances, which is, according to Van Ginneken (2009), achieved through (1) the interactions taking place between a large number of similar and free moving "agents," which (2) react autonomously and quickly toward one another and their surrounding, resulting in (3) the development of a collective new entity and a coherent larger unity of higher order.

So if we want a system to reach a higher adaptive capacity, offering better possibilities to deal with wicked problems, we need to encourage swarm behavior. As demonstrated in the design of buildings, Oosterhuis (2006, 2011) increases the adjustability of the building by programmatic labeling and tagging of building elements, enabling buildings to customize temporary desires or changing demands, hence encouraging swarm behavior. The aforementioned "ingredients" form the basis for developing the theory of swarm planning.

8.5.4 Proposition: Swarm Planning

The objective is to present the first contours and basic elements of swarm planning, which ultimately aims to increase the potential of a landscape or city to deal with wicked problems, such as climate change. Elaborating the aforementioned, this means that if the landscape could perform swarm behavior, it increases its capacity to deal with uncertainty, complexity, and change (e.g., to deal with wicked problems). Therefore, a planning theory that enables swarm behavior to occur supports landscapes to reach higher levels of adaptive capacity. This planning theory, swarm planning, needs to take at its core the dual complexity of the landscape and therefore to combine complex behavior of the elements of the system and the complex adaptive behavior of the system as a whole. And thus, swarm planning needs to actively intervene on both levels (the whole and its parts) of the *dual complex* landscape.

8.5.4.1 Intervention

At the level of the whole, an intervention needs to take place in order to start the swarm to behave in the first place. In current theory, tipping points are identified after they have occurred (Gladwell, 2000) or identify the patterns that announce these points (Scheffer, 2009), but they are not planned. In essence, it describes the process of an evolving system, becoming unstable, ends up in a crisis, "tips," and transforms through self-organization to another stable state. However, in the case of climate change, this system change preferably anticipates the actual change. Hence, an early intervention must allow the system to be able to "flip." We need to actively intervene in the system to start self-organizing processes to anticipate the wicked problem. Hence, this requires an intervention to get things started.

Obviously, the difficulty is to identify the location, the type, and the actor to intervene. As demonstrated elsewhere (Roggema et al., forthcoming), network theory holds the key to identifying the location. The type of intervention cannot be otherwise determined than through the local context (existing landscape combined with specific wicked problem). The actor identifies the point where and the type of intervention. The person or institution most eligible to decide upon this is the problem owner, not necessarily the government.

8.5.4.2 Freedom to Emerge

The second level is the level of the parts (the elements in the landscape). At this level, the components of the system need to make use of their joint capabilities to perform as a system as a whole. Only then the system is able to produce swarm behavior and achieve a higher adaptive capacity. Therefore, interacting relationships need to be provided with the qualities allowing them to develop emergent properties, to self-organize, and to change (Manson, 2001). The hypothesis is that if the landscape elements are attributed with the capabilities as described earlier, they will support swarm behavior of the whole system. Once the individual components are attributed with these capabilities and free self-organization takes place, the system will strive for the most optimal stable state (in general, the mountaintop in the fitness landscape), which represents the highest adaptive capacity.

This theoretical proposition requires further research on the question of how individual landscape elements can be attributed with qualities to allow them to perform emergent behavior, self-organize, and change. The first attempts to answer this question have been undertaken in the work of Kas Oosterhuis (2006, 2011), who attributed swarm characteristics to building elements, and by linking complex adaptive system properties to landscape entities (Roggema, 2011). However, further research is required in this area.

The proposition of swarm planning combines directive steering, in the form of an active design intervention (system level), with the freedom of individual landscape elements to shape (and self-organize) the system. The outcome of this process is fundamentally unpredictable, but this does not mean that we cannot be confident that the system, when performing swarm behavior, reaches a higher adaptive capacity (or to speak with the terms of complexity theory: reaches the top of the fitness landscape).

When swarm planning and the proposed definition of sustainable energy landscapes—"a physical environment that can be sustained on the basis of locally available renewable energy sources without compromising…"—are linked, swarm planning can be seen as the way sustainable energy landscapes can be planned and realized. What is unclear in the proposed definition however is how locally available energy sources are connected at a systemic level. Emphasis could be put on the location where abundance of renewable energy sources are found to be most productive (the strategic or focal points in the system), linked with locations where the demand for energy is largest. Identification of these crucial points allows the system to adjust itself toward higher adaptive capacities.

The second level of swarm planning, the individual elements allowed to shape the system, can be easily linked with the locally available renewable energy sources referred to in the definition of sustainable energy landscapes. However, this part of the definition could be adjusted with emphasis on the self-organizing capabilities of the individual elements, being the locally available energy sources (e.g., the freedom to increase the use of wind power in case the wind increases on a long-term change basis or, preferably, on the basis of immediate changeable configurations, numbers, and patterns).

Part of a planning theory must be, in my opinion, besides a theoretical basis as presented earlier, a practical strategy and real-world applications. The theoretical basis has been used and translated into a practical approach (Roggema et al., 2012) with the five-layer strategy as the centerpiece, in which the first two layers identify the point of intervention, layer three arranges and defines the freedom to emerge, and layers four and five allow for the individual components to self-organize. The first practical applications of this theory have also been identified. In the work of Massoud Amin, the principle of self-organization in order to reach higher levels of agility in the energy network (Massoud Amin and Horowitz, 2007; Massoud Amin, 2008a,b, 2009) can be explained as an early form of swarm planning "avant la lettre." A second body of knowledge has been developed, designing "swarm" landscapes for regional climate adaptation (Roggema, 2008a,b; Roggema and Van den Dobbelsteen, 2008).

8.6 Conclusion

In this chapter, it has been demonstrated that current planning discourses are strongly focused on the government as major actor and rely mainly on existing rules, regulations, and established procedures. Research on the current academic discourse, as represented in four international planning journals, illuminates the scarcity of articles focusing on strategies and practices that focus on interventions, incorporate a changing and dynamic future, and emphasize a paradigm shift. There exists a gap in planning practice and theory that will lead to suboptimal solutions with respect to climate change, both in developing sustainable energy landscapes and climate adaptive landscapes. This is seen as very risky, because, even if decision makers should decide now that planning practice should implement swarm planning immediately, such a change will take much time. This time lag might imply that it becomes too late to implement the required changes in due time.

Complexity theory might open the opportunity to integrate the characteristics of wicked problems in spatial planning and, as demonstrated in this chapter, has been subject of debate to link complexity and city and geography. Unfortunately, complexity theory is mainly used in mathematical modeling to better understand self-organizing processes in cities and not to identify design interventions and to increase the capability of cities and landscapes to deal with wicked problems. This leads to suboptimal preparation of communities in mitigating and adapting to climate change.

Therefore, a proposition is launched in this chapter to develop a planning approach, which can integrate complexity theory and planning. "Swarm planning" is an attempt to do so, and, learning from the fact that cities have been attributed with a dual complexity (Portugali, 2000), it identifies two major levels of intervention: the whole system—a level at which a strategic intervention is required—and the level of the individual components—to which the properties of complex adaptive systems need to be attributed in order to allow free emergence. Both levels in conjunction are able to perform swarm behavior, which improves resiliency through lessening the impact of uncertainties, complexity, and change.

Compared with the way spatial planning is practiced in many institutions, thinking in points, enforcing change and free emergence, is the opposite of current practice. Generally, tipping points are not sought, but comprehensive developments are seen as the interventions, and these comprehensive interventions are planned in great detail and for entire areas, not allowing them to develop freely. The aversion against tipping points, and surprises, and the willing to paternalize the entire planning process, including its detailed execution, is grounded in the political culture in many countries where risk has to be avoided and uncertainties or "uncontrollabilities" must be abandoned. However, pursuing the existing (and historical) path-dependent political pathways will lead to "more of the same" policy, which, and this is for certain, will not produce the spatial interventions that are required to deal with wicked problems.

The results of swarm planning are, partly, unpredictable, and this is, especially to the responsible decision makers, a danger, but it is also a *conditio sine qua non*. Because in swarm planning, the new state of the system is undefined, and not possible to define either, there always is the danger of ending up with the "wrong outcome." However, continuing on the same pathway of not adjusting will end in repetition of history, and this will certainly not bring the answers to fundamental different problems of the future. Having said this, there is a lack of understanding of what the future system, planned through swarm planning, may be, and more research can be carried out in this field. However, given the unpredictable future state of complex adaptive systems, it can be questioned whether more understanding will shine brighter lights on the actual future of the system.

This leaves alone the potential of swarm planning to be used in the design of sustainable energy landscapes and cities. As has been demonstrated earlier (Stremke et al., 2011; Van den Dobbelsteen et al., 2011), the use of sustainable energy principles, such as *exergy* and *locality* in the design of landscapes and neighborhoods, leads to more sustainable energy use and supply and, moreover, the design of attractive and sustainable energy landscapes.

Finally, swarm planning is positioned in a revised paradigm framework (Figure 8.6). On the axes of inside–outside government and regulatory–interventionist, swarm planning can be positioned at the upper end of "interventionistic" approaches. It has potentially a widespread range of applications, both within governmental contexts and among individual institutions outside the government. One of the biggest challenges in swarm planning will be to practice an interventionist approach within governmental policy making.

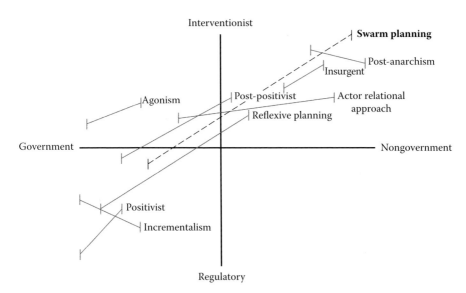

FIGURE 8.6
Swarm planning positioned in the paradigm scheme.

References

Allen, P.M. (1996) *Cities and Regions as Self-Organising Systems, Models of Complexity.* London and New York: Taylor & Francis.

Allmendinger, P. (2001) *Planning in Postmodern Times.* London, U.K.: Routledge.

Allmendinger, P. (2002a) Towards a post-positivist typology of planning theory. *Planning Theory* **1** (1): 77–99.

Allmendinger, P. (2002b) *Planning Theory.* New York: Palgrave.

Amin, A. (2004) An institutionalist perspective on regional economic development. In: T.J. Barnes, J. Peck, E. Sheppard, and A. Tickell (Eds.) *Reading in Economic Geography.* Oxford, U.K.: Blackwell.

Banfield, E.C. (1973) Ends and means in planning. In: A. Faludi (Ed.) *A Reader in Planning Theory. Urban and Regional Planning Series,* Vol. 5. Oxford, New York, Toronto, Sydney, Paris, Frankfurt: Pergamon Press.

Batty, M. (2005) *Cities and Complexity, Understanding Cities with Cellular Automata, Agent-Based Models, and Fractals.* Cambridge, Mass, London: The MIT press.

Beauregard, R. (1996) Between modernity and post-modernity: The ambiguous position of US planning. In: S. Campbell and S. Fainstein (Eds.) *Readings in Planning Theory.* Oxford, U.K.: Blackwell.

Beck, U. (1992) *Risk Society: Towards a New Modernity.* (trans. M. Ritter). London, U.K.: Sage.

Beck, U. (1994) The reinvention of politics. In: U. Beck, A. Giddens, and S. Lash (Eds.) *Reflexive Modernization: Politics, Tradition and Aesthetics in the Modern Social Order,* pp. 1–55. Cambridge, U.K.: Polity.

Beck, U., W. Bonss, and C. Lau (2003) The theory of reflexive modernization: Problematic, hypotheses and research programme. *Theory, Culture & Society* **20**: 1–33.

Beck, U., A. Giddens, and S. Lash (1994) *Reflexive Modernization: Politics, Tradition and Aesthetics in the Modern Social Order.* Cambridge, U.K.: Polity.

Belin, H. (2008) The Rifkin vision. We are in the twilight of a great energy era. *European Energy Review,* special edition, June 2008.

Boelens, L. (2010) Theorizing practice and practising theory: Outlines for an actor-relational—Approach in planning. *Planning Theory* **9** (1): 28–62.

Campbell, C.J. (1999) *The Imminent Peak of World Oil Production.* Presentation to a House of Commons All-Party Committee, July 7, 1999. URL: http://www.hubbertpeak.com/campbell/commons.htm. Accessed: May 10, 2010.

Campbell, C.J. (2002a) *Peak Oil: An Outlook on Crude Oil Depletion.* www.mbendi.com/indy/oilg/p0070.htm#27. Accessed: May 10, 2010.

Campbell, C.J. (2002b) *Forecasting Global Oil Supply 2000–2050.* M. King Hubbert Center for Petroleum Supply Studies, Newsletter # 2002/3.

Campbell, C.J. and J.H. Laherrrère (1998) The end of cheap oil. Global production of conventional oil will begin to decline sooner than most people think, probably within 10 years. *Scientific American,* 1998–03.

Cates, C. (1979) Beyond muddling: Creativity. *Public Administration Review* **39** (6): 527–532.

Cohen, J. and I. Stewart (1994) *The Collapse of Chaos: Discovering Simplicity in a Complex World.* London, U.K.: Penguin Group Ltd.

Crawford, T.W., J.P. Messina, S.M. Manson, and D. O'Sullivan (2005) Guest editorial. *Environment and Planning B* **32**: 792–798.

Dam F. Van and en K.-J. Noorman (Eds.) (2005) *Grounds for Change: Bridging Energy Planning and Spatial Design Strategies.* Charrette Report. Groningen, the Netherlands: Grounds for Change/IGU.

Davy, B. (2008) Plan it without a condom! *Planning Theory* **7** (3): 301–317.

Dobbelsteen A. van den, S. Broersma, and S. Stremke (2011) Energy potential mapping for energy-producing neighborhoods. *International Journal of Sustainable Building Technology and Urban Development* **2** (2): 170–176.

Dos Santos, F.T. and M.R. Partidário (2011) SPARK: Strategic planning approach for resilience keeping. *European Planning Studies* **19** (8): 1517–1536.

Dror, Y. (1964) Muddling through-'science' or inertia? *Public Administration Review* **24**: 154.

Dror, Y. (1973) The planning process: A facet design. In: A. Faludi (Ed.) *A Reader in Planning Theory. Urban and Regional Planning Series*, Vol. 5. Oxford, New York, Toronto, Sydney, Paris, Frankfurt: Pergamon Press. (Reprinted, original publication in: *International Review of Administrative Sciences* **29** (1) (1963): 46–58).

Farmer, J. (1993) A poststructuralist analysis of the legal research process. *Law Library Journal* **85**: 391–404.

Fisher, L. (2009) *The Perfect Swarm, the Science of Complexity in Everyday Life.* New York: Basic Books.

Forester, J. (1989) *Planning in the Face of Power.* Berkeley, CA: University of California Press.

Ginneken J. van (2009) *De kracht van de zwerm.* Amsterdam/Antwerpen: Uitgeverij Business Contact.

Gladwell, M. (2000) *The Tipping Point.* New York: Little, Brown and Company, Time Warner Book Group.

Gleick, J. (1987) *Chaos, Making a New Science.* Harmondsworth, England: Penguin Books Ltd.

Grin, J. and R. Weterings (2005) Reflexive monitoring of system innovative projects: Strategic nature and relevant competences. Paper prepared for the sixth Open Meeting of the Human Dimensions of Global Environmental Change Research Community, University of Bonn, Bonn, Germany, October 2005.

Gunder, M. (2011) Fake it until you make it, and then... *Planning Theory* **10** (3): 201–212.

Gunder, M. and J. Hillier (2009) *Planning, in Ten Words or Less.* London, U.K.: Ashgate.

Habermas, J. (1987) *The Theory of Communicative Action*, Vol. 2. Cambridge, U.K.: Polity Press.

Habermas, J. (1993) *Justification and Application: Remarks on Discourse Ethics.* Cambridge, U.K.: Polity Press.

Healey, P. (1997) *Collaborative Planning: Shaping Places in Fragmented Societies.* London, U.K.: Palgrave.

Healey, P. (2009) In search of the "strategic" in spatial strategy making. *Planning Theory and Practice* **10** (4): 439–457.

Hillier, J. (2003) Agon'ising over consensus: Why Habermasian ideals cannot be 'real'. *Planning Theory* **2** (1): 37–59.

Hillier, J. (2007) *Stretching Beyond the Horizon. A Multiplanar Theory of Spatial Planning and Governance.* Aldershot, U.K.: Ashgate.

Hillier, J. and K. Cao (2011) Enabling Chinese strategic spatial planners to paint green dragons. *Planning Theory* **10** (4): 366–378.

Innes, J. (2004) Consensus building: Clarifications for the critics. *Planning Theory* **3** (1): 5–20.

Innes, J. and D. Booher (1999) Consensus building and complex adaptive systems— A framework for evaluating collaborative planning. *APA Journal*, Autumn **65** (4): 412–423.

Innes, J. and D. Booher (2004) Reframing public participation: Strategies for the 21st century. *Planning Theory* **5** (4): 419–436.

Jencks, C. (1987) *Post-Modernism, The New Classicism in Art and Architecture*. London, U.K.: Academy Editions.

Johnson, N. (2007) *Simply Complexity, a Clear Guide to Complexity Theory*. Oxford, U.K.: Oneworld Publications.

Johnson, S. (2001) *Emergence. The Connected Lives of Ants, Brains, Cities and Software*. New York, London, Torontoi, Sydney: Scribner.

Jones, R. (2010) A risk management approach to climate change adaptation. In: R.A.C. Nottage, D.S. Wratt, J.F. Bornman, and K. Jones (Eds.) *Climate Change Adaptation in New Zealand: Future Scenarios and Some Sectoral Perspectives*, pp. 10–25. Wellington, New Zealand: Climate Change Centre.

Kauffman, S.A. (1993) *The Origin of Order: Self-Organisation and Selection in Evolution*. New York: Oxford University Press.

Kauffman, S. (1995) *At Home in the Universe. The Search for Laws of Self-Organisation and Complexity*. New York, Oxford: Oxford University Press.

Langton, C.G., C. Taylor, J.D. Farmer, and S. Rasmussen (1992) *Artificial Life II. Studies in the Sciences of Complexity. Proceedings*, Vol. 10. Redwood City, CA: Santa Fe Institute.

Lansing, J.S. and J.N. Kremer (1993) Emergent properties of Balinese water temple networks: Co-adaptation on a rugged fitness landscape. *American Anthropologist* **95** (1): 97–114.

Lewin, R. (1993) *Complexity, Life at the Edge of Chaos*. London, U.K.: JM Dent Ltd.

Lietaer, B. and S. Belgin (2008) Big change. In: *Of Human Wealth: New Money for a New World*. Boulder, CO: Citerra Press.

Lindblom, C.E. (1959) The Science of "Muddling Through". *Public Administration Review* **19** (2): 79–88.

Lissandrello, E. and J. Grin (2011) Reflexive planning as design and work: Lessons from the port of Amsterdam. *Planning Theory and Practice* **12** (2): 223–248.

Lombardi, M. (1994) L'evoluzione del distretto industriale come sistema informativo: Alcuni spunti di riflessione. *L'Industria* **15** (3): 523–535.

Manson, S.M. (2001) Simplifying complexity: A review of complexity theory. *Geoforum* **32**: 405–414.

Massoud Amin, S. (2008a) Resilience and self-healing challenges: Present-possible futures. *CRITIS'08, Third International Workshop on Critical Information Security*. Frascati-Rome URL: http://critis08.dia.uniroma3.it/pdf/CRITIS_08_9.pdf. Accessed: November 8, 2011.

Massoud Amin, S. (2008b) The smart self-healing electric power grid: Challenges in security and resilience of energy infrastructure. *Proceedings of the 2008 IEEE Power Engineering Society General Meeting*. Pittsburgh, PA, July 2008. URL: http://panda.ece.utk.edu/w/images/e/e7/Amin_IEEE_PES_GM_July_2008.pdf. Accessed: November 8, 2011.

Massoud Amin, S. (2009) *Smart Grid: Opportunities and Challenges—Toward a Stronger and Smarter Grid*. Keynote at the Smart Grid Workshop sponsored by Sandia National Laboratories at the MIT Energy Conference Cambridge, Massachusetts, March 6, 2009. URL: http://panda.ece.utk.edu/w/images/9/92/2009_SNL_Amin_MIT_Energy.pdf. Accessed: November 8, 2011.

Massoud Amin, S. and B. Horowitz (2007) Toward agile and resilient large-scale systems: Adaptive robust national/international infrastructures. In: *Flexibility with Business Excellence in the Knowledge Economy*, pp. 247–265, November 2007, ISBN: 81-903397-7-X. URL: http://panda.ece.utk.edu/w/images/1/1a/Amin_Agile_Resilient_Systems.pdf. Accessed: November 8, 2011.

McLoughlin, B. (1969) *Urban and Regional Planning: A Systems Approach*. London, U.K.: Faber and Faber.

Miller, J.H. and S.E. Page (2007) *Complex Adaptive Systems. An Introduction to Computational Models of Social Life*. Princeton, Oxford: Princeton University Press.

Miller, P. (2010) *The Smart Swarm*. New York: The Penguin Group.

Miraftab, F. (2009) Insurgent planning: Situating radical planning in the Global South. *Planning Theory* **8** (1): 32–50.

Mitchell Waldrop, M. (1992) *Complexity. The Emerging Science at the Edge of Order and Chaos*. New York, London, Toronto, Sydney: Simon and Schuster Paperbacks.

Mouffe, C. (1993) [2005 edition] *The Return of the Political*. London, U.K.: Verso.

Mouffe, C. (1999) Deliberative democracy or agonistic pluralism. *Social Research* **66** (3): 745–758.

Mouffe, C. (2000) *The Democratic Paradox*. London, U.K.: Verso.

Mouffe, C. (2005) *On the Political*. New York: Routledge.

Newman, S. (2011) Postanarchism and space: Revolutionary fantasies and autonomous zones. *Planning Theory* **10** (4): 344–365.

Northrop, R.B. (2011) *Introduction to Complexity and Complex Systems*. Boca Raton, London, New York: CRC Press, Taylor & Francis Group.

Oosterhuis, K. (2006) Swarm Architecture II. In: K. Oosterhuis and L. Feireiss (Eds.) *Game, Set and Match II, On Computer Games, Advanced Geometries and Digital Technologies*. Rotterdam, the Netherlands: Episode Publishers.

Oosterhuis, K. (2011) *Towards a New Kind of Building, A Designer's Guide to Nonstandard Architecture*. Rotterdam, the Netherlands: NAi Uitgevers.

O'Sullivan, D. (2004) Complexity science and human geography. *Transactions of the Institute of British Geographers* **29**: 282–295.

O'Sullivan, D., S.M. Manson, J.P. Messina, and T.W. Crawford (2006) Guest editorial. *Environment and Planning A* **38**: 611–617.

PBL, KNMI, and WUR (2009) *News in Climate Science and Exploring Boundaries, a Policy Brief on Developments Since the IPCC AR4 Report in 2007*. Bilthoven, the Netherlands: PBL Publication number 500114013.

Pløger, J. (2004) Strife: Urban planning and agonism. *Planning Theory* **3** (1): 71–92.

Portugali, J. (2000) *Self-Organisation and the City*. Berlin, Heidelberg, New York: Springer-Verlag.

Portugali, J. (2006) Complexity theory as a link between space and place. *Environment and Planning A* **38** (4): 647–664.

Portugali, J. (2008) Learning from paradoxes about prediction and planning in self-organising cities. *Planning Theory* **7** (3): 248–262.

Prigogine, Y. and I. Stengers (1984) *Order Out of Chaos. Man's New Dialogue with Nature*. New York: Bantam Books, Inc.

Ramirez, R., J.W. Selsky, and K. van der Heijden (Eds.) (2008) *Business Planning for Turbulent Times, New Methods for Applying Scenarios*. London, Sterling, VA: Earthscan.

Richardson, K., W. Steffen, H.J. Schellnhuber, J. Alcamo, T. Barker, D.M. Kammen, R. Leemans, D. Liverman, M. Munasinghe, B. Osman-Elasha, N. Stern, O. Wæver (2009) *Climate Change—Synthesis Report: Global Risks, Challenges and Decisions, Copenhagen 2009*. Copenhagen, Denmark: University of Copenhagen.

Rifkin, J. (2002) *The Hydrogen Economy: The Creation of the World-Wide Energy Web and the Redistribution of Power on Earth*. New York: Penguin Group (USA) Inc.

Rittel, H. and M. Webber (1973) Dilemmas in a general theory of planning. *Policy Sciences* 4: 155–169. Amsterdam, the Netherlands: Elsevier Scientific Publishing Company, Inc., 1973. (Reprinted in Cross, N. (Ed.) (1984) *Developments in Design Methodology*, pp. 135–144. Chichester, U.K.: John Wiley & Sons.

Roggema, R. (2008a) The use of spatial planning to increase the resilience for future turbulence in the spatial system of the Groningen region to deal with climate change. *Proceedings UKSS -Conference*. Oxford, U.K.

Roggema, R. (2008b) Swarm Planning: A new design paradigm dealing with long term problems associated with turbulence. In: R. Ramirez, J.W. Selsky, and K. Van der Heijden, *Business Planning for Turbulent Times, New Methods for Applying Scenarios*, pp. 103–129. London, U.K.: Earthscan.

Roggema, R. (2011) Swarming landscapes, new pathways for resilient cities. *Proceedings 4th International Urban Design Conference 'Resilience in Urban Design'*. Surfers Paradise, Australia.

Roggema, R. and A. van den Dobbelsteen (2008) Swarm planning: Development of a new planning paradigm, which improves the capacity of regional spatial systems to adapt to climate change. *Proceedings World Sustainable Building Conference* (sb08), Melbourne, Australia.

Roggema, R., A. van den Dobbelsteen, and P. Kabat (2012) Towards a spatial planning framework for climate adaptation. *SASBE* 1(1).

Roggema, R., T. Vermeend, and A. van den Dobbelsteen (forthcoming) Incremental change, transition or transformation? Optimising change pathways for climate adaptation in spatial planning. *Environment, Development and Sustainability* (in review).

Roo, G. de (2006) *Understanding Planning and Complexity—A Systems Approach*. AESOP-working group complexity and planning; IIIrd meeting; Cardiff. URL: http://www.aesop-planning.com/Groups_webpages/COMPLEX/cardiff/DeRoo.pdf. Accessed: November 8, 2011.

Roo, G. de and G. Porter (2007) *Fuzzy Planning: The Role of Actors in a Fuzzy Governance Environment*. Aldershot, U.K.: Ashgate.

Sabel, C.F. (1994) Learning by monitoring: The institutions of economic development. In: N. Smelser and R. Swedberg (Eds.) *Handbook of Economic Sociology*. Princeton, NJ: Princeton University Press.

Scheffer, M. (2009) *Critical Transitions in Nature and Society*. Princeton and Oxford: Princeton University Press.

Scheinkman, J.A. and M. Woodford (1994) Self-organised criticality and economic fluctuations. *American Economic Review* 84 (2): 417–421.

Schieve, W.C. and P.M. Allen (Eds.) (1982) *Self-Organisation and Dissipative Structures: Applications in the Physical and Social Sciences.* Austin, TX: University of Texas Press.

Schwank, T. (1965) *Sensitive Chaos, the Creation of Flowing Forms in Water and Air.* Forest Row, England: Rudolf Steiner Press.

Sergeev, V., R. Roggema, V. Artrushkin, W. Mallon, and E. Alekseenkova (2009) What is the price to use gasoline in contemporary society? The Groningen experience. In: R. Roggema (Ed.) *INCREASE 2, International Conference on Renewable Energy Approaches for the Spatial Environment, Conference Proceedings.* Groningen, the Netherlands: Province of Groningen.

Seymour R. (2011) Towards a new model commune. Lenin's tomb blog: http://lenin-ology.blogspot.com/2011/03/towards-new-model-commune.html. Accessed November 8, 2011.

Simon, H.A. (1999) *The Sciences of the Artificial.* Cambridge, U.K.: MIT Press.

Sommerkorn, M. and S.J. Hassol (Eds.) (2009) *Arctic Climate Feedbacks: Global Implications.* Oslo, Norway: WWF International Arctic Programme.

Stremke, S., A. van den Dobbelsteen, and J. Koh (2011) Exergy landscapes: Exploration of second-law thinking towards sustainable landscape design. *International Journal of Exergy* **8** (2): 148–174.

Tin, T. (2008) *Climate Change: Faster, Stronger, Sooner, an Overview of the Climate Science Published Since the UN IPCC Fourth Assessment Report.* Brussels, Belgium: WWF European Policy Office.

Youssefmir, M. and B.A. Huberman (1997) Clustered volatility in multiagent dynamics. *Journal of Economics Behavior and Organization* **31** (1): 101–118.

9

Planning Sustainable Energy Landscapes: From Collaborative Approaches to Individuals' Active Planning

Claudia Basta, Wim van der Knaap, and Gerrit J. Carsjens

CONTENTS

9.1 Introduction

In this chapter, we depart from recalling early writings on the epochal transition toward a renewable energy era that seminal scholars envisioned to occur at the turn of the past two centuries (Hayes 1977, Van Til 1979). These groundbreaking writings initiated the discussion on the future perspectives of urban planning in relation to the transition toward a nonfossil era, providing in so doing a precious source of reflection for the current research in the field. The main intuition of these early authors was that a planning system designed to support a global energy transition would have required the combination of major involvement by the side of individuals and strong leadership by the side of decision makers, both based on a strong sense of commitment to the making of a more sustainable future (Van Til 1979).

This notion of commitment is the starting point of this contribution, proposing a perspective on the planning of future energy landscapes based on the centrality of individuals' responsibility to take action. This perspective owes to recent climate ethics literature (Gardiner 2004, Gardiner et al. 2010) and constitutes a development from the collaborative planning paradigm, based upon a constructivist notion of knowledge and its transposition into shared planning objectives by means of societal involvement (Healey 2003), to an active planning approach, based upon a normative conception of the intergenerational responsibility of individuals to collaborate to the achievement of evidence-based sustainability objectives. On these bases, we envision a planning framework within which individuals ought to collaborate in energy transition processes by choosing among equivalent forms of contributions on the ground of a fair distribution of burdens and benefits. The resulting planning practice would be bent toward the promotion and facilitation of individuals' active participation to, particularly, the siting of renewable energy technologies, the subscription of energy-saving programs, and generally the active involvement of each individual in the body of interventions required to enhance a sustainable energy transition.

To elaborate this perspective, this chapter builds upon interdisciplinary literature, particularly climate ethics and planning theory literature, and the lessons learned from selected projects developed in the framework of the Landscape Architecture and Planning MSc program (in the following: LAR program) of Wageningen University, the Netherlands. The chapter is structured as follows: Section 9.2 introduces the main theories of reference at length and elaborates the theoretical framework of the following discussion on planning. Section 9.3 discusses the more practical implications of the siting of renewable energy installations and provides some important clarifications regarding the relevant spatial implications. In Section 9.4, some case studies that highlight the centrality of individual agency to the making of sustainable energy landscapes are briefly illustrated. In Section 9.5, we finally draw some conclusions on the planning approach emerging from the combination of the proposed theoretical perspective and the illustrated empirical investigations, so to finally indicate some possible research horizons.

9.2 New Centrality of Individual Agency: Toward an Active Planning Paradigm

Despite it taking decades to root onto solid scientific ground, the awareness of the cause–consequence relationship between anthropogenic global warming and climate change has penetrated virtually all disciplinary domains, escaping the cautious fences of academic speculation to finally penetrate public discussions worldwide. Nowadays, "[It] has become an issue about which it is virtually impossible to listen to any of the broadcast media without hearing mention (…). Images of the plight of (cuddly) famished polar bears clearly make good copy, but it is symptomatic of the transition of debate about climate change from the scientific and academic to the popular" (Campbell 2006, p. 201).

In front of these images of crumbling ecosystems, global society becomes aware of two complementary realities: the enormity of its cumulative impact on the environment and the correspondingly infinitesimal power of individuals in front of the urgency of its abatement. Considering the featuring component of collective agency that characterizes the "making of plans," this is likely the reason why prominent authors wondered, provocatively, whether the question of climate change is indeed "too big" for spatial planning to address (Campbell 2006). Notably, authoritative attempts to bridge climate change as a problem to spatial planning as devices of solutions were made in the past years (Meyer et al. 2010, Schmidt-Thomé 2006, Wilson and Piper 2010 among others). Nevertheless, the general impression is that the planning discipline, rather than having taken early intellectual responsibility of arguing on the *causes* of anthropogenic global warming, entered the debate on climate by arguing on possible solutions to the manifestations of its consequences. In simple words, spatial planners seem to reason more in terms of *adaptation* than of *prevention* (Füssel 2007, Füssel and Klein 2004, Wilson 2007). This is a welcomed contribution as it mirrors the way in which the phenomenon of climate change is being framed within most other disciplines. However, by addressing the question of the new spatial planning perspectives needed to advance the transition toward a nonfossil energy era, this contribution tries to extend the role of spatial planning toward devising preventive measures rather than mitigation measures only.

Whereas this might seem a deviation from mainstream planning research, it has to be noted that, to the contrary, in a somehow premonitory article of 1979 Van Til indicated in an energy crisis, likely to occur at the end of the past century, the most urgent call the planning discipline should have responded to by means of preventive adaptation, anticipating, in so doing, many of the considerations at the core of the contributions collected in this volume. In his work, Van Til recalls the work of Hayes on the future of energy supply, wherein the latter had rightfully asserted how the future of energy lies in renewable energy sources like wind, water, biomass, and direct sunlight (Hayes 1977). According to Hayes, a supply up to 75% of the expected energy demand in 2025 could have been realized by renewables, provided sufficient saving and efficiency policies would have been promoted in parallel.* Departing from this argument, Van Til suggested that "the prudent planner should consider it when making decisions and recommendations that will involve energy use and conservation for the coming decades and centuries" (Van Til 1979, p. 321). His main conclusion was that any *responsible* form of planning should have taken into consideration the possible spatial implications of these energy scenarios well in advance, suggesting to elaborate

---

* These figures were quite optimistic; according to the International Energy Agency in 2004, the global renewable energy supply was of 13.1% (OECD/IEA 2007) and, realistically, this share is not likely to peak to 75% in the next 13 years.

consistent urban plans preventively and with a strategic horizon penetrating the twentieth century and beyond.

Leaving his detailed elaborations apart, what we feel like retaining of Van Til's well-grounded recommendations is, first and foremost, the spirit that animates them: "The transition to an energy-short future will involve one of the greatest shifts in cultural and individual expectations known in history (...) To succeed in the transformation will require early warning, brilliant planning, great leadership, and a rare sense of commitment and compassion among all citizens" (Van Til 1979, p. 328). In particular, we intend to dwell onto Van Til's reference to the notions of *commitment* and *compassion* among citizens in relation to what he rightfully envisioned as an epochal, globally relevant, and incontrovertible energy transition. The call into play of these notions is remarkable in the context of a scientific writing of more than 30 years ago, when the sense of urgency to act against global climate change was not yet fueled by images of perishing polar bears and catastrophic climate events. Remarkably, what Van Til captures through them is the essence of what the sustainability paradigm would have shortly identified as the future crucial mandate of policy-making, that is, the enhancement of societal commitment and participation to the making of a common sustainable future. A future that, note, calls to action first of all individuals, and whose realization is only achievable through each individual intent to convert cumulative societal impacts into sustainable impacts. Arguably, this semantic nuance of "individual intent" is implicit to a sustainable energy transition in all respects as it underlies both individuals' support to the siting of renewable technologies and the voluntary renunciation to indiscriminate energy consumption through mobility, housing, and, generally, life styles.

What *commitment* and *compassion* among individuals indicate in both Van Til's reflections and in the sustainability paradigm in general is therefore an individual *sense of belonging* to the human community, a following *sense of responsibility* for its future together with that of its future members and a consequent *intent to act* for transiting toward a sustainable energy horizon. That is why these notions, and the individual agency they subtend, should be central to the formulation of any planning approach pointing at the horizon.

The underlying normative character of this vision is resonant with recent literature on climate ethics (Gardiner 2004, Gardiner 2006, and Gardiner et al. 2010 in particular), and, as such, it is worth some further clarification. Gardiner in particular argues on the centrality of both individual and collective agency to the effective mitigation of climate change, and in support of his vision, he clarifies recurrent misconceptions of, among others, the notions of risk and uncertainty (Gardiner 2004). On the basis of a careful examination of climate literature, he argues that what permeates the science on climate is the notion of risk, that is, the probability of the consequences of global warming, and not, as often claimed, the notion of uncertainty, implying ignorance regarding such probability and eventually of the very consequences. This clarifies misleading interpretations of the science on climate and allows him to conclude that on the ground of current evidence there is no legitimate justification *for not acting*.

Gardiner's view has important planning implications. The notion of individual responsibility that underpins his motivated viewpoint confounds with that of *duty*, more precisely, the duty of taking what could be called both *negative* and *positive* actions. These may entail, respectively, not opposing the installation of large-scale renewable technologies on unmotivated grounds and installing a small windmill in the own backyard to achieve household energy sufficiency. In both cases, note, actions are in clear support of sustainable energy landscape development. The conception of individual responsibility is therefore not to be understood as a discretional form of participation to the planning process, but rather as a *due* form of collaboration within it.

In theoretical terms, this is a significant advance as it promotes a development from the predominant collaborative planning paradigm, based on a constructivist notion of *knowledge*, to what might be called an active planning paradigm, based on a normative perspective on individuals' duty to collaborate, actively, to the making of sustainable (energy) plans. This is not in contradiction but rather in continuity with Healey's most mature work on collaborative planning. In particular, it fully embraces a vision of planning as "form of collaborative process [that] may have the potential to be transformative, to change the practices, cultures and outcomes of 'place governance'" and as "processes [that] could be made more socially just, and, in the context of the multiplicity of urban social worlds, more socially inclusive" (Healey 2003, p. 107, 108). Nevertheless, in most of Healey's discourse, the normative accent is on transposing a multiplicity of worldviews into governed realities by means of interactive, both institutionalized and "spontaneous," forms of communicative rationality; it is therefore up to the planning process to open up to the collaboration of individuals in "constructing" both the institutional and informal objectives that could be incorporated into inclusive plans. But, as discussed, when introducing the work of Gardiner abating climate change is already *per se* a moral objective; it is a *duty* in front of which there is no legitimate justification for not acting.

Somehow here the objective "preexists" the plan: it is neither to be constructed nor susceptible of conflicting interpretations. The constructivist notion of knowledge that permeates the collaborative discourse and the consequent moral equivalence to be recognized to contrasting perceptions of the priorities and objectives to be mirrored by collaborative plans, in essence, risks to weaken individuals' capacity to assume, both individually and collectively, the responsibility to act on climate change with due cohesion and sense of urgency.

This observation derives from the evidence that the spatial implications of a renewable energy transition are undeniably problematic. They entail dramatic changes of landscapes, penetrate private before public spaces, and create unknown forms of proximity to site-specific energy technologies. This latter aspect in particular is discussed in more detail in the following section.

9.3 Siting Renewable Energy Technologies: An Overview of Criteria and Spatial Challenges

Wind farms and CO_2 underground disposals, to mention two representative examples, are different yet equally important "transition" technologies called to replace more established ones (such as coal and nuclear power plants) in providing sustainable energy. According to an established classification, such technologies are also called second-generation technologies (IEA 2007) as they replace, in fact, the first generation of fossil-based technologies.

In addition to a "greener" nature, most of these second-generation technologies contribute to abate the global risk of climate change while creating impacts and posing risks on localities. That is why the matter of their siting received increasing attention in the past years (e.g., Boholm and Lofsted 2004, Hayden Lesbirel and Shaw 2005, Menoni and Margottini 2011). The dilemmas that accompany virtually any siting process have been discussed from the perspective of different disciplinary domains, ranging from risk analysis to spatial planning up to applied ethics (e.g., Amendola 2004, Boholm 2004, Hermansson 2007, Peterson and Hansson 2004). Despite the fact that planning literature dedicated more attention to the so-called site-specific hazardous technologies rather than to renewable energy technologies (RET), many of the relevant reflections are fully applicable to the latter, particularly in regard of the criteria and principles of fair siting processes. Some of them are, therefore, discussed in the following.

Wind turbines, photovoltaic panels, biogas plants, hydrogen plants, and CO_2 underground disposals are all site-specific technologies with remarkably different characteristics and, consequently, siting implications. Their different scales, impacts, risks, and, finally, the uncertainty concerning their potential effects on man and the environment are sometimes remarkable and do not allow for generally valid considerations. Nevertheless, any locational study should generally identify the following:

- Visual and place-identity impacts
- Environmental and safety impacts
- Environmental and safety risks

Perhaps, the most interesting aspect of RET is, generally, the smaller scale and consequent major proximity to end users in comparison to established energy technologies. A main implication though is the major visual impact onto the landscape. The stigmatizing effects of these impacts on self- and place-identities are discussed in an established literature (Cass and Walker 2009, Satterfield and Gregory 2002, Simmons and Walker 2004, Slovic et al. 1994, Walker 1995). These studies put forward a notion of technological *impact* that include both tangible (e.g., fumes, noise, lightning, and pollution) and intangible components, in particular the perceptive component of altered identities of places and people. These intangible impacts are often found to be central to so-called Not in My Backyard (NIMBY) controversies (Boholm 2004, Hermansson 2007, also see Chapter 3 in of this volume by Sijmons and Van Dorst), showing how the sustainability of RET may not suffice in enhancing their acceptance by the side of localities (see among others Basta 2012).

For some of these technologies, like wind turbines, the problem seems to relate mostly to their aesthetics impact, to the point of giving rise to a specialized literature (Breukers and Wolsink 2007, Pasqualetti et al. 2002, Wüstenhagen et al. 2007). For others, like CO_2 underground disposals and biomass plants, the problem seems to relate to more "classic" matters of impacts, risks, uncertainty, and ignorance. The understanding of these latter notions in particular is control to individuals' attitude towards the relevant technologies. Some guiding definitions are therefore provided:

- With *impacts* there could be identified all operational standard effects of the technology under considerations on the surrounding environment
- With *risk* the probability of such operational standards to be disrupted and leading to major unwanted consequences
- With *uncertainty* the impossibility to estimate such probability
- With *ignorance* the lack of knowledge regarding what these consequences could be

What gets lost from the first to the last notion is the possibility of estimating the likelihood of given effects up to the effects themselves; in more simple words, the capacity to assess, with a given degree of certainty, *what* may happen *when*. These distinctions are crucial in approaching the siting of RET as for some of them empirical evidence has not yet cleared the ground from major uncertainties, like in the case of CO_2 disposals (see, e.g., Holloway 1997). For other technologies such as wind turbines, it has evidently been done so (see Desholm and Kalhert 2005, Mangwell et al. 2002). Whereas for some technologies the matter of uncertainty is therefore a major factor of rejection, for others it is (or should be) an issue of minor relevance.

It is therefore somehow surprising, to mention a significant example, that the siting of wind turbines, which in light of their long technological history are *de facto* the icon of renewable energy transition, keeps being periodically opposed in several European regions on the ground of aesthetic judgments, especially when thinking that some of these regions are familiar with the landmarks of windmills in the landscape since centuries and that they are regarded as pillars of cultural identity and hence undeletable elements of any landscape development perspective (Vos and Meekes 1999).

These problematic aspects of the siting of RET call for the definition of solid guiding notions regarding the principles that should govern the siting of technologies, regardless of their scale, in a view of intergenerational fairness. The theoretical premise of our discourse is that individuals ought to collaborate to plans designed to facilitate renewable energy transition; this does not exclude the necessity to further identify the criteria of legitimacy of such collaboration and of eventual, equally legitimate, oppositions.

First of all, with all due caution, it could be said that controversies marked by a specific and identifiable spatial dimension (the *hic* dimension) are likely to regard technologies about which there is minor uncertainty but that cause major visual, environmental, and hence place-identity impacts. Differently, controversies that lose their initial spatial localization to evolve up to an overall rejection of a particular technology at regional or even national scale are likely to regard installations about which there are perceived major uncertainty and ignorance (Basta 2011). Somehow, "localized" controversies oppose the siting; "unlocalized" rejections oppose the technology as a whole. Arguably, to be legitimate, siting oppositions of the former type are to be motivated on the ground of concrete place-identity impacts, unacceptable safety risks, or major uncertainty, whereas oppositions of the second type should not be framed as NIMBY opposition in the first instance (Basta 2011, Basta 2012, Boholm 2004) but discussed in a more ample political arena.

This may seem subtle but it becomes a fundamental guiding principle when brought together with the premises of our discourse. As discussed, what we defined as the moral duty of individuals to participate to energy technology transitions entails both positive and negative actions, among which the acceptance of a visual evolution of landscapes' toward *energy* landscapes. When such transition regards RET like wind turbines and photovoltaic panels that do not pose neither environmental nor safety risks of documented unacceptable relevance, it is questionable whether the aesthetics argument alone could suffice to motivate rejection. Nevertheless, it is surely relevant to individuals' acceptance, and, as such, it becomes a planning theme of clear importance. Arguably, restricting the space of liberty of individuals to the active support of sustainable energy transition processes does not imply imposing any form of RET in the landscape, regardless of individuals' perceptions. As argued in the introduction of this chapter, such space of liberty should therefore include the option of mutually equivalent forms of contribution, in so doing allowing individuals to choose, freely, the type of action they wish to take.

This is defendable in theory but much harder to realize in practice though. Technological and spatial constraints impose in fact a restricted range of interventions depending on the features taken into consideration and the type of active contribution individuals can provide. Together with the centrality of individual agency to the making of sustainable energy landscapes, these more operational aspects are discussed in the following section, wherein two cases of, respectively, a small-scale and a large-scale energy transition study are illustrated in detail.

9.4 Planning Energy Landscapes: Some Experiences

The past two editions of the Landscape Architecture and Planning studio at Wageningen University provided valuable perspectives on the challenges posed by transiting toward sustainable energy landscapes (Helm et al. 2010, Stremke et al. 2011a). The relevant projects saw the involvement of commissioners from various institutions and were therefore aligned with the concrete perspectives of development of the respective areas. Some of these projects are of clear support to our discussion and are therefore discussed in more detail later.*

The focus of those studies was the City Region of Arnhem-Nijmegen (the Netherlands) and the islands of Schouwen-Duivenland and Goeree-Overflakkee (the Netherlands). Additional inputs were provided by a master thesis, following the 2010 edition of the studio, elaborating a comparative study of the islands of Texel, Samsø, and Gotland (the Netherlands, Denmark, and Sweden, respectively).

All areas were approached from the perspective of renewable energy transition and object of detailed scenario analyses. The analyses considered different spatial-policy and socioeconomic scenarios resulting, per region, in four different visions. Each vision provided the framework for the development of specific projects. Several lessons of general relevance, partially discussed in recent literature (Dobbelsteen et al. 2011, Stremke et al. 2011a,b), can be derived from these studies as follows:

1. The effective implementation of RET depends on the different energy sources available at local level, on the establishment of clear energy targets, and on the promotion of energy-saving programs within what could be defined as "integrated spatial-energy plans."

2. A key implementation factor is the will of individuals to support the installation of RET in both private space and the landscape and to change energy consumption behaviors. Analyzing the local conditions of, respectively, the availability of local renewable energy sources and of the local acceptance of spatial and behavioral changes should be therefore a precondition to the elaboration of integrated spatial-energy plans.

---

* For more information on the setting and general assignment of the studios, please refer to Chapter 20 authored by De Waal and her colleagues from Wageningen University.

3. From an economic perspective, renewable energy sources are more easily converted into usable energy and less dispersed along the supply chain when collected and converted at short distances from final users. Such technological requirements may constitute the premise of proximity between settlements and RET and should therefore be given special attention during the planning and design processes.

4. Diversified energy system is indeed more robust but it is more difficult and expensive to collect large quantities of energy. Furthermore, daily and seasonal fluctuations pose the problem of ensuring constant availability. In its first stage of development, a credible energy transition should therefore rely on the supply of a smart combination of fossil and nonfossil energy sources and on the parallel massive investment in renewable energy infrastructures.

The first point—that is the point of the individuals' acceptance of new installations in the landscapes and of the behavioral changes required by new forms of provision and consumption of energy—interrelates strongly with the point of proximity between RET and settlements. As mentioned earlier, two projects are focused on this aspect in particular.

The first consisted of the analysis of the island landscapes of Texel, Samsø, and Gotland. Specifically, of how individuals and groups engaged into different aspects of the sustainable transition promoted in the islands and were prone to accept RET together with the related behavioral adaptations. It was observed that, in relation to RET, an initial enthusiasm had not necessarily led to long-term commitment. This seems to confirm the study by Geels and Schot (2007) describing the "speed of changes" related to transition processes, describable as three different levels of innovation with different timeframes. An important aspect of the transition along the three levels of innovation is the leadership exerted by key actors acting with determinacy and fuelling collective motivation (Geels and Schot 2007). In line with this finding, the determinant factors of lasting commitment to the energy transition of, particularly, the island of Samsø were identified in the charismatic leadership exerted by the promoters of the initiative, and in the possibility for involved citizens to benefit from the installation of off-shore wind turbines by owning shares. The possibility of accessing such benefits in terms of reduced electricity costs, improved supply, and even profit was recognized as the crucial factor of individuals' lasting commitment to the overall transition and the consequent long-term perspective of an energy-sufficient future for the island. In conclusion, the study showed that the transition in Samsø would have failed without strong leadership, the transparency of the related goals and benefits, and the gradual absorption of the process of innovation by the side of inhabitants through their early involvement in and constant information about the phases of implementation (Andela and Willems 2011).

The second project focused on the analysis and promotion of energy-saving solutions in the heart of the historical city center of Zierikzee, capital of the island of Schouwen-Duivenland in the southern Dutch province of Zeeland (Panait 2011). The project was developed in the framework of the Global Economy scenario. The mapping of renewable energy potentials in the initial phase of the studio led to the division of the island into energy sinks and sources. The former are the areas wherein energy is mostly consumed, whereas the latter the areas wherein renewable energy provision exceeds consumption.* Historical centers are considered energy sinks and are therefore indicated as areas wherein the contribution of inhabitants to energy transition should take place in the form of energy saving. The project discussed here did focus on the strategies for implementing energy-saving solutions in protected historical centers.

A detailed analysis of the amount of energy that could be saved by means of basic interventions (e.g., improved isolation, secondary glazing, and optimized use of domestic appliances) shows how individual savings at household scale could accumulate into major savings at the urban scale. Through a smart combination of technical devices within dwellings and simple behavioral changes (e.g., moderate temperature during the day and making best use of solar irradiation), it is calculated that up to 60% of the yearly energy consumption in the city center could be saved (Panait 2011). It is argued that the implementation of such technical and behavioral adaptations, particularly in considerations of the private nature of residential housing, could be promoted through both "hard" and "soft" measures. The former would include a series of mandatory requirements with regard to energy consumption to be applied to both existing and new buildings, whereas the latter would include fiscal incentives for technological adaptation and energy-saving targets. Individuals' active involvement would be facilitated through establishment of mandatory requirements and the possibility of accessing direct benefits.

In both studies, despite their different scale and location, what emerges with clarity is the crucial relevance of individuals' active participation to the aspects of the renewable energy transition that penetrate, shape, and hence "resignificate" the daily experienced living environment. In light of the often private nature of the spaces intruded by RET, without a firm individual commitment onto which grounding sometimes uncomfortable forms of adaptation, sustainable energy transition seems to be difficult to implement and would likely miss to meet the objective of responding to climatic changes with due urgency. These are the main conclusions we discuss in the last section of this contribution.

---

* For more information on the concepts of sources and sinks, please refer to Stremke and Koh (2011).

9.5 Planning Sustainable Energy Landscapes: Some Conclusions

This chapter started with retracing the seminal writings of early planning scholars who reflected on the epochal energy transition envisioned to occur at the end of the past century, following the crisis of supply of fossil fuels and the consequences on the global environment (Van Til 1979). Here, the main intuition was that any successful energy transition would have required active forms of commitment and support by the side of individuals on the basis of the adhesion to the moral mandate of preventing a global environmental catastrophe. This literature was then put in relation with more recent writings of climate ethics scholars who reflected on the normative implications of such individual commitment, arriving to conclude that individuals ought to act in front of the threat of climate change without any legitimate justification for not doing so (Gardiner 2004).

Following these theoretical premises and the general lessons learned from the cases illustrated in this chapter, the main perspective put forward by this contribution is that a sustainable energy transition is a moral objective calling for a planning approach based upon, and focusing on, the promotion and empowerment of a constellation of individual actions bent toward cumulating durable positive impacts on the global environment. This should find correspondence in integrated spatial-energy plans whose driving forces are individuals actively supporting, and therefore gradually adapting to, the new forms of self- and place-identities shaped by renewable energy technologies. In this view, individuals' support and adaptation are to be seen as active components of integrated spatial-energy plans to be regulated through a virtuous combination of both mandatory and discretional forms of contribution to the achievement of transparent energy targets.

As shown in the case studies, sustainable energy transition enters both private and public spaces and, as such, calls for the active participation of individuals both in terms of concrete and specific actions and in terms of more general support. This suggests that a coherent planning system must show clarity both in regard of achievable energy targets and in regard of individuals' duty to collaborate to their achievement. Integrated spatial-energy plans that combine these objectives and provide the roadmaps for their achievement seem, therefore, a desirable contemporary spatial planning instrument. Such integrated plans should assume the role of means of dissemination of renewable energy targets and of the pathways to their achievement in an intergenerational view. More in practice, "hard" and "soft" implementation measures, such as mandatory requirements and incentives for energy savings, respectively, should become integrant parts of local land use and development plans.

In the light of the theoretical premises of our discourse and the general findings of the illustrated case studies, what we envision is, in essence, an active planning process to which individuals *ought to collaborate*. This is the main innovative point of our vision, but at the same time, the most problematic point of our argumentation. Contrary to what is supported by predominant planning paradigms regarding the need of extending public involvement to the definition of planning objectives, what is defended here is that mitigating climate change is *per se* a moral objective that "preexists" and hence determines the plan. As mentioned earlier, the involvement of individuals may indeed regard the collaborative definition of the means and timelines of mitigation but not of its desirability, overall necessity, and duty to support it. How this duty could be concretely imposed through planning instruments, particularly in Western democracies where planning systems have little capacity to impose measures that overcome the boundaries of private space, remains an open research question though. However, as Van Til intuited more than 30 years ago, energy transition calls for a planning system capable to impose forms of collaboration when necessary and to activate individuals' participation when possible, that is, after all, what the sustainability challenge demands us for the good of the current and future generations.

References

Amendola A. (2004), Management of change, disaster risk, and uncertainty: An overview. *Journal of Natural Disaster Science* 26 (2), 55–61.

Andela M. and Willems R. (2011), Landscape is energy; analysis of sustainable energy transition in renewable energy regions. MSc thesis Landscape Architecture and Spatial Planning, Wageningen University, Wageningen, the Netherlands.

Basta C. (2011), Siting technological risks: Cultural approaches and cross-cultural ethics. *Journal of Risk Research* 14 (7), 799–817.

Basta C. (2012), Risk and spatial planning. In Roeser S. (Editor in Chief) et al., *Handbook of Risk Theory*. Springer, Berlin, Germany.

Boholm A. (2004), What are the new perspectives on siting controversies? *Journal of Risk Research* 7 (2), 99–100.

Boholm A. and Lofsted A.R. (2004), *Facility Siting: Risk, Power and Identity in Land Use Planning*. Earthscan, London, U.K.

Breukers S. and Wolsink M. (2007), Wind power implementation in changing institutional landscapes: An international comparison. *Energy Policy* 35 (5), 2737–2750.

Campbell H. (2006), Is the issue of climate change too big for spatial planning? *Planning Theory and Practice* 7 (2), 201–230.

Cass N. and Walker G. (2009), Emotion and rationality: The characterization and evaluation of opposition to renewable energy projects. *Emotion, Space and Society* 2 (1), 62–69.

Desholm M. and Kalhert J. (2005), Avian Collision risk at an offshore wind farm. *Biological Letter* 1 (3), 296–298.

Dobbelsteen A. van den, Broersma S., and Stremke S. (2011), Energy potential mapping for energy-producing neighborhoods. *International Journal of Sustainable Building Technology and Urban Development* 2 (2), 170–176.

Füssel H.M. (2007), Adaptation planning for climate change: Concepts, assessment approaches, and key lessons. *Sustainable Science* 2, 265–275.

Füssel H.-M. and Klein R.J.T. (2004), Conceptual frameworks of adaptation to climate change and their applicability to human health. PIK report no. 91, Potsdam Institute for Climate Impact Research, Potsdam, Germany.

Gardiner S.M. (2004), Ethics and global climate change. *Ethics* 114, 555–600.

Gardiner S.M. (2006), A perfect moral storm: Climate change, intergenerational ethics and the problem of moral corruption. *Environmental Values* 15 (3), 397–413.

Gardiner S.M., Caney S., Jamieson D., and Shue H. (Eds.) (2010), *Climate Ethics: Essential Readings*. Oxford University Press, New York.

Geels F. and Schot J. (2007), Typology of socio-technical transition pathways. *Research Policy* 36 (3), 399–417.

Hayden Lesbirel H. and Shaw D. (2005), *Managing Conflict in Facility Siting*. Edward Elgar, Cheltenham, U.K.

Hayes D. (1977), *Rays of Hope: The Transition to a Post-Petroleum World*. W. W. Norton, New York.

Healey P. (2003), Collaborative planning in perspectives. *Planning Theory* 2 (2), 101–123.

Helm S. van der, Knaap W. van der, and Telgen S. van (2010), *Sustainable Energy Landscapes; Biomass and Other Opportunities for an Energy Neutral City Region Arnhem Nijmegen. Overview of the Atelier*. Wageningen University, Wageningen, the Netherlands.

Hermansson H. (2007), The ethics of NIMBY conflicts. *Ethical Theory and Moral Practice* 10 (1), 23–34.

Holloway S. (1997), An overview of the underground disposal of carbon dioxide. *Energy Conversion and Management* 38, 193–198.

Mangwell J.F., Mc. Gowan J.G., and Rogers A.L. (2002), *Wind Energy Explained: Theory, Design and Application*. John Wiley & Sons, New York.

Menoni S. and Margottini S. (Eds.) (2011), *Inside Risk: A Strategy for Sustainable Risk Mitigation*. Springer, Berlin, Germany.

Meyer B.C., Rannow S., and Loibl W. (2010), Climate change and spatial planning. *Landscape and Urban Planning* 98 (3), 139–140.

OECD/IEA. (2007), *Renewables in Global Energy Supply*. Paris, France.

Panait I. (2011), *Energy Saving Solutions in Protected Urban Areas*. Wageningen University, Wageningen, the Netherlands.

Pasqualetti M.J., Gipe P., and Righter W.R. (2002), *Wind Power in View: Energy Landscapes in a Crowded World*. Academic Press, San Diego, CA.

Peterson M. and Hansson S.O. (2004), On the application of right-based moral theories to siting controversies. *Journal of Risk Research* 7 (2), 269–275.

Satterfield T. and Gregory R. (2002), The experience of risk and stigma in community context. *Risk Analysis* 22, 347–358.

Schmidt-Thomé P. (2006), Integration of natural hazards, risk and climate change into spatial planning practices. PhD thesis no. 193 of the Department of Geology, University of Helsinki, Helsinki, Finland.

Simmons P. and Walker G. (2004), Living with technological risk: Industrial encroachment on sense of place. In Boholm A. and Lofsted R. (Eds.), *Facility Siting Risk. Power and Identity in Land-Use Planning*. Earthscan, London, U.K., pp. 90–106.

Slovic P., Flynn J., and Gregory R. (1994), Stigma happens: Social problems in the siting of nuclear waste facilities. *Risk Analysis* 14 (5), 773–778.

Stremke S., Dobbelsteen A. van den, and Koh J. (2011b), Exergy landscapes: Exploration of second-law thinking towards sustainable landscape design. *International Journal of Exergy* 8 (2), 148–174.

Stremke S., van Etteger R., de Waal R., de Haan H., Basta C., and Andela M. (2011a), *Envisioning Desired Futures for two Sustainable Energy Islands in the Dutch Delta Region.* Wageningen University, Wageningen, the Netherlands.

Stremke S. and Koh J. (2011), Integration of ecological and thermodynamic concepts in the design of sustainable energy landscapes. *Landscape Journal* 30 (2), 194–213.

Van Til J. (1979), Spatial form and structure in a possible future: Some implications of energy shortfall for urban planning. *Journal of the American Planning Association* 45 (3), 318–329.

Vos W. and Meekes H. (1999), Trends in European cultural landscape development: Perspectives for a sustainable future. *Landscape and Urban Planning* 46, 3–14.

Walker G. (1995), Renewable energy and the public. *Land Use Policy* 12, 49–59.

Wilson E. (2007), Adapting to climate change at the local level: The spatial planning response. *Local Environment* 11 (6), 609–625.

Wilson E. and Piper J. (2010), *Spatial Planning and Climate Change.* Taylor & Francis, New York.

Wüstenhagen R., Wolsink M., and Bürer M.J. (2007), Social acceptance of renewable energy innovation: An introduction to the concept. *Energy Policy* 35 (5), 2683–2691.

10

Integrated Optimization of Spatial Structures and Energy Systems

Gernot Stoeglehner and Michael Narodoslawsky

CONTENTS

10.1 Introduction

Spatial planning decisions about settlement and infrastructure development largely determine not only the energy demand of cities and regions, but also the potentials to supply certain areas with renewable energies. Both aspects are important to achieve not only emission targets concerning climate change, but also long-term availability of secure and affordable energy in a post-fossil and post-nuclear society and are key issues for shaping sustainable societies. According to the importance of the topic, literature coverage is already rich. Also, planning principles to optimize the interrelation between spatial planning and energy systems are readily available. However, with the exception of a few good practice examples, planning practice is far from implementation of energy transition. An analysis of the spatial planning literature concerning the interrelation with energy systems reveals the following dominant lines of the discourse (based on Stoeglehner et al. 2011a,b):

1. Many initiatives and visions for urban and regional planning point out that multifunctional, densely populated settlement structures that are based on proximity concerning walking/biking/public transport not only reduce energy demand but also increase quality of life for the inhabitants (see, e.g., Brunner et al. 1999, CNU 1996, Dittmar and Ohland 2004, Farr 2008, Gaffron et al. 2005, 2008, Lerch 2007, Lineau 1995, Motzkus 2002, Newman and Jennings 2008, Register 2002, Schriefl et al. 2009).

2. A further line of the discourse reflects that choosing sites for new settlements (e.g., topography, exposition, microclimate) as well as building schemes and the detailed planning of buildings (e.g., the volume–surface relation) determine the energy demand of settlements and the potential active and passive use of solar energy (see, e.g., Heinze and Voss 2009, Neufert 2009, Prehal and Poppe 2003, Stremke and Koh 2010, Tappeiner et al. 2002, Treberspurg 1999).

3. A third line of debate draws upon the relation of spatial structures and potential options for (renewable) energy supply, as energy provision causes land use demands, land use conflicts, and the necessity to protect the use of resources and, therefore, needs spatial planning. Potential effects of certain energy sources are assessed; methods and processes for (local and regional) energy planning are developed (see, e.g., Dobbelsteen et al. 2011, Girardin et al. 2010, Halász et al. 2006, Joanneum Research 2001, Kanning et al. 2009, Krotscheck and Narodoslawsky 1996, Mandl et al. 2008, Narodoslawsky and Stoeglehner 2010, Narodoslawsky et al. 2010, Rode and Kanning 2010, Späth 2007, Stoeglehner 2003, Stoeglehner and Narodoslawsky 2008, 2009, Stremke 2010, Tillie et al. 2009).

4. A fourth line of debate, for example, in Austria, appraises the effectiveness of spatial planning concerning the potentials to save energy and to supply renewable energy, mainly with the result that energy efficient spatial structures are mostly obtained in good practice, while a wide exhaustion of the potentials cannot be detected (see, e.g., Dallhammer 2008, Steininger 2008, Stoeglehner and Grossauer 2009). In this respect, much potential for the mitigation of climate change, for the development of sustainable energy systems, and for the strengthening of local and regional economies remain untouched.

Drawing from the aforementioned literature, we define "energy-optimized spatial structures" as structures that allow for energy efficiency by the organization of spatial functions (e.g., to reduce energy demand of buildings via siting or choosing certain densities, before energy efficiency of buildings is taken into account) on the one hand and to allow for sustainable energy supplies based on renewable resources on the other hand. The aim of this chapter is to frame the theoretical background for integrated spatial and energy planning to achieve energy-optimized spatial structures on the local and regional level from a systems perspective.

This leads to the following key questions: (1) Which system elements can be identified to jointly steer spatial structures as well as energy supply and consumption in order to achieve overall sustainability? (2) How can spatial archetypes be characterized from the perspective of a sustainable energy system? (3) What are the tasks of the spatial archetypes in a sustainable resource economy?

The chapter is structured accordingly and as follows. First, a systems approach for integrated spatial and energy planning is shown, defining system elements of spatial planning and energy planning and mapping the interrelations. Hence, new approaches can be derived concerning visions on the one hand, and practical methods on the other hand (Section 10.2). Second, generic sustainable characteristics for four spatial archetypes—cities, suburban areas, rural small towns, and rural areas—concerning integrated spatial and energy planning are derived (Section 10.3). Third, the role of active system elements in steering the system is elaborated (Section 10.4). Finally, in Section 10.5, we draw some brief conclusions.*

10.2 Systems Approach toward Integrated Spatial and Energy Planning

The interrelations between spatial structures and energy demand as well as energy supply options are manifold and highly complex. The increased importance of renewable resources will have a fundamental impact on landscape and urban development and will happen because of market forces. If this takes place without proactive consideration in planning and design processes, it may lead to unsustainable degradation of natural resources and loss of biodiversity and landscape diversity as well as unsustainable spatial structures. In order to allow planners to understand the complex interrelations between spatial structures and energy supply, we present a systems[†]

---

* The research presented in this chapter was elaborated during the project, "PlanVision–Visions for energy optimized spatial planning" (also see Stoeglehner et al. 2011a).

† Systems are approaches to explain reality. They define the interrelations of phenomena and impact factors from a certain perspective, which means that the generation of systems is highly dependent on the aim of the survey and bound to the human perception of (highly) complex issues (Heitzinger 1995).

approach, which was elaborated to find out which system elements offer the best options for effective interventions in the system in order to achieve energy-optimized spatial structures. The method for the system analysis builds upon the approach of Vester (2007, 1980, 1976) and adapts the "paper computer" to comprehend complex systems and to depict the effects and impacts of the system elements within the system. Vester distinguishes four types of system elements concerning their impacts within systems and their potentials for steering a system:

1. *Active elements* strongly affect other system elements, but are weakly impacted by other elements. These elements are well suited to steer, respectively, change systems as changes of these system elements have large impacts on the one hand, but unexpected rebound, follow-up, and side effects are to be expected on lower probability levels or are easier to detect and to be considered in planning interventions.

2. *Passive elements* impair other elements weakly but are influenced by others to a huge extent. To change passive elements means to achieve relatively small effects in the overall system.

3. *Critical elements* affect other elements and are also impacted by many elements. Changes take strong effects but may destabilize the system, as unpredictable follow-up, side, and rebound effects might occur more likely as with active elements.

4. *Buffering elements* have little effect on other elements and are little affected. They are stabilizing systems but are not very suitable if system changes shall be reached.

In order to determine the steering potential for the system of spatial planning versus energy supplies, 34 relevant system elements were identified via literature surveys and structured brainstorming in the project team. They were subsequently arranged in a symmetric matrix (see Figure 10.1). For each cell of the matrix, we assessed, in a series of interdisciplinary expert discussions drawing on the expertise of 19 researchers from different fields,* if a system interrelation is present. In order to avoid methodological objections against the paper computer according to Vester, the assessment and aggregation rules were changed in PlanVision. It was first determined if an interrelation is affected (0 = no impact; 1 = impact). Counting the frequencies of interrelations identifies the active and passive sums of each system element. A graph was generated with the *x*-axis containing the passive sums and the

---

* Experts with a background in landscape planning, spatial (urban and regional) planning, environmental planning, engineering, environmental systems science, climate change research, economy, planning law, and energy law.

y-axis containing the active sums. The quadrants of the graphs between the extreme values represent the appraisal: the up-right quadrant with high active and passive sums characterizes critical elements, the low-left corner buffering elements with low active sums and passive sums, etc.

In the PlanVision-matrix, 19 system elements of spatial planning and 15 elements of energy supplies were brought together to define the system, and assessed concerning their role in the system (Figure 10.1). From the system elements of spatial planning, a mix of spatial functions,[*] density of settlements, and site[†] were identified as active; from the side of energy supplies, only the useable resources are considered as active elements (Figure 10.1).

Although the results are not very surprising, the analysis reveals that shaping sustainable energy landscapes can be based on well-known and well-established system elements of spatial planning. Moreover, the change of the resource base actively determines the sustainability and efficiency of energy systems. Starting from here, visions for energy-optimized spatial structures can be drafted to show the direction in which change should be implemented. In the following section, we specify our vision for the four spatial archetypes (i.e., urban centers, suburban areas, rural small towns, and rural areas) and explain the target state of each active system element.

10.3 Generic Sustainable Characteristics of Spatial Archetypes

In order to draft the direction of system change, the functions of each spatial archetype (urban centers, suburban areas, rural small towns, and rural areas) have to be defined. We use sustainable characteristics for energy-optimized spatial structures in these archetypes to achieve this. The sustainable characteristics are appropriate to the situation, as spatial structures are long lasting, highly resistant to change because of massive material and energy investments as well as financial means. Land ownership is included too, as it influences and introduces new aspects concerning energy and resource flows, where considerable gaps between the state and the target state can be detected in actual structures and recent developments. The following subsections rest upon these sources. Figure 10.2 presents an overview for the generic sustainable characteristics.

---

[*] The basic spatial functions normally used in spatial planning are housing, working, supply and disposal, food supply, recreation, communication, and transport (Lienau 1995).
[†] Consisting of topography, location, and exposition.

System matrix spatial planning-energy supply

The matrix cross-references 19 spatial planning factors (rows) against 34 system factors (columns, grouped under *Spatial planning* and *Energy supply*). Column headers and the summarising "AS role in system" assessment are listed below.

Row / column factor groups

Spatial planning

- (de-)centrality mischung
  1. Mix of spatial functions
  2. Nearness
  3. Cluster development
  4. Mix of industrial sectors
  5. Combination of destinations
- Accessability
  6. Means of transport
  7. Distance
  8. Travel time
- Density
  9. Population density
  10. Employment density
  11. Resource density
  12. Technological density (e.g. energy density)
- Land consumption
  13. Sealing of soil
  14. Preliminary land use (brownfield/greenfield)
  15. Quality and shape of buildings
- Site
  16. Topography
  17. Location
  18. Exposition
  19. Shape of the surroundings

Energy supply (columns only)

- Regional resource potential
  20. Raw materials
  21. Residues
  22. Energy cascades
  23. Dynamics of energy production
- Location of energy production facilities
  24. Land use conflicts
  25. Environmental impacts
  26. Demand for location factors
- Technological options
  27. Applied resources
  28. Energy conversion technologies
  29. Energy distribution
- Energy consumption
  30. Room heating and cooling
  31. Process energy
  32. Light and power
  33. Mobility
  34. Dynamics of energy demand

AS role in system (per column factor)

#	Factor	AS	Role
1	Mix of spatial functions	22	Active
2	Nearness	17	Critical
3	Cluster development	14	Critical
4	Mix of industrial sectors	9	Buffering
5	Combination of destinations	6	Buffering
6	Means of transport	11	Passive
7	Distance	12	Passive
8	Travel time	6	Passive
9	Population density	22	Active
10	Employment density	21	Critical
11	Resource density	10	Buffering
12	Technological density (e.g. energy density)	4	Passive
13	Sealing of soil	6	Passive
14	Preliminary land use (brownfield/greenfield)	13	Buffering
15	Quality and shape of buildings	20	Critical
16	Topography	24	Active
17	Location	28	Active
18	Exposition	14	Active
19	Shape of the surroundings	11	Passive

			PS	11	13	15	11	9	16	13	15	10	13	11	16	14	4	15	0	6	0	14	12	14	12	18	16	22	25	26	21	18	6	0	5	14	19	AS	Role
Energy supply	Regional resource potential	Raw materials	20	0	0	1	1	0	1	0	0	0	0	0	0	1	1	0	1	0	0	1	1	1	1	1	1	1	1	1	0	0	0	0	0	0	0	17	Active
		Residues	21	0	0	1	1	0	1	0	0	0	0	0	1	1	0	1	1	0	0	1	1	1	1	1	1	1	1	1	0	0	0	0	0	0	0	13	Passive
		Energy cascades	22	0	0	1	1	0	0	0	0	0	0	0	0	1	0	0	1	0	0	1	1	1	1	1	0	1	1	1	0	0	0	0	0	0	0	11	Buffering
		Dynamics of energy production	23	0	0	0	0	0	0	0	0	0	0	0	0	1	0	0	0	0	0	0	1	0	1	0	0	1	1	1	0	0	0	0	0	1	0	7	Passive
	Location of energy production facilities	Land use conflicts	24	1	1	0	0	0	1	1	1	0	0	0	0	0	0	0	0	0	0	0	0	0	0	1	1	1	1	1	1	0	0	0	0	0	0	10	Passive
		Environmental impacts	25	0	0	1	1	0	0	0	1	1	0	0	1	1	0	0	0	0	0	0	1	1	1	1	1	1	1	1	1	0	0	0	1	0	0	14	Critical
		Demand for location factors	26	0	1	0	0	1	1	0	0	1	1	0	0	0	1	0	0	0	0	0	0	1	0	1	1	1	1	1	1	0	0	0	0	0	0	10	Passive
	Technological options	Applied resources	27	1	1	1	1	1	1	1	0	1	1	0	0	1	0	1	0	0	0	1	1	1	1	1	1	1	1	1	1	1	0	0	0	1	0	22	Critical
		Energy conversion technologies	28	0	0	1	0	0	0	0	0	1	0	1	1	1	1	0	1	0	0	1	1	1	1	1	1	1	1	1	1	1	0	0	0	1	0	14	Critical
		Energy distribution	29	1	1	0	0	1	0	0	0	0	0	0	0	1	0	0	1	0	1	0	0	1	0	1	1	0	0	1	1	0	0	0	0	0	0	6	Passive
	Energy consumption	Room heating and cooling	30	0	0	0	0	0	0	0	0	0	0	0	0	0	0	0	0	0	0	0	0	1	1	0	1	0	1	0	0	1	0	0	0	0	1	7	Buffering
		Process energy	31	0	0	0	0	0	0	0	0	0	0	0	0	0	1	0	0	0	0	0	0	0	1	0	1	1	1	1	1	1	0	0	1	0	1	9	Buffering
		Light and power	32	0	0	0	0	0	0	0	0	0	0	0	0	0	0	0	0	0	0	0	1	0	0	1	0	1	1	1	1	1	1	0	0	0	1	10	Buffering
		Mobility	33	0	0	0	0	0	0	0	0	0	0	0	0	0	0	0	0	0	0	0	0	0	0	1	0	0	1	1	0	0	0	0	0	0	0	6	Passive
		Dynamics of energy demand	34	0	0	0	0	0	0	0	0	0	0	0	0	0	1	0	1	0	0	1	1	1	1	0	1	1	1	1	0	1	0	0	0	0	0	8	Passive

FIGURE 10.1

Matrix of the system spatial planning—energy supply (From Stoeglehner, G. et al., In: PlanVision—Visionen für eine energieoptimierte Raumplanung, Projektendbericht, Gefördert aus Mitteln des KlimaE- und nergiefonds, Wien, Vienna, Austria, 2011a, https://www.boku.ac.at/fileadmin/_/H85/ H855/materialien/planvision/Endbericht_PlanVision.pdf, last accessed: May 21, 2012.)

Basic Function	Urban Center		Suburban Area		Small Town/Rural Center		Rural Area	
	Basic Function	Goal	Basic Function	Goal	Basic Function	Goal	Basic Function	Goal
Living space for majority of people	Living space for majority of people	Highest quality of living Sufficient leisure time opportunities High environmental quality Comprehensive provision of • Goods • Education (up to tertiary level) • Social services (health/care) • Cultural services • Research	Spatial reserve for urban center	Highest logistic efficiency for people and goods High environmental quality Basic provision of • Goods (daily consumption) • Education (primary level) • Social services • Cultural services	Attractive living space for de-centralized industrial society	High quality of living Excellent leisure time opportunities Highest environmental quality Advanced provision of • Goods • Education (up to secondary level) • Social services (health/care) • Cultural services • Research	Sufficient population density for primary production and sustainance Recreational space	Basic provision of • Goods (daily consumption) • Education (primary level) • Social services • Cultural services Highest environmental quality Sufficient touristic infrastructure

			Resource conversion	Sustainable resource provision
Main energy/resource consumer	Highest efficiency of use; Lowest pressure in energy provision/utilization; Lowest resource consumption; Highest resource conversion efficiency; Strong societal interaction; International interconnectedness	Highest utilization efficiency; Lowest pressure in energy provision/utilization; Lowest resource consumption; Highest resource conversion efficiency; Highest efficiency in space utilization; Max. long-term yield per area	Lowest pressure in resource conversion/utilization	Highest efficiency in space utilization; Max. long-term yield per area; Stable eco systems
Provider of complex (industrial) goods and services		Space reserve for provision of complex goods; Provision of fresh goods for urban center	Highest conversion efficiency; Linking the distribution grids	Highest logistical efficiency for renewable resources and by-products of conversion processes; No resource import

FIGURE 10.2

Generic sustainable characteristics for spatial archetypes. (Based on Stoeglehner et al. 2011a, In: PlanVision—Visionen für eine energieoptimierte Raumplanung, Projektendbericht, Gefördert aus Mitteln des KlimaE- und nergiefonds, Wien, Vienna, Austria, 2011a, https://www.boku.ac.at/fileadmin/_/H85/H855/materialien/planvision/Endbericht_PlanVision.pdf, last accessed: May 21, 2012.)

10.3.1 Urban Centers

In urban centers, the large demand for energy can be explained, on the one hand, because of high population densities and, on the other hand, because urban centers (and their suburban surroundings) are the main location for the production of industrial and consumer goods as well as the provision of services.

The potentials for renewable energy provision are limited because the high share of sealed soil and potential land use conflicts. Therefore, the main task for urban centers is to increase energy and resource efficiency as much as possible, so that resource input can be minimized. Furthermore, we propose that energy provision based on urban resources can be mainly based on thermal use of nonrecyclable waste and the exploitation of solar energy, both thermal and photovoltaic. This calls for ambitious energy efficiency programs concerning thermal and process energy, electricity, and mobility as well as conscious energy and material flow management and the utilization of energy cascades (from energy uses with higher exergy to those with lower exergy, e.g., low temperature heating).*

We propose to plan urban energy systems "backward," up the energy cascade, in order to achieve maximum efficiency of the overall system: first, the amount of low temperature demand has to be determined. This shall be supplied as waste heat from high temperature process energy and electricity provision. So the heat demand in residential and commercial areas influences the technological choices for industrial and energy provision purposes. This also leads to system integrations of energy source areas, industrial areas, residential areas, and agricultural structures like glasshouses. District heating systems become an important role in energy-optimized structures as they allow for networking the different user profiles and allow for the maximum exhaustion of the exergy input in the spatial structures. Furthermore, the energy demand of low exergy energy (e.g., low temperature heat) determines the efficiency targets for high exergy energy provision (e.g., process energy and, in particular, electricity) as well as the choice of technological options. This implies, for example, the priority of "flameless" electricity provision (e.g., wind, hydropower, or solar energy) in spatial structures where heat consumption is low. This calls for new methods to carry out these optimization processes, and these methods have to integrate both energy and spatial data.

10.3.2 Suburban Areas

For suburban areas, our ideas and the reality are differing considerably. We propose that suburban areas are the "spatial reserve" for urban centers. They should provide fresh foods with low durability to urban settlers and suburbanites. Production should, however, comply with environmental capacity limits.

---

* More implications of thermodynamic concepts such as exergy for planning are discussed in Stremke et al. 2011.

Suburban areas may also provide energy, especially using waste materials from the production of fresh goods, other wastes, and solar (if applicable, wind) energy. Furthermore, land has to be provided for industrial production and residential areas where urban areas are not able to accommodate such processes. We propose that suburban areas do only provide for basic supply, suburban shopping centers or hypermarkets shall be redeveloped as agricultural or industrial sites as most of them violate principles of energy and resource efficiency. Efficiency in suburban areas also includes logistic efficiency for persons and goods, the use of public transport, as well as creating nodes between regional and supraregional grids (for energy and transport) and locating at least industrial facilities on these nodes.

10.3.3 Rural Small Towns

For rural small towns, we propose totally new functions. In a sustainable and renewable resource economy, not only energy will be provided but also raw materials produced for industries. Especially the latter will result in an increase of biomass use. Therefore, we suggest that the mere use of biomass for energy provision will only be a step on the way for the multi-shift usage of plants: first, the production of the so-called commodities* in green biorefineries; second, the production of electricity from the sludge of biorefineries via biogas; and third, the use of low temperature heating in urban structures. We propose that the rural small towns become the centers of commodity production and therefore the nodes between energy provision, production of biomass-based industrial raw materials, transport networks, and energy grids for electricity and low temperature heat. By implementing this vision, rural small towns become focal points for quality of life in a decentralized industrial society. We also suggest bringing commodity-related research, development, and education to the rural small towns, so that they develop into innovation centers.

10.3.4 Rural Areas

Rural areas have to supply the remaining amount of biomass-based resources and renewable energies for the whole society—the amount that cannot be produced within the other three spatial archetypes. Basic functions are agricultural and forest production, the securing of daily supply for the population as well as recreation. Main challenges are to balance production levels with environmental capacity, to protect biodiversity, and to bring back nutrients to the soil. Efficiency in rural areas also means to delegate the processing of raw materials to the rural small towns, to protect as much unsealed soil as possible from settlement and infrastructure development, and to prevent sprawl.

---

* Commodities are intermediate products to transfer biomass primary products with low durability and low transport density in intermediate products with high durability and transport density for processing in the centers of industrial production (Narodoslawsky et al. 2010).

10.4 Planning Energy Landscapes via Active System Elements

In this section, we depict how the four active system elements—*mix of spatial functions*, *density*, *site*, and *resources*—shall be considered in spatial planning processes in order to achieve energy-optimized spatial structures. Therefore, this section also is normative.

10.4.1 Mix of Spatial Functions

How to generate a mix of spatial functions in order to enhance quality of life and to reduce environmental pressures is a long-lasting debate in spatial planning and illustrated in many visions for urban design.* Concerning energy transition, the mix of spatial functions is mainly seen as means to reduce transport demands. We add the notion that mixing of functions is important to level out dynamics of energy supply and demand in grid-bound energy supplies so that their efficiency can be increased.

Mix of spatial functions has different meanings in the four spatial archetypes characterized earlier. In *urban centers*, high levels of public services (education, culture, specialized supply, etc.) on public transport nodes as well as basic supply, jobs etc. in walking distance shall be reached. Because of the high population densities, multifunctional residential quarters shall be developed near the center (where public services, research, offices, and big-sized shopping facilities are primarily located). The multifunctional residential quarters also provide outdoor recreation, ecological functions, shopping, services, gastronomy, low-emission or zero-emission companies (mainly small and medium sized), and partly education and research. Residential and center functions are linked to industrial and commercial areas mainly through powerful public transport grids. A limited primary production may also take place, which contributes to the food supply. A target of the domestic production should be determined, and production areas protected by spatial planning. These provisions support energy efficiency, reduce transport, and increase the chance to make use of energy cascades.

In *suburban areas*, a mix of functions means to provide for ecological compensation areas, to provide agricultural production supporting the urban population, and to keep land reserves for industrial and commercial sites that cannot be accommodated in the urban centers. As suburban areas are directed toward the urban centers, we believe that basic daily supply is enough. Residential areas should be interlinked with working and supply functions as well as high-capacity public transport lines within walking distance. In order to achieve this goal, we believe that it is necessary to reverse and withdraw certain developments especially concerning shopping facilities as well as residential sprawl e.g., when opportunities arise because

---

* See, for example, New Urbanism (CNU 1996) and Ecocities (Newman and Jennings 2008).

of shrinking population or the emergence of brownfields. We acknowledge that such visions might face enormous implementation barriers, but we also believe that energy and resource scarcity (e.g., expressed in increased energy prices, or loss of property values in declining areas) might change the perception of such proposals.

In the *rural small town*, a mix of spatial functions has similar tasks than in urban areas and guarantees structural efficiency by interlinking different supply structures. By mix of spatial functions, all functions including more specialized supply can be reached in walking and biking distance. Also, the linking of the small town with its rural surroundings for transport of materials, residues, and persons is important. Rural small towns will be developed as important nodes between different energy and resource supply functions and, therefore, will provide new job opportunities at all levels of education if this vision is implemented.

In *rural areas*, a mix of spatial functions means to focus on different kinds of primary production. Rural areas will be the most important source for energy and resources in a renewable economy. That is why local and regional planning has to support these functions as well as ecological compensation, recreation, and tourism functions. The basic daily supply has to be oriented on these functions. Specialized supply is granted in the rural small towns or in the urban centers. Residential functions should be connected to public transport to the small towns and urban centers.

10.4.2 Density

Density has to be similarly treated in all spatial archetypes. It means to provide land use efficiency of all kinds of settlements in order to protect areas for primary production, ecological compensation, recreation, and tourism, on the one hand, and to provide for structural energy efficiency and supply efficiency from both a resource and energy flow perspective as well as an economic perspective, on the other hand. Density is an inherent precondition for the economic efficiency of energy networks, the economic feasibility of supply structures and transport systems. Therefore, we propose to define minimum densities for all spatial archetypes. In addition, especially for urban areas, maximum densities have to be defined to keep land use pressure both for nature and recreation areas within acceptable limits.

10.4.3 Siting

Choosing sites for physical interventions and projects largely determines if aforementioned planning objectives can be met. Preference should be given to sites that allow for the use of residual heat and renewable energy sources. The site largely determines, by systems interrelations, the energy demand of buildings or the active and passive usability of solar energy. Aspects of topography, exposition, and the building environment can be treated similarly in all spatial archetypes as funda-mental physical interrelations between solar irradiation, microclimate, etc. Concerning the choice of locations, one has to differentiate between the spatial archetypes.

In *urban centers* and *rural small towns,* location factors are likely relatively good to optimize spatial structures from an energy perspective. Therefore, the most important factor is to choose a site in the urban center or rural small town. Because of the multifunctional structure and the (relatively) high densities, energy optimization has good preconditions.

In *suburban areas*, shopping areas and services should be related to exist-ing settlement cores and high-performance public transport. Facilities for industrial productions shall be located close to high-performance energy and transport grids.

In *rural areas*, siting shall be oriented toward village cores and public trans-port. We propose to formulate objectives for the increase of production and ecological compensation areas as well as zones for the long-term removal of functions and infrastructures, which do not comply with the primary tasks of rural areas (primary production, ecological compensation, recreation).

10.4.4 Resources

Resource situations and subsequent production tasks are very different for each of the four spatial archetypes. Energy efficiency and savings are important aspects to reduce pressures on natural resources and maintain bioproductive ecosystems, especially when food, raw material, and energy provision are considered from a systems perspective.

For *urban centers*, energy efficiency and energy-saving targets have to be determined in the first place. Furthermore, via indicators like per capita solar thermal and photovoltaic-area, and the degree of waste utilization for recy-cling and energy provision, the urban contribution to a sustainable energy system has to be defined. Full autonomy of urban centers is not realistic under current technological conditions.

In *suburban areas*, such measures have to be completed with measures for biomass-based raw material production. The maximum withdrawal of resources (taking into account environmental friendly production tech-niques) has to be defined, as well as minimum return rates of nutrients to the agricultural land (e.g., as kg/ha).

In *rural small towns*, measures like energy efficiency, energy savings, defini-tion of solar use, targets for renewable energy provisions, as well as material and energetic performance rations for residues have to be applied. Furthermore, commodity production has to be regionally differentiated depending on the production conditions and possibilities for crop cultivation in the surround-ing rural areas. Moreover, the return rate of nutrients and residues from rural small towns to the surrounding rural areas has to be defined.

In *rural areas*, energy efficiency, energy saving, definition of solar use, targets for renewable energy provisions, and minimum return rates of nutrients to agricultural land (e.g., as kg/ha) have to be applied. In addition, it has to be determined that rural areas utilize 100% renewable energy sources. In the other spatial archetypes, we assume that in the foreseeable future a certain amount of fossil resources will still constitute a part of the energy supply (at least via waste burning).

10.5 Conclusions

The system approach to depict the relation between spatial structures and energy supplies reveals that the "classical" steering elements of spatial planning (i.e., mix of spatial functions, density, siting) and long-desired normative principles concerning the direction of steering would also allow for the creation of energy-optimized spatial structures. The use of renewable energy sources will impose challenges for shaping sustainable landscapes, as there is a direct link between energy consumption, energy provision, and the pressures and changes imposed on landscapes and ecosystems. There is a definite danger that overuse of renewable sources will overshoot environmental capacities and lead to unsustainable development of the environment, the society, and the economy. Hence, sustainability must be a criterion throughout the planning, the design, and the development of sustainable energy landscapes (this book).

Spatial planning has considerable potentials to save energy and increase energy efficiency. Spatial planning can also influence the potentials to supply renewable energy efficiently and environmentally friendly. Furthermore, the consideration of energy in spatial planning adds further notions to the planning principles that might make amendments necessary and help to further argue for desired spatial visions.

The approach presented here offers a new view on the role of spatial archetypes within a sustainable resource economy. Defining the functions of these archetypes within the energy system goes a long way for finding the right approaches to planning and designing sustainable energy-optimized spatial structures. Even though the archetypes are generic and not new, the line of argument reveals that in a sustainable resource economy, the particular characteristics will have to differ considerably from the recent state. The approach presented here points out, for each archetype, in which direction they have to be developed. A massive effort is needed to achieve sustainability, even if the change from fossil to renewable sources is further facilitated by resource scarcities. Planners and designers have to be aware of these issues in order to prevent wrong decisions and to avoid false investments. If the active system elements identified in this chapter are developed in the directions specified for each spatial archetype, an optimal energy system can be achieved within a larger sustainable spatial structure both on local and regional levels.

Acknowledgments

The research presented here was conducted in the research project "PlanVision—Visions for energy optimized spatial planning." This project was funded by the Austrian Climate and Energy Fund and carried out within the program "NEUE ENERGIEN 2020" (grant Number 818916). We thank our fellow researchers Michael Weiss, Hermine Mitter, Georg Neugebauer, and Gerlind Weber (University of Natural Resources and Life Sciences Vienna); Karl-Heinz Kettl, Michael Eder, and Nora Sandor (Graz University of Technology); Horst Steinmüller, Barbara Pflüglmayer, Beatrice Markl, Andrea Kollmann, Christina Friedl, Johannes Lindorfer, and Martin Luger (Johannes Kepler University Linz); as well as Karl Steininger and Veronika Kulmer (Karl-Franzens-University Graz).

References

Brunner, O., Gutmann, R., Lung, E. et al. (1999). Forderungen für die Realisierung einer ausgewogenen Nutzungsmischung bei der Siedlungsentwicklung. In: Wachten, K., Brunner, O., and Kaiser, H.-J., [Hrsg.]: Kurze Wege durch die Nuztungsmischung. Grundlagen für eine nachhaltige Siedlungsentwicklung. Linzer Planungsinstitut, Linz, Austria.

Congress for the New Urbanism (CNU) (1996): Charta des New Urbanism. http:// www.cnu.org/sites/www.cnu.org/files/Charta_deutsch.pdf (accessed on May 21, 2012) (letzter Zugriff: März 3, 2011).

Dallhammer, E. (2008). Verkehrsbedingte Treibhausgase. Die Verantwortung der Siedlungspolitik. *RAUM* 71/2008, 37–39.

Dittmar, H. and Ohland, G. (2004). *The New Transit Town: Best Practices in Transit-Oriented Development.* Island Press, Washington, DC.

Dobbelsteen, A., Broersma, S., and Stremke, S. (2011). Energy potential mapping for energy-producing neighborhoods. *Sustainable Building Technology and Urban Development* 2(2): 170–176.

Farr, D. (2008). *Sustainable Urbanism: Urban Design with Nature.* John Wiley & Sons, Hoboken, NJ.

Gaffron, P., Huismans, G., and Skala, F. (2005). *Ecocity Book I: A Better Place to Live.* Facultas Verlags- und Buchhandels AG, Vienna, Austria.

Gaffron, P., Huismans, G., and Skala, F. (2008). *Ecocity Book II: How to Make It Happen.* Facultas Verlags- und Buchhandels AG, Vienna, Austria.

Girardin, L., Marechal, F., Dubuis, M. et al. (2010). EnerGis: A geographical information based system for the evaluation of integrated energy conversion systems in urban areas. *Energy* 35: 830–840.

Halász, L., Zachhuber, C., and Narodoslawsky, M. (2006). Beyond processes—Application of process synthesis to the optimisation of regional economies. *CHISA 2006*, August 27–31, 2006. Prague, Czech Republic.

Heinze, M. and Voss, K. (2009). Ziel Null Energie. Erfahrungen am Beispiel der Solarsiedlung Freiburg am Schlierberg. *Deutsche Bauzeitschrift*, 57(1): 72–74.

Heitzinger, P. (1995). System integrierter Umweltschutz und Saubere Technik—Ein komplexes Phänomen, Dissertation an der TU Graz, Austria.

Joanneum Research. (2001): *Handbuch für kommunale und regionale Energieplanung (KREP 2000) [Handbook for Municipal and Regional Energy Planning]*. Joanneum Research, Graz, Austria.

Kanning, H., Buhr, N., and Steinkraus, K. (2009). Erneuerbare Energien—Räumliche Dimensionen, neue Akteurslandschaften und planerische (Mit) Gestaltungspotenziale am Beispiel des Biogaspfades. *Raumforschung und Raumordnung* 67(2): 142–156.

Krotscheck, C. and Narodoslawsky, M. (1996). The Sustainable process index: A new dimension in ecological evaluation. *Ecological Engineering* 6: 241–258.

Lerch, D. (2007). Post Carbon Cities: Planning for energy and climate uncertainty. A guidebook on peak oil and global warming for local governments. Post Carbon Institute. http://postcarboncities.net

Lienau, C. (1995). *Die Siedlungen des ländlichen Raumes. 2.* Auflage Westermann Schulbuchverlag GmbH, Braunschweig, Germany.

Mandl M. et al. (2008). Projektbericht des Projektes KOMEOS, Program Energy systems of tomorrow, http://www.energiesystemederzukunft.at/

Motzkus, A.-H. (2002). Dezentrale Konzentration—Leitbild für eine Region der kurzen Wege? Auf der Suche nach einer verkehrssparsamen Siedlungsstruktur als Beitrag für eine nachhaltige Gestaltung des Mobilitätsgeschehens in der Metropolregion Rhein-Main, Bonner Geographische Abhandlungen, Band 107.

Narodoslawsky, M., Eder, M., Niemetz, N. et al. (2010). Durchführbarkeit von nahchaltigen Energiesystemen in INKOBA-Parks. Studie durchgeführt im Auftrag der INKOBA-Region Freistadt, des Klima- und Energiefonds sowie des Landes Oberösterreich.

Narodoslawsky, M. and Stoeglehner, G. (2010). Planning for local and regional energy strategies with the ecological footprint. *Journal of Environmental Policy and Planning* 12(4): 363–379.

Neufert, E. (2009). *Bauentwurfslehre. Grundlagen, Normen, Vorschriften über Anlage, Bau, Gestaltung, Raumbedarf, Raumbeziehungen, Maße für Gebäude, Räume, Einrichtungen, Geräte mit dem Menschen als Maßund Ziel. Handbuch für den Baufachmann, Bauherrn, Lehrenden und Lernenden. 39.* Aufl. Vieweg + Teubner | GWV Fachverlage GmbH, Wiesbaden, Germany.

Newmann, P. and Jennings, I. (2008). *Cities as Sustainable Ecosystems*. Island Press, Washington, DC.

Prehal, A. and Poppe, H. (2003). Siedlungsmodelle in Passivhausqualität. Berichte aus Energie- und Umweltforschung 1/2003. Bundesministerium für Verkehr, Innovation und Technologie, Wien, Vienna, Austria.

Register, R. (2002). *Ecocities: Building Cities in Balance with Nature*, Berkeley Hills Books, Berkeley, CA.

Rode, M.W. and Kanning, H. (2010) (Hrsg.). *Natur- und raumverträglicher Ausbau energetischer Biomassepfade.* Ibidem-Verlag: Stuttgart, Germany.

Schriefl, E., Schubert, U., Skala, F. et al. (2009). Urban development for carbon neutral mobility. *World Transport Policy and Practice* 14(4): 25–35.

Späth, P. (2007). *EnergieRegionen: Wirksame Leitbildprozesse und Netzwerke zur regionalen Gestaltung sozio-technischen Wandels.* BMVIT Berichte aus Energie- und Umweltforschung 29/2007. Wien, Vienna, Austria.

Steininger, K.W. (2008). Raumplanung als Emissionsbremse. Großes Potenzial in der Theorie, wenig Effizienz in der Praxis. *RAUM* 71(2008): S22–S26.

Stoeglehner, G. (2003). Ecological footprint—A tool for assessing sustainable energy supplies. *Journal of Cleaner Production* 11(2003): 267–277.

Stoeglehner, G. (2009). Energieversorgung als Aufgabe der Raumplanung— Perspektiven aus Österreich. 15. Thüringer Regionalplanertagung, SEPT 24–25, 2009, Bad Langensalza. www.arl-net.de (last accessed: December 22, 2010).

Stoeglehner, G. and Grossauer, F. (2009). Raumordnung und Klima. Die Bedeutung der Raumplanung für Klimaschutz und Energiewende. *Wissenschaft & Umwelt— Interdisziplinär*, 12(2009): 137–142.

Stoeglehner, G. and Narodoslawsky, M. (2008). Implementing ecological footprinting in decision-making processes. *Land Use Policy* 25: 421–431.

Stoeglehner, G. and Narodoslawsky, M. (2009). How sustainable are biofuels? Answers and further questions arising from an ecological footprint perspective. *Bioresource Technology* 100: 3825–3830.

Stoeglehner, G., Narodoslawsky, M., Steinmüller, H. et al. (2011a). In: PlanVision— Visionen für eine energieoptimierte Raumplanung. Projektendbericht. Gefördert aus Mitteln des Klima- und Energiefonds. Wien: Vienna, Austria. https://www.boku.ac.at/fileadmin/_/H85/H855/materialien/planvision/ Endbericht_PlanVision.pdf (last accessed: May 21, 2012).

Stoeglehner, G., Narodoslawsky, M., Baaske, W. et al. (2011b). ELAS—Energetische Langzeitanalyse von Siedlungsstrukturen. Projektendbericht. Gefördert aus Mitteln des Klima- und Energiefonds. Wien: Vienna, Austria. www.elas-calculator.eu (last accessed: May 21, 2012).

Stremke, S. (2010). Designing sustainable energy landscapes. *Concepts, Principles and Procedures*. Thesis, Wageningen University, Wageningen, the Netherlands.

Stremke, S., van den Dobbelsteen, A., and Koh, J. (2011). Exergy landscapes: Exploration of second-law thinking towards sustainable landscape design. *International Journal of Exergy* 8(2): 148–174.

Stremke, S. and Koh, J. (2010). Ecological concepts and strategies with relevance to energy-conscious spatial planning and design. *Environment and Planning B: Planning and Design* 37: 518–532.

Tappeiner, G. et al. (2002). Heimwert. Ökologisch-ökonomische Bewertung von Siedlungsformen. Berichte aus der Energie- und Umweltforschung 25/2002. Bundesministerium für Verkehr, Innovation und Technologie, Wien, Vienna, Austria.

Tillie, N., van den Dobbelsteen A., Doepel, M., Jobert, M., and Jager, W. (2009). Towards CO_2-neutral urban planning—Presenting the Rotterdam energy approach and planning (REAP). 45. ISOCARP Congress 2009. www.isocarp. net/paper/case_studies/1488.pdf (last accessed: January 25, 2012).

Treberspurg, M. (1999). Neues Bauen mit der Sonne. Ansätze zu einer klimagerechten Architektur. 2. Aufl. Springer-Verlag, Wien, New York.

Vester, F. (1976). Ballungsgebiete in der Krise: Vom Verstehen und Planen menschli-cher Lebensräume. Deutscher Taschenbuch Verlag, München, Germany.

Vester, F. (1980). Neuland des Denkens. Deutsche Verlags-Anstalt GmbH, Stuttgart, Germany.

Vester, F. (2007). Die Kunst vernetzt zu denken. dtv, 6. Aufl. 2007.

11

Employing Exergy and Carbon Models to Determine the Sustainability of Alternative Energy Landscapes

Sven E. Jørgensen

CONTENTS

11.1 Introduction

There is a clear need for indicators to assess sustainability, to be able to give a quantitative answer to the question: Is the proposed development of, for instance, a landscape sustainable? This chapter presents two indicators that could and should be used to answer this question quantitatively: eco-exergy (the work energy capacity) and a model for the complete carbon cycling in the landscape. To completely answer the question raised, in most cases, it is probably necessary to use supplementary indicators, but to be able to give a complete answer, the calculations of eco-exergy and the erection of a carbon cycling model is compulsory under all circumstances, because of the following:

1. Energy can be divided into exergy or work energy and anergy, which cannot do work. It is therefore crucial not only to set up an energy balance but also an exergy balance, because our focus is of course on the capacity to do work. We use energy in various forms because we want to perform work, for instance, by the use of vehicles for transportation and different machinery for various tasks in the industries. It is therefore important to know how much of the energy can be used to do work and how much is lost as anergy. By calculations of the work capacity (exergy), it will be possible to know how much work energy is available, which is the core question. So, exergy calculations will give

us the efficiency of our energy use. The exergy balance proposed will be able to indicate which energy applications have a low efficiency, and therefore where it is possible to get more work energy out of the energy. Increased efficiency means, of course, energy can be saved and that the corresponding emission of greenhouse gases is eliminated.

2. It is agreed that sustainable development requires that the emission of greenhouse gases (mainly carbon dioxide and methane) is reduced to a minimum, because these accumulate in the atmosphere and thereby change our climate, which is a completely accepted theory today. The main source of greenhouse gases is our increasing use of fossil fuel, but there are also other sources of greenhouse gases (for instance, methane emission from wetlands or carbon dioxide emission from soil) and several important sinks for the greenhouse gases (first of all photosynthesis). These processes must of course be taken into account, when the sustainability of a certain development is evaluated.

These two important indicators are presented in the following to give the reader a clear understanding of the definition of the indicators, of their application in landscape planning, and how much the indicators tell us about the sustainability of various plans for further landscape development.

11.2 Exergy and Eco-Exergy

Exergy is defined as the work the system can perform when brought into thermodynamic equilibrium with the environment, considering the differences in temperature (heat energy), pressure (pressure or expansion energy), altitude (mechanical potential energy), voltage (electrical energy), and chemical potential (chemical energy), to mention the most applied energy forms (Table 11.1). The work energy is found as a gradient in an intensive variable times the extensive variable. For instance, expansion work is the difference in

TABLE 11.1

Different Forms of Energy and Their Intensive and Extensive Variables

Energy Form	Extensive Variable	Intensive Variable
Heat	Entropy (J/K)	Temperature (K)
Expansion	Volume (m^3)	Pressure (Pa = $kg/s^2 \cdot m$)
Chemical	Moles (M)	Chemical potential (J/mol)
Electrical	Charge (Ampere second = A.s)	Voltage (volt = V)
Potential	Mass (kg)	Gravity * height (m^2/s^2)
Kinetic	Mass (kg)	0.5 * (velocity)^2 (m^2/s^2)

pressure times volume, electrical work is the difference in voltage times the charge, and potential energy the difference in altitude times mass and times the gravity constant.

Potential and kinetic energy is denoted as mechanical energy.

This form of exergy is denoted as technological exergy. Technological exergy is not practical to use in the ecosystem context, because it presumes the environment is the reference state, which means an ecosystem for an ecosystem. Since the work energy embodied in the organic components and the biological structure and information contributes far most to the exergy content of an ecosystem, there seems no reason to assume (minor) temperature and pressure difference between the ecosystem and the reference environment. Eco-exergy (application of exergy or work energy in ecological context) is defined (Figure 11.1) as the work the ecosystem can perform

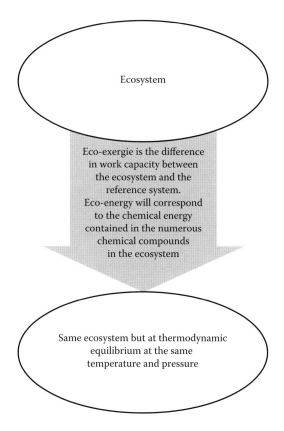

Ecosystem

Eco-exergie is the difference in work capacity between the ecosystem and the reference system. Eco-energy will correspond to the chemical energy contained in the numerous chemical compounds in the ecosystem

Same ecosystem but at thermodynamic equilibrium at the same temperature and pressure

FIGURE 11.1
Definition of eco-exergy. Eco-exergy is the work capacity in the ecosystem in the form of chemical energy of the many different and complex chemical compounds relative to the reference system. The reference system is the same ecosystem but at thermodynamic equilibrium, that is, a homogeneous system without life. All the chemical compounds are inorganic and there are no gradients.

relative to the same ecosystem at the same temperature and pressure but at thermodynamic equilibrium, where there is no gradient and all components are inorganic at the highest possible oxidation state. Under these circumstances, we can calculate the exergy, which has been denoted as eco-exergy to distinguish it from the technological exergy, as coming entirely from the chemical energy of the many biochemical compounds in the ecosystem. Eco-exergy has successfully been used to develop structurally dynamic models (see Jørgensen [2002] and Jørgensen and Fath [2011]), as a holistic ecological indicator (Jørgensen [2006] and Jørgensen et al. [2007]) and as an important variable to describe ecosystem dynamics (Jørgensen 2012).

Eco-exergy represents the nonflow of biochemical exergy. It is determined by the difference in chemical potential ($\mu_c - \mu_{co}$) between the ecosystem and the same system at thermodynamic equilibrium. This difference is determined by the activities (approximated by the use of the concentrations) of the considered components in the system and in the reference state (thermodynamic equilibrium), as is the case for calculations of all chemical processes. We can measure or determine the concentrations in the ecosystem, but the concentrations in the reference state (thermodynamic equilibrium) are more difficult to find. Nonetheless, it is possible to find good estimations, as will be shown later. Eco-exergy is a concept close to Gibb's free energy but opposing it: Eco-exergy has a different reference state from case to case (from ecosystem to ecosystem) and it can furthermore be used far from thermodynamic equilibrium, while Gibb's free energy, in accordance to its exact thermodynamic definition, is a state function close to thermodynamic equilibrium. In addition, eco-exergy of organisms is mainly embodied in the information content (see also the more detailed discussion in Jørgensen et al. [2010b]).

As ($\mu_c - \mu_{co}$) can be found from the definition of the chemical potential replacing activities by concentrations, we get the following expressions for the exergy:

$$Ex = RT \sum_{i=0}^{i=n} C_i \ln \frac{C_i}{C_{i,o}} \qquad (11.1)$$

where

R is the gas constant (8.317 J/K.mol = 0.08207 L.atm/K.mol)

T is the temperature of the environment

C_i is the concentration of the ith component expressed in a suitable unit

$C_{i,o}$ is the concentration of the ith component at thermodynamic equilibrium

n is the number of components

$C_{i,o}$ is of course a very small concentration (except for i = 0, which is considered to cover the inorganic compounds), due to a very low probability of forming complex organic compounds spontaneously in an inorganic soup at thermodynamic equilibrium. $C_{i,o}$ is even lower for the various organisms, because the probability of forming the organisms is very low with their embodied information, which implies that the genetic code should be correct.

By using this particular exergy based on the same system at thermodynamic equilibrium as reference, the eco-exergy becomes dependent only on the chemical potential of the numerous biochemical components.

In Equation 11.1, it is possible to distinguish between the contribution to the eco-exergy from the information and from the biomass. We define p_i as c_i/A, where

$$A = \sum_{i=1}^{n} c_i \tag{11.2}$$

is the total amount of matter density in the system. With the introduction of this new variable, we get

$$EX = ART \sum_{i=1}^{n} p_i \ln \frac{p_i}{p_{io}} + A \ln \frac{A}{A_o} \tag{11.3}$$

As $A \approx A_o$, eco-exergy becomes a product of the total biomass A (multiplied by RT) and Kullback measure

$$K = \sum_{i=1}^{n} pi \ln \left(\frac{p_i}{p_{io}} \right) \tag{11.4}$$

where p_i and p_{io} are probability distributions, a posteriori and a priori to an observation of the molecular detail of the system. It means that K expresses the amount of information that is gained as a result of the observations. For different organisms that contribute to the eco-exergy of the ecosystem, the eco-exergy density becomes c RT $\ln(p_i/p_{io})$, where c is the concentration of the considered organism. RT $\ln(p_i/p_{io})$, denoted as β, is found by calculation of the probability to form the considered organism at thermodynamic equilibrium, which would require that organic matter is formed and that the proteins (enzymes) controlling the life processes in the considered organism have the right amino acid sequence. These calculations can be seen in Jørgensen and Svirezhev (2004) and in Jørgensen (2012). In the latter reference, the latest information about the β values for various organisms is presented (see also Table 11.2). For humans, the β value is 2173, when the eco-exergy is expressed in detritus equivalent or 18.7 times as much or 40,635 kJ/g if the eco-exergy should be expressed as kJ and the concentration by unit of mass divided by unit of volume or area. The β value has not surprisingly increased as a result of the evolution. A few β values from Table 11.2 are as follows: bacteria 8.5, protozoa 39, flatworms 120, ants 167, crustaceans 232, mollusks 310, fish 499, reptiles 833, birds 980, and mammals 2127. The evolution has, in other words, resulted in an ever more effective transfer of what we could call the classical work capacity to the work capacity of the information. A β value of 2.0 means

TABLE 11.2

ß-Values = Exergy Content Relative to the Exergy of Detritus

Organisms	Plants		Animals
Detritus		1.00	
Viroids		1.0004	
Virus		1.01	
Minimal cell		5.0	
Bacteria		8.5	
Archaea		13.8	
Protists (algae)		20	
Yeast		17.8	
		33	Mesozoa, Placozoa
		39	Protozoa, amoebe
		43	Phasmida (stick insects)
Fungi, molds		61	
		76	Nemertina
		91	Cnidaria (corals, sea anemones, jelly fish)
	Rhodophyta	92	
		97	Gastrotricha
Porifera, sponges		98	
		109	Brachiopoda
		120	Plathyhelminthes (flatworms)
		133	Nematoda (round worms)
		133	Annelida (leeches)
		143	Gnathostomulida
	Mustard weed	143	
		165	Kinorhyncha
	Seedless vascular plants	158	
		163	Rotifera (wheel animals)
		164	Entoprocta
	Moss	174	
		167	Insecta (beetles, flies, bees, wasps, bugs, ants)
		191	Coleoidea (sea squirt)
		221	Lepidoptera (butterflies)
		232	Crustaceans
		246	Chordata
	Rice	275	
	Gymnosperms (incl. pinus)	314	
		310	Mollusca, bivalvia, gastropodea
		322	Mosquito

TABLE 11.2 (continued)

ß-Values = Exergy Content Relative to the Exergy of Detritus

Organisms	Plants		Animals
	Flowering plants	393	
		499	Fish
		688	Amphibia
		833	Reptilia
		980	Aves (Birds)
		2127	Mammalia
		2138	Monkeys
		2145	Anthropoid apes
		2173	Homo sapiens

Source: Jørgensen, S.E. et al., *Ecol. Model.*, 185, 165, 2005.
ß-values = Eco-exergy content relative to the eco-exergy of detritus.

that the eco-exergy embodied in the organic matter and the information are equal. As the β values are much bigger than 2.0 (except for a virus, where the β value is 1.01—slightly more than 1.0), the information eco-exergy is the most significant part of the eco-exergy of organisms.

In accordance with Equations 11.3 and 11.4 and the aforementioned interpretation of these equations, it is now possible to find the eco-exergy density for a model as

$$\text{Eco-exergy density} = \sum_{i=1}^{i=n} \beta_i c_i \tag{11.5}$$

The eco-exergy due to the "fuel" value of organic matter (chemical energy) is about 18.7 kJ/g (compared with coal: about 30 kJ/g and crude oil: 42 kJ/g). Notice that for these calculations, the same value is found for technological exergy and eco-exergy. The chemical energy of detritus, coal, and oil can be transferred to other energy forms, for instance, mechanical work directly, and be measured by bomb calorimetry, which however requires destruction of the sample (maybe an organism). The information eco-exergy is equal to $(\beta - 1)*$biomass or density of information eco-exergy is equal to $(\beta - 1) *$concentration. The information eco-exergy is taking care of control and functioning of the many biochemical processes. The ability of the living system to do work is contingent upon its functioning as a living dissipative system. Without the information, the organic matter could only be used as fuel similar to fossil fuel. But due to the information eco-exergy, organisms are able to make a network of the sophisticated biochemical processes that characterize life. The eco-exergy (of which the major part is embodied in the information) is a measure of the organization (Jørgensen and Svirezhev 2004). This is the intimate relationship between energy and organization that Schrödinger (1944) was struggling to find.

The eco-exergy is a result of the evolution and of what Elsasser (1981, 1987) calls recreativity to emphasize that the information is copied and copied again and again in a long chain of copies where only minor changes are introduced for each new copy. The energy required for the copying process is very small, but it requires of course a lot of energy to come from the "mother" copy through evolution, for instance, from prokaryotes to human cells. To cite Margalef (1977) in this context: the evolution provides for cheap—unfortunately often "erroneous," i.e., not exact—copies of messages or pieces of information. The information concerns the degree of uniqueness of entities that exhibit one characteristic complexion that may be described.

The application of eco-exergy is based on what could be considered a translation of Darwin's Law to thermodynamics. Biological systems have many possibilities for moving away from thermodynamic equilibrium, and it is important to know along which pathways among the possible ones a system will develop. This leads to the following hypothesis sometimes denoted as the ecological law of thermodynamics (ELT) (Jørgensen 2002, Jørgensen 2006, Jørgensen et al. 2007, Jørgensen 2012): If a system receives input of exergy (free energy, for instance, from solar radiation), then it will utilize this exergy to perform work. The work performed is first applied to maintain the system (far) away from thermodynamic equilibrium whereby exergy is lost as anergy by transformation into heat at the temperature of the environment. If more exergy is available than needed for maintenance, the system will be moved further away from thermodynamic equilibrium, reflected in growth of gradients. If more than one pathway is offered to depart from equilibrium, the one yielding the highest eco-exergy storage (denoted Ex) will tend to be selected. Expressed differently, among the many ways for ecosystems to move away from thermodynamic equilibrium, the one maximizing dEx/dt under the prevailing conditions will have the propensity to be selected.

This hypothesis is supported by several ecological observations and case studies (Jørgensen et al. 2000, Jørgensen 2002, Jørgensen et al. 2007, Jørgensen 2012). Survival implies maintenance of biomass, and growth means increase of biomass and information. It costs exergy to construct biomass and gain information; biomass and information therefore possess exergy. Survival and growth can therefore be measured by use of the thermodynamic concept of eco-exergy, which may be understood as the work capacity the ecosystem possesses. The eco-exergy of a landscape is found as the sum of the eco-exergy of the ecosystems in the landscape.

By development of an exergy balance, it is possible to use eco-exergy because technological exergy and eco-exergy is the same for fossil fuel, chemical energy of any form, and electrical exergy, except for expansion exergy and exergy of heat due to a temperature difference. For these two energy forms, the technological exergy should be applied, because the eco-exergy does not consider differences in temperature and pressure as it is defined.

The work capacity or exergy balance is an important supplement to an *energy balance*, because the work capacity balance considers only the energy that can do work (useful energy) and exclude in the balance the waste energy or anergy—heat energy released to the environment. The balance is used to indicate the activities that are not giving a satisfactory use of the energy to do work and how it would be possible to improve the work capacity balance by increasing the work energy efficiencies of these activities. As the work capacity is a measure for the sustainability (see Jørgensen [2006] and Jørgensen [2010]), the development of sustainability is also determined. To express it differently, the work capacity balance determines together with the energy balance not only the use of energy but also the efficiency of this use. Moreover, the work capacity considers also the changes of work energy in a landscape and in nature due to the management of nature and agriculture. It implies that an exergy (work capacity) balance is made for a landscape—the annual gains and losses of work energy including eco-exergy are found. When the overall balance shows a loss of exergy (work capacity), the landscape is not sustainable, and if on the other side the work capacity of the landscape is increasing or not changed, the landscape is sustainable.

The described use of an eco-exergy balance for the assessment of sustainability was demonstrated by Jørgensen (2006) and will also be illustrated in Chapter 18. Eco-exergy has been widely used as a holistic ecological indicator; see Jørgensen et al. (2010a). It is an application for ecosystem planning and management and as landscapes consist of ecosystems, an extrapolation of the application to landscapes is obvious; see for instance, the application in Stremke et al. (2011).

11.3 Modeling the Carbon Cycling

The carbon compounds causing the greenhouse effect not only come from the use of fossil fuel but are also adsorbed and emitted from soil, land-fills, incineration of waste, wetlands, agricultural land, and the entire nature. The carbon compounds include carbon dioxide and methane and other possible carbon-containing greenhouse compounds. So, control of climate change requires that we consider the entire carbon cycle.

A carbon cycle model is set up for an area, for instance, a landscape. The model will describe the pools of carbon in the area and how these pools exchange carbon and how much they release or take up carbon to and from the environment. The carbon pools included in the model—the state variables of the model—are parks, green areas and other natural vegetation (bushes, grass, and so on), other biomass pools, waste, food, and feed for pets and domestic animals, as well as the carbon of the soil (this should probably be divided into, for instance, five or more pools covering different soil types in different parts of the landscape, including asphalt, concrete, and different types of soil). The carbon cycle is furthermore influenced by the exchange with the environment: import and consumption of fossil fuel, photosynthesis (nature, parks, and green areas), respiration of humans and animals, carbon dioxide emission by incineration (for instance, biomass used for heating), export and import of carbon (for instance, in vegetables), emission and uptake of carbon by nature, and emission and uptake from soil. The state variables are connected by transfer processes and are connected to the environment by the exchange processes mentioned previously. Figure 11.2 shows a conceptual

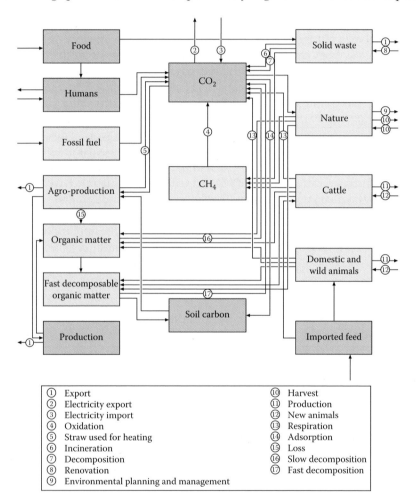

FIGURE 11.2
Conceptual diagram of the proposed carbon cycle model. The boxes indicate the state variables that describe the cycle, and the arrows indicate the processes either between two state variables or the influence from the environment. When the model is used as a management tool, it can be used to answer questions as, for instance, if we change this and this process (green field during the winter, incineration of waste instead of landfill, or more forest to give three concrete examples), how will it change the carbon dioxide emission? The corresponding changes in the work capacity balance are also easily calculated. The soil carbon pool may be divided into two or more boxes according to the different soil characteristics.

diagram of a carbon model erected for the Danish island of Samsø. The carbon model that is going to be developed for a specific landscape will of course be different and the carbon model will never be the same for two different cities. The model must be changed according to the available information. The state variables are indicated as boxes in the figure, the processes as thick arrows (both the transfer between boxes and the processes exchanging carbon with the environment) and the thin arrows indicate transfer of information within the model. It should be stressed that all these three components in the conceptual model will be different from landscape to landscape and of course different from the conceptual diagram showed, which is presented in this chapter to illustrate the idea behind the erection and application of a carbon model. The mathematical formulation of the processes and the corresponding coefficients (parameters) are known from various sources including general ecology. The model should as always be calibrated and validated by the use of the collected data, which are considered very important tests of the model. The model results can be translated to carbon-ecological footprints, expressed in g carbon dioxide equivalents per ha and per year. Methane has 23 times higher greenhouse effect than carbon dioxide but is decomposed by a biological half-life time of 7 years in the atmosphere. These properties of methane are taken into account by the translation of methane emission to carbon dioxide equivalents. The carbon cycle model is applied to overview all the carbon sources and sinks. It is thereby possible to assess the processes that should and could be changed to improve the carbon cycle and reduce the net carbon emission. The model can be used to answer questions such as: How much would the net carbon emission be influenced if we erected 5 ha more park, restored a wetland of 3 ha, and used as recreational area or changed the waste treatment system from use of landfills to incineration followed by use of the generated heat? In this context, the release and uptake of carbon by soil and vegetation are important processes. An increased accumulation of carbon in the soil is considered beneficial for the fertility of the soil, for instance, for parks and green areas.

11.4 Conclusions

When a sustainability analysis of a landscape is developed, it is recommended to produce a manual to explain the background knowledge that has been applied to develop the analysis and how the available information has been applied to obtain the exergy balance and the carbon model. If the analysis is supplemented with other indicators that indeed can be recommended, for instance, by the use of biodiversity, the additional indicators must of course also be explained in the manual. Moreover, a manual explaining the application of the developed tools in all details and also demonstrating how to use the tools in as many details as possible is very beneficial. The manual should of course also demonstrate how the work capacity calculations and the carbon cycle model could be applied as powerful tools for designing, planning, and developing of sustainable landscapes and, not the least, to reduce the carbon dioxide emission of the focal area.

In the analysis of the sustainability of a landscape, it is often beneficial to make a complete flow diagram containing (a) the sources to solid waste, (b) the transfer processes for the collected waste, and (c) the various treatments of the solid waste. The result of the flow diagram should of course be included in the carbon model. The possibilities of recycling reuse and changes of the waste and the water can be elucidated by the application of eco-exergy and the carbon cycle.

References

Elsasser WM. 1981. A form of logic suited for biology? In: Rosen R (ed.), *Progress in Theoretical Biology*, Vol. 6. Academic Press, New York, pp. 23–62.

Elsasser WM. 1987. *Reflections on a Theory of Organisms. Holism in Biology.* John Hopkins University Press, Baltimore, MA, p. 160.

Jørgensen SE. 2002. *Integration of Ecosystem Theories: A Pattern.* Kluwer, Dordrecht, the Netherlands, 386pp.

Jørgensen SE. 2006. *Eco-Exergy as Sustainability.* WIT, Southampton, NY, 220pp.

Jørgensen SE. 2010. Ecosystem Services, sustainability and thermodynamic indicators. *Ecological Complexity* 7, 311–313.

Jørgensen SE. 2012. *Fundamentals of Systems Ecology.* CRC Press, Boca Raton, FL, 320pp.

Jørgensen SE, Costanza R, and Xu F-L. 2010a. *Handbook of Ecological Indicators for Assessment of Ecosystem Health.* 2nd edn. CRC press, Boca Raton, FL, 482pp.

Jørgensen SE and Fath B. 2011. *Fundamentals of Ecological Modelling.* 4th edn. Elsevier, Amsterdam, the Netherlands, 400pp.

Jørgensen SE, Fath B, Bastianoni S, Marques JC, Mueller F, Nielsen SN, Patten BC, Tiezzi E, and Ulanowicz RE. 2007. *A New Ecology.* Elsevier, Amsterdam, Oxford, 276pp.

Jørgensen SE, Ladegaard N, Debeljak M, and Marques JC. 2005. Calculations of exergy for organisms. *Ecological Modelling* 185, 165–176.

Jørgensen SE, Ludovisi A, and Nielsen SN. 2010b. The free energy and information embodied in the amino acid chains of organisms. *Ecological Modelling* 221, 2388–2392.

Jørgensen SE, Patten BC, and Straškraba M. 2000. Ecosystems emerging: 4 Growth. *Ecological Modelling* 126, 249–284.

Jørgensen SE and Svirezhev Y. 2004. *Toward a Thermodynamic Theory for Ecological Systems.* Elsevier, Amsterdam, the Netherlands, 366pp.

Margalef R. 1977. *Ecologia.* Omega, Barcelona, CT, p. 951.

Schrödinger E. 1944. *What Is Life? The Physical Aspect of the Living Cell.* Cambridge University Press, Cambridge, U.K., 90pp.

Stremke S, Dobbelsteen AV, and Koh J. 2011. Exergy landscapes: Exploration of second-las thinking towards sustainable landscape design. *International Journal of Exergy* 8, 148–174.

Part III

Case Studies

12

Energy-Conscious Design Practice in Asia: Smart City Chengdu and the Taiwan Strait Smart Region

Raoul Bunschoten

CONTENTS

12.1 Energy Landscape as Incubator

A dynamic human environment needs energy to keep it going. This energy is the source of creativity, generating change and visions for new cultures and societies. But energy can be dangerous. The energy driving most of society comes from the earth and is now the source of one of the greatest threats humanity has ever faced: climate change. This leads to problems such as rising temperatures; scarcity of water; lack of raw materials to feed and provide comfort for a rapidly increasing population; the poisoning of air, soil, and oceans with chemical substances, heavy metals, and waste; and the immanent energy wars.

The term energy landscape, much in use currently to indicate the relationship between spatial planning and energy infrastructures, combines the essence of the two skins of the Earth. As introduced in my book *Urban Flotsam* (010 Publishers), published in 2011, the two skins of the Earth define the "real" space that humans operate in. The first skin of the Earth is nature in its totality. The second skin is whatever we, humanity, have created on top, within, and underneath the first. This second skin is an intensely interconnected and dynamic construction lightly interweaving between the crust of the Earth and the atmosphere. This "weaving," however, is constantly a balancing act between nature and human action and is becoming overstressed. Instead of coexisting in a planetary biotope, people are beginning to influence the balance of this biotope, mainly through changes in energy behavior, to an extent that will not endanger, as is commonly being said, the Earth ("save the planet" is the faint plea in all hotel rooms to reuse your towels), but humanity itself—not least because it is rapidly expanding in numbers. The unbalanced interweaving within this dynamic skin has become the strongest feature of the global energy landscape people plan and live in and where humanity projects its possible futures (there is no escape: the next habitable planet is around 600 light-years away). These futures cannot be traded by the hedge bankers for quick profits; they are real life choices. In a sustainable energy landscape, this balance is managed successfully and for the sake of future generations.

But how do we prepare for these decisions? Can we trial different futures? There is a long tradition of constructing secluded test beds or ideal gardens (paradise as a reflection of heaven) reflecting ideals. Utopias describe ideal futures, and there is a history of realized utopian fragments that act as test beds. Taking inspiration from the business, agricultural, and medical communities, we aim to create test beds as incubators. In these incubators, we try to nurture a multitude of alternative energy behaviors and therefore landscapes, which would lead to a rebalancing of the dynamics of the skins, of the Earth. These incubators are stages for urban choreographies that lead to new forms of energy behavior that can prolong our existence on Earth. An incubator is something like a performance space as well as test bed and curatorial space, nurturing new relationships and new life forms.

As in the tradition of the original paradise, these incubators are political and scientific ideals—utopias but at the same time playful and inspiring. Four main aspects help incubators mitigate and adapt to climate change: choreography, coevolution, cybernetic management, and curatorship (Figure 12.1). The first two are forms of creation, and the second two forms of management. Cybernetics help to manage the choreographies into complex dynamics of an urban society; curatorship animates and coaches order out of the chaotic evolutionary processes that define a rapidly growing and highly volatile interconnected world.

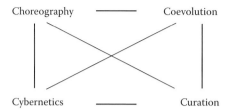

FIGURE 12.1
Four main aspects help incubators mitigate and adapt to climate change: choreography, coevolution, cybernetic management, and curatorship (CHORA schema).

Our projects link extremes of scientific programming with artistic license, or to reflect on the role of Faust, harnessing knowledge as well as playing with the factors of reality. Both Marlowe's and Goethes Faust make a pact with the devil to do this. Our pact may have to be with the energy and smart city development sector—some corporations such as Siemens have already initiated this sector. Our profession, urban planners, architects, geographers, and others, forms the platform between this sector and the urban population.

12.2 Emission Reduction Targets and Smart Cities

Within the second skin, cities and their policymakers have a major role to play in changing energy use and meeting national and global targets of greenhouse gas emission reductions. The mitigation of humanity's damage and defacement of the first skin is a huge challenge. As most of the global population now lives in an urban environment, planners and architects have a greater responsibility to change the energy structure of the metropolis. This means changing the interactive dynamics between the skins, in which cities become the main providers of energy, and where a change in energy consumption is most likely to have an effect. This contradicts the more traditional view of contemporary energy production and consumption, which is a rural-to-urban, source-to-sink relationship. In order for an urban incubator to become a low-carbon or even energy-producing entity, it has to become a "smart city," which means, in effect, connecting as many systems and processes as possible with the aim of improving energy efficiency and promoting renewable energy use.

What is the context for the creation of large-scale incubators and smart cities? China certainly has raised the stakes. In March 2011, the People's Republic of China issued its 12th 5 year plan. In this, climate change played a key role and it defined serious targets to which its provinces should adhere to. The plan outlines the goal to source 11.4% of primary energy consumption from nonfossil fuel sources by 2015. Combined with a 17% reduction in CO_2 emissions per unit of Gross Domestic Product (carbon intensity) by 2015, the country seems well on its way to reaching the 40%–45% drop in carbon intensity set out in Copenhagen in 2009. These kinds of environmental objectives, combined with the forecast growth rates (over 9%), generate the requirement of low-carbon projects at an urban scale. Preceding this is an infrastructure that can accommodate and nourish such projects, allowing them to grow from a business plan to a physical object with an impact on the urban environment. Since this is uncharted territory, the previously mentioned incubators are needed to test these infrastructures and projects in various combinations and under differing circumstances.

Two key CHORA incubator proposals are presented here that attempt to address the void between strenuous international climate change targets and agreements, and the practical realization of low-carbon policies in cities (Figure 12.2). The first project is titled Smart City Chengdu, a recent installation by CHORA at the 2011 Chengdu Biennale. Chengdu's plans for the development of its fifth circumnavigating transport ring provided a unique opportunity for CHORA to demonstrate its vision for Chengdu as a "Smart City," as instigated by this development ring. The second project is the Taiwan Strait Smart Region—a project that aims at transforming the geopolitical complexity of the region that links Mainland China and Taiwan into that of a special low-carbon zone. The project started out as an atlas project

FIGURE 12.2
Map of CHORA incubators in Asia.

produced with award-winning graphic designer Joost Grootens and has now become a combined atlas and manual (as well as other potential materials), which is set to be published in 2012. The manual will guide and inform urban planners and city officials in the region on how to implement a low-carbon incubator. It does this by linking components of a CHORA-designed operating system, entitled the "Urban Gallery," to the sampling of maps. Within the maps, the operating system is used to create pilot projects, policies, and the cross-pollination of action. The Urban Gallery acts as a cybernetic management tool used to choreograph action and coevolution and curate results.

12.3 Technical Methodology/The Urban Gallery

Managing the construction of an energy landscape as incubator is an experiment in cybernetics. Cybernetics was invented during the Second World War and is about the management of dynamic complex and interconnected process arrays. The only time cybernetics has been used in urban planning has been in the CyberSyn project initiated by the Allende government in Chile during the 1970s. The experiment was led by Stafford Beer, one of the founders of the cybernetics movement, but was cut short by the military coup that ended Allende's government and life. Cybernetics aims at curating information and choreographing action in a context where constant change takes place. An incubator is a cybernetic device. It connects various policies, strategies, technologies, etc., and it creates feedback to manifestations on the ground, in the form of local information and conditions. The Urban Gallery is the operating system CHORA is developing for urban incubators. It is based on cybernetic principles, but it is unique in the sense that there are no precedents. This purpose of the system is to demonstrate how targets can be met in urban environments. As the incubator is an exhibition of possibilities, the Urban Gallery is a planning support system aimed at curating this information and choreographing action. CHORA has used this methodology in the creation, among other things, of an energy plan for Taichung government (Figure 12.3). Its operation is split across four layers, which are as follows:

1. *Database*: gathering and disseminating the necessary local information related to climate change, energy, water, waste, and other urban planning and social and cultural issues that relate to communities. Like the prototypes, this has a dynamic influence to the plan as it represents a real understanding of the dynamic local environment.

2. *Prototype technologies and design methods*: introducing a pool of specific means with which to achieve concrete results. Such a taxonomy of options would be live and represents some of the most advanced technology and ideas available.

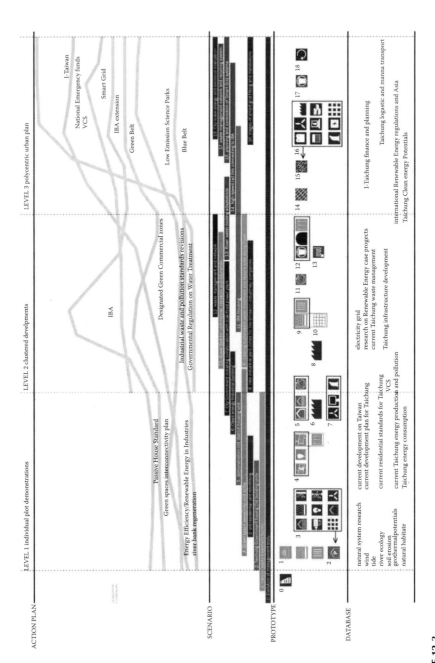

FIGURE 12.3
Timeplan for the Taichung energy plan by CHORA.

3. *Scenario creation and stakeholder management*: sustaining cooperation between actors, agents, and different projects through the mediation of regular congregation of innovative groups of partners.

4. *Action planning*: aimed at realizing projects, policies, and strategies. These milestones and targets of planning may be set, and adjusted, by a task force who will manage the planning of the low-carbon region.

The Urban Gallery is a live instrument that changes and reacts over time. Figure 12.3 shows a timeplan of the Urban Gallery for CHORA's Taichung Energy Plan, a project commissioned by the local government.

12.4 Smart City Chengdu

12.4.1 Conditions for Progress

The Chengdu Biennale is an art, design, and architecture biennale and is the largest local contemporary art event in the history of the city. In 2011, the theme of the fifth biennale was Changing Vistas: Creative Duration and was divided into three sub-exhibitions: Pure Views: Contemporary Art Exhibition, The Solutions: International Design Exhibition, and Holistic Realm: International Architecture Exhibition, which CHORA took part in.

The city of Chengdu, capital of the Sichuan province, is one of the several official development zones outlined in the latest 5 year plan. In 2009, Chengdu established a long-range objective of constructing a "Modern International Garden City"—the core ideas being natural beauty, a just society, and an organic integration of urban and rural areas. Currently it is a favorite for investment from major international enterprises and is a powerhouse of innovation and production. Intel, Cisco, Sony, Siemens, Foxconn (makers of Apple products, including 1.2 million iPads a month in Chengdu), and Toyota hold assembly, manufacturing, and corporate bases there, as well as Motorola, Ericsson, and Microsoft having R&D centers in Chengdu. Dell plans to open its second major China operations center in 2011 in Chengdu, while AMD (Intel's rival in chip manufacturing) aims to set up an R&D center in the city. There is also national investment into the biopharmaceutical sector there as well as high tech industrial parks for software, space, and aviation technology.

The Bureau of Commerce aimed for the city to grow in the tertiary sector by 13.5% in 2011 as well as maintaining an 18% increase in consumer sales and 44% increase in exports. It has recently achieved such accolades as the world's fastest growing city in the next decade by Forbes and China's happiest city by Oriental Outlook. By October 2011, Chengdu produced 100 million computers, accounting for 20% of global output.

Historically, Chengdu's expansion has been achieved through concentric growth, in which the city expands in increasingly large rings. Currently there are four; however, a fifth is under way. The first two encase the main urbanized districts of the city, as well as the location of the original city wall built in 311 BC—they densely house approximately 960,000 people. Beyond that is the third city ring, currently used as an urban mobility tool, and the fourth ring, which acts as a green belt. As the population within the city limits recently reached 14,000,000, demand for a high-speed fifth ring motorway has sparked the debate of the future of Chengdu's planning. Traffic throughout the inner rings has reached a point where cars now alternate days of access (based on odd or even number plates), as an attempt to alleviate circulatory traffic.

12.4.2 Fifth Ring as a Low-Carbon Incubator

CHORA's aim is to transform the construction of the fifth ring, and all the secondary development that follows, into a "smart" ring. The term "smart ring" refers to the shape of the physical manifestation as a ring form. In theory, this is merely a fragment of the smart city, as in time, it may grow to affect the entirety of Chengdu. It will become an incubator of low-carbon projects and will adapt to new technologies and projects in the future. The development of this fifth ring is an opportunity. It may allow Chengdu to shift toward a low-carbon trajectory and away from static, heavy, and inflexible infrastructures.

The construction of a motorway of this scale today must consider the targets set out by the state in regard to climate change. Soon we will be deep into the transition to electric personal transportation, and potentially even intelligent guiding systems. This implies a smart infrastructure that will easily adapt over time without becoming redundant. An adaptive, additive, and expansive infrastructure will allow the stakeholders involved in the progression of the city to scale up—as larger agents approach, the infrastructure can be modified to accommodate them.

A truly smart strategy will address carbon reduction targets not only through technology, adaptability, and policy but also through the integration of this with the newly emerging lifestyle in China today. This lifestyle is the result of the growth and development of Chengdu and China that was previously mentioned. Such GDP (per capita) increases create a new Chinese society; one that allows its citizens to enjoy a new lifestyle, which in turn, has the potential to facilitate a new type of energy behaviour. A smart city allows this type of growth (environmentally sensitive and sustainable) by integrating all systems, cultures, and societies and is achieved with a smart infrastructure.

This is a holistic strategy where the fifth ring can lead to such integration (nature and city, urban and rural, high and low tech, inside and outside, etc.) as it is a liminal body (a threshold space) between these fields. Nature can be seen as a resource to urban dwellers psychologically (recovery and leisure), genetically (biodiversity, carbon sequestering, etc.), and technically (biomass, water purification, etc.). The transformation of the fifth ring forms an ideal incubator by adapting the behavior within the second skin. Its aspects imply an ideal, or utopian, society, the architecture of which has to lead the way by creating cultural, social, and economic identity, one that the people can relate to. Design is at the heart of this utopian society, not just planning.

12.4.3 Low-Carbon Planning, Utopia, and Dramatic Necessity

Today, the definition of an urban utopia has shifted toward a perfectly working low-carbon plan. This is an exciting idea to aspire to; however, there are many unknowns and variables preventing the immediate success of such a development. To lower carbon emissions as dramatically as China is proposing to is a difficult step to take. It begins by starting with a specific project—a large-scale pilot—through which we can demonstrate the possibility to create a smart city. It acts as a start-up and initiates the learning curve.

In order for the fifth ring to succeed as a smart city (fragment), and become a prototype for China, it must satisfy four main objectives: (1) sufficient branding, (2) occupancy of a clearly marked space, (3) possession of a set of effective flows (traffic, energy, etc.), and (4) it must have clear management structures in place. These criteria relate directly to the layers of the operating system used by CHORA and can act as indicators for the testing of pilot schemes.

CHORA proposes to use the fifth ring as a pilot, which can both become the site of a large range of renewable, efficient, and highly technological measures and become core, adaptable infrastructure that can eventually serve the heart of the city and its high-speed traffic, while also expanding outward.

In a recent interview with Harvard University, conducted during the opening session celebrating the launch of the book *Ecological Urbanism* (Lars Muller Publishers), I explained the four levels of the Urban Gallery with these four terms. In this manner, these terms connect the aforementioned indicators with the generic and technical methodology discussed earlier:

1. *Pressure*: The pressure is caused by the driving forces at play. Smart City Chengdu focuses on the pressures of expansion and growth, the need for increased transport capacity, climate change targets, and the drive to become a global pilot project.

2. *Toolbox*: The toolbox provides the main ingredients of a smart city; it is what a city has to apply to become low carbon. The toolbox of Smart City Chengdu is a varied one and can be technical, political, biological, etc. A Smart City activates these ingredients in the right frequency and density.

3. *Spin*: The spin is the story behind the plan, how it is told and to whom. It represents the relevant stakeholders such as city government, provincial government, local communities, new inhabitants, local investors and project developers, and foreign investors and experts. Vital to the project's success are the corporate developers who stake a claim in the plan and encourage it to be adapted.

4. *Onto the stage*: This is where the incubator takes place. In this case, it is the fifth ring of Chengdu. It will be divided into planning sectors, which become receptacles for the toolbox. This allows us to begin addressing the targets and relieving the pressure, thus satisfying the stakeholders.

If operated correctly, this method can aid Chengdu in becoming an internationally recognized pilot incubator, pioneering the Urban Gallery as its operating system to manage its layers.

12.4.4 Chengdu Biennale Installation

CHORA's installation visualizes and simulates the Smart City Chengdu incubator and its effect on energy use and carbon emissions. The following are the components of the installation.

12.4.4.1 Table

The fifth ring of Chengdu is represented by a tabular surface divided into sectors representing districts and based on studies of various utopian projects of the twentieth century. It illustrates the fifth ring as a potential incubator for smart developments and demonstrates the need for a widespread and versatile application of a low-carbon planning and construction toolbox. The table itself is a timber jigsaw of plates (representing development parcels or districts) attached to a metal frame. A high-quality print of the fifth ring utopian plan, whereby each sector is represented by a utopian plan (Figure 12.4a), is laminated onto the timber and protected by glass plates. Legs of varying diameter are bolted to the metal frame at intersections. This allows the structure underneath the table's surface to suggest a network of interconnected sites of varying scales, mapped by the legs. This is amplified by the use of different sized steel sections for primary and secondary supports, mimicking the motorway and its arterial offshoots (Figure 12.4b).

12.4.4.2 Dice

The toolbox of prototypes is presented in a playful manner as a vast quantity of 2000 custom-made dice, which carry a selection of icons relating to a low-carbon urban toolbox. Dice are thrown across the table to represent the provision of various prototypes throughout the fifth ring. The fall is accidental, it is not important which specific projects are realized or where, what is important is the variation of projects (complexity) and the critical mass of their frequency (scale) within the fifth ring. Two thousand dice were initially placed on the table top and this gradually dwindled as the exhibition progressed and visitors took home a CHORA prototype dice as a souvenir.

(a)

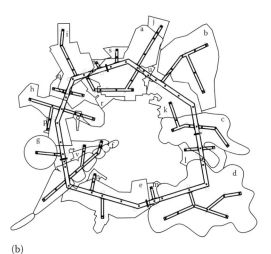

(b)

FIGURE 12.4
(a) Collective design of the utopian segments of the fifth ring of Chengdu by students at Düsseldorf University of Applied Sciences. (b) (Infra)structure of the table and its completed state.

(continued)

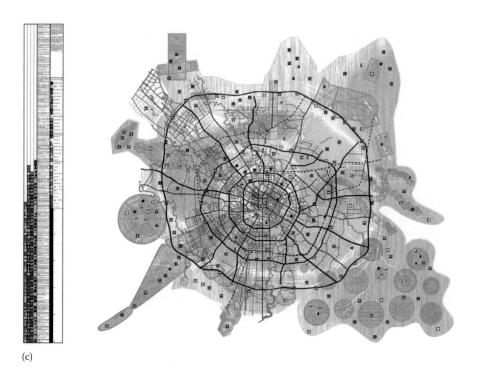

(c)

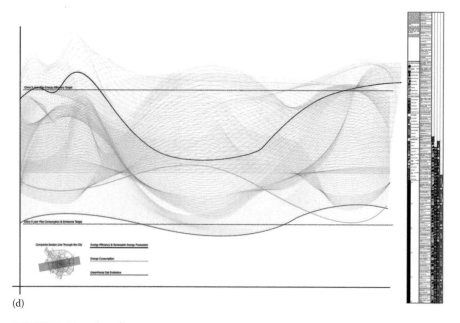

(d)

FIGURE 12.4 (continued)
(c) Scenario mapping animation. (d) Graph animation of climate change indicators.

(e)

FIGURE 12.4 (continued)
(e) CHORA's Smart City Chengdu installation.

12.4.4.3 Animations

The impact and result of these strategies, technologies, policies, etc., is important—we need to know the result of such strategies in reference to climate change indicators.

This principle of cause and effect is illustrated by graphic animations projected onto the walls. On one wall is an animation showing the rise and fall of prototypes in frequency, density, type, and organization throughout the city, over a map (Figure 12.4c). It shows constantly changing scenarios of stratagem in plan using the icons of the prototype dice. At the same time, an animated graph shows the changes caused by these scenarios on various climate change criteria as a result (Figure 12.4d). This graph is essentially a section through a three-dimensional (3D) variable net of statistical data that sits over the entire city, illustrating the changes in these criteria across the space. It attempts to visualize the normally abstract concepts of carbon emissions and energy consumption and highlights the direct influence of a Smart City incubator on these measurements. The three indicators nominally represented are energy efficiency and renewable energy production, energy consumption, and greenhouse gas emissions.

In order to catalog and register the scenarios, a data sidebar is placed on the ends of both animations. This bar contains a running list of every active prototype, a bilingual scenario summary as well as an index of prototype icons and timer. This tool helps synchronize the animations and cements the relationship between cause (prototype proliferation change) and effect (impact on climate change indicators throughout the city).

The aim of the animations is to demonstrate the need for complexity and scale of projects. The optimum situation occurs when a large number of diverse projects exist throughout the city. At this point, on average, the renewable energy production is at its peak, as is energy efficiency, and the consumption of energy is at its lowest. Of course, this is also affected by the distribution of projects within the rings of the city. For example, the dense city center may not house large renewable energy farms that push the production graph up, but may instead test home improvement kits or solar water heaters that divert the energy efficiency graph upward at the center.

The animated graph is a section through a 3D net of climate change data, which reacts to the implementation of prototypes, shown in plan on the map. The sidebar records the scenario in time as a short description and a log of every prototype active at that point in time.

The overall installation is an enclosed space whereby visitors can manually distribute prototypes throughout the smart city using the dice while observing potential changes in climate change indicators as a projection of theoretical scenarios (Figure 12.4e).

The Chengdu Smart City Incubator acts as a decision-making tool for stakeholders of various scales to encourage their capital growth with the aid of low-carbon technologies, allowing the city to achieve its goals of carbon intensity reduction, which is monitored throughout the city.

A total of 300,000 people were registered while passing through the doorway of the Architecture Biennale. Many of these, as visible in the image (and demonstrated by the films available on Youtube and Vimeo), were attracted to the table and played with the dice—using them mostly as building blocks, piling them into towers, laying them into geometric configurations, curvy lines, etc. In this manner, they become unwitting stakeholders and designers of the fifth ring incubator. The sight of whole families including young children, older ladies, etc., playing with the dice at the table is the kind of collective participation and mass interaction an incubator needs.

12.4.5 Conclusion: The Future of Smart City Chengdu

An earlier version of the Smart City Chengdu installation employed "Smart Dice" as opposed to regular plastic ones. In this instance, each dice had a built-in microchip that would register the orientation of the dice as it was rolled. This information (location on table and the prototype that was rolled) would be processed by an on-site computer and would have a direct impact on a "live" graph and map projected onto the walls. Effectively, this would be a live, interactive calculator of low-carbon planning—a simplified, digitalized model of Stafford Beer's CyberSyn.

The benefits of this optimum installation would be the translation of this eager interactivity seen by all visitors, to a more intuitive "cause–effect" relationship (between rolling several dice and seeing the animations morph in response).

Unfortunately, this was financially prohibitive, and the number of dice would be dramatically reduced—turning the whole plan into more of a digital toy (the dice would have to be fixed to the table to some degree) as opposed to a lively plethora of colors and options.

Pleasingly, many of the visitors who were intrigued by the objects took time to read the bilingual description, and would hopefully better understand the impact of every role.

12.5 Taiwan Strait Smart Region

Regardless of the policies reviewed at COP17, in Durban, we will need large-scale urban incubators to test cities' contribution toward global emission targets—this is especially poignant as the majority of people now live in them. The Taiwan Strait is a complex geopolitical region; in effect, it is a very large urban agglomeration containing an inner sea, a natural liminal body. It separates Fujian Province from Taiwan (Figure 12.5a). Taiwan claims sovereignty but is not recognized by most of the world, including the United Kingdom. China sees Taiwan as one of its provinces and is maintaining a constant pressure. This makes mapping the Taiwan Strait, as an emergent urban region, new and unique.

12.5.1 Components

The research aims at completing two parts of the Taiwan Strait Smart Region, with the potential for additional materials/tools in the future:

1. *Atlas*: Mapping of emergent conditions of the Taiwan Strait describing it as a potential incubator territory (Figure 12.5b).
2. *Manual*: This is the application of the low-carbon planning operating system on up to 100 sample sites within the region, as well as two demonstration masterplans for Xiamen and Taichung as pilots for the incubator region (Figure 12.5c).

These two ingredients of the TSSR form the material for a vision of the Taiwan Strait as Smart City Incubator. The function of the TSSR Atlas and Manual is, as a curatorial device, not only to create a new cultural and political environment with the focus on energy and carbon management cooperation but

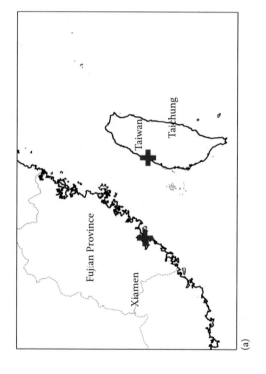

FIGURE 12.5

(a) Map showing location of Xiamen and Taichung, cities facing each other across the strait.

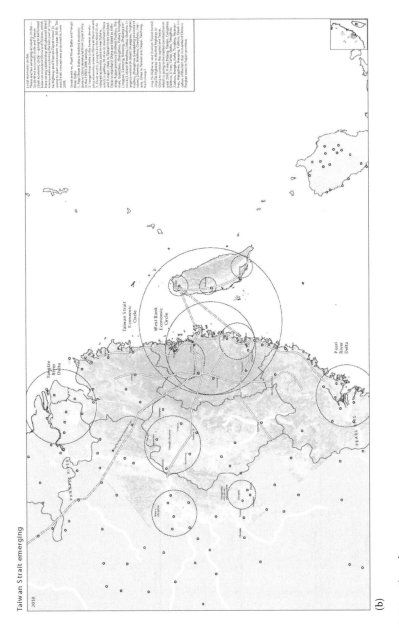

(continued)

FIGURE 12.5 (continued)
(b) Sample spread from the Atlas of the TSSR.

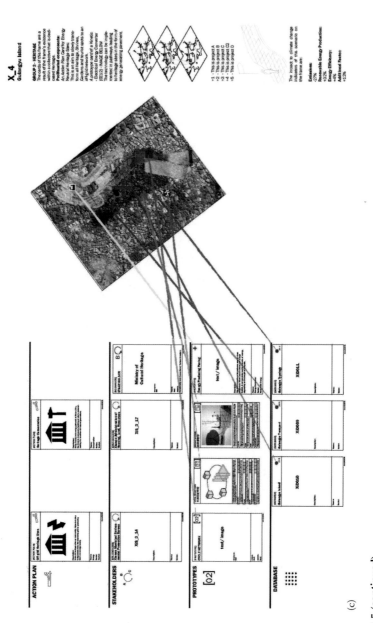

FIGURE 12.5 (continued)
(c) Sample spread from the Manual of the TSSR.

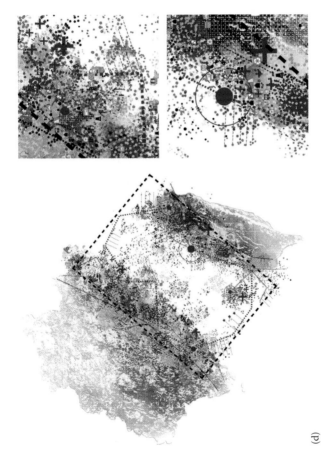

(d)

FIGURE 12.5 (continued)
(d) Taiwan Strait Smart Region, toward a low-carbon incubator.

also to enhance an urban society by democratizing its action to achieve the overall aims and eventually, to meet global targets.

Within the context of climate change, cities and their urban policies have to play a role in changing energy use and meeting national and global targets of greenhouse emissions reductions. The cities Xiamen and Taichung, facing each other across the Taiwan Strait, are keen to promote low-carbon plans, acting as catalysts for the Taiwan Strait region as a whole. Climate change is a universal topic; it focuses the mind, elicits public opinion, and it democratizes. Sustainability and energy efficiency are shared values. Cooperative projects on low carbon fit the policies of governments on Beijing and Taipei alike. The Atlas has the potential to become a curatorial device in the achievement of the low-carbon incubator, providing a guide to the necessary stepping stones in this process.

12.5.2 Local Pressure

Climate change is a daily feature of life in China. With much of its coastal wetland, arable farmland, and urbanized districts lying only 2 m above sea level, there is plenty at stake. This portrays parts of China as being at the mercy of climate change, creating obvious societal and humanitarian difficulties. The hazards of climate change in Fujian Province alone are apparent; for example, the 1000 year old plantations of the tea industry are encountering droughts and floods, which impact the socioeconomic activity. There is an increase in storms and floods, increased intensity of rainfall, hotter temperatures and heat waves, more "super typhoons," and increasing landslides. Additionally, 4 of the worlds 10 worst earthquakes since 1900 occurred in China, including Fujian province.

Unlike Mainland China, Taiwan struggles to contribute to global policy; Taiwanese leaders do not get invited to international negotiations on climate change. As a result, Taiwanese companies and cities cannot lower their emissions through the framework of CDM (UN regulated Clean Development Mechanism, discussed again in Durban, with matters unclear regarding their continuity) or in conjunction with Chinese action/agents. Also, the targets set by the Chinese government, such as the 5 year plans, do not affect the Taiwanese population, and that the three billion yuan pledged by the Chinese to tackle climate change has not seen by the Taiwanese.

Ecologically, however, Taiwan is possibly more susceptible to climate change effects than most other Asian countries due to it being an island, on the edge of the Philippine Plate, facing the Pacific Ocean. Earthquakes are frequent, and the island is interspersed with steam vents and sulfur springs. It is a subtropical island and vulnerable to rising sea levels and changing temperatures as most of its mass is a rugged mountain range, pushing the roughly 22 million inhabitants to the coasts. Furthermore, its industry is booming, contributing a large quantity of carbon to the surrounding environment—its CO_2 emissions per capita is three times the world average, and growing at a groundbreaking rate. However, as it is not recognized by most of the UN nations, it has no pressure to make national commitments to carbon emission reductions. Some cities within Taiwan have signed the ICLEI climate agreements, a global conglomeration of local governments and authorities. Taiwan's efforts are largely led by grassroots organizations such as the Green Formosa Front and the Homemakers Union and Foundation.

12.5.3 Approaching the Taiwan Strait Smart Region

The aim of the Taiwan Strait Smart Region is to demonstrate a joint strategy of climate change mitigation. This incubator will in turn spawn future low-carbon projects, technologies, and policies that will cumulatively create a low-carbon region. The use of the *Urban Gallery* in this context creates an interconnected public domain or metaspace that rises above political and geographic separation and tension. The Taiwan Strait is an area of global significance as a long-time potential conflict zone of Chinese and United States of America interests.

Creating a metaspace in the form of public domain, a kind of integrated carbon market creates bridges by circumventing political systems, thereby making a contribution to Beijing's much promoted "social harmony" and to world peace. The process of searching for low-carbon planning methods on the other hand is in itself a process that brings out local voices and self-expression of communities.

But this low-carbon incubator has the potential to enhance commercial activity and trade as well, which is also in Western interests. This emerging marketplace creates the conditions for a place of innovation. The density of projects and cultural nature of the project can be described as a geographic metaspace.

12.5.4 Cybernetic Experiment and Curatorial Device

When such a geographic metaspace (real projects, connected culturally or numerically) acts as a planning tool, it is like an incubator, as the Urban Gallery operating system choreographs and curates projects. The cross-pollination of stakeholders and policies across sites with similar agents nurtures projects to grow and coevolve in a complex manner (Figure 12.5d). Traditionally, when confronted with national emission targets, local governments tend to pass much of the duty (and expense) to the private sector through legislation and directives, forcing developers to independently meet these demands using the quickest and cheapest (not necessarily the best) method. This may often be coupled with public investment into infrastructure, like public transport (such as Curitiba's Bus Rapid Transit System) or recycling initiatives. This however, is limited in its reach and does not publicly deal with issues of energy behavior, creation, efficiency, dedicated pilot projects, or more importantly, smart energy networks (the coevolution of multiple projects into complex new systems). In order for this to happen naturally, the conditions for such projects to emerge on their own, or gradually, have to be set up—this is the function of an incubator.

Incubators are where ideas are born and start-ups are tested and proved. Ones that succeed are stimulated, ones that fail are discarded. The key thing is that they share space and resources (in this case both public and private), within which there can be cross-pollination, mutual influence, integration, and development. A competitive nature exists while allowing for the accidental discovery of connections and inter-relationships—these inter-relationships are key to incubators of any type. Within the realm of low-carbon planning, the incubator is the city-wide framework for future planning, whereby stakeholders are gathered to address the pressure of targets by implementing and testing prototypes in relation to local conditions. One typical example of this is the production of clean energy for the city. By using low-carbon planning of this nature, production is not restricted to publicly funded super-projects, which have their problems. Critical mass can be reached through policy and stakeholders in a region pursuing alternative solutions, such as the proliferation of micro production centers at consistent site conditions. In this case, the basis of an incubator is the territory (or liminal body), as opposed to a business park or training framework. The territory supplies all the resources for the individual projects to grow, allowing for cross-pollination, etc., while at the same time adapting itself over time to suit more frequent and new types of project.

It is also important to maintain sight of the principle of cause and effect at this scale. This means that as the strategy becomes "smarter," its monitoring becomes more intelligent and responsive, allowing for an immediate sustainability perspective of the entire city in question at the touch of a button (as was demonstrated in the animations of the Smart City Chengdu installation). Other urban design incubator projects include the Energy Incubator, Tempelhof, Berlin. Tempelhof intended to be a space where technologies and nature merge, where energy is produced by a series of different technologies and a space for learning and experiments. It would reflect cities which not only consist of low energy structures, but also of energy producing systems and smart networks.

12.5.5 Future of the Taiwan Strait Smart Region

Creating a new low-carbon identity for this region as an energy landscape is of enormous significance for both China as well as the whole Pacific region. The material of the Taiwan Strait Smart Region currently includes the book (atlas + manual); however, it is intended to progress into a live website, mobile app, and game, that all express either the interactivity or the dynamics of this landscape. The immediate aim is to encourage two task forces, one in Xiamen, one in Taichung to act as initiators. A further aim is to create a third, cross-strait, task force that will act as choreographer of the development of this incubator. The result will be a physical and chronological plan that can act as a resource for a public investment strategy. This will also satisfy the international need for a successful low-carbon incubator, which could be invested in and tested by all. Seen by the entire world, this low-carbon incubator will have a strong impact on how Europe and the West strive to reach their goals.

12.6 Conclusion

The idea of testing potential futures (that are nurtured by energy behavior in an interactive landscape) has already been modeled, in the form of an online digital calculator for the United Kingdom's climate change targets. Professor David MacKay, chief scientific adviser to the Department of Energy and Climate Change in the United Kingdom and Professor in the Department of Physics in the University of Cambridge has developed The DECC 2050 Pathways Calculator (Figure 12.6). This calculator enables you to test, in detail, the ways in which the United Kingdom can reach its carbon targets; it is also demonstrated in the form of a game to better experience the dynamics of changing energy behavior. It is accessed online via 2050-calculator-tool. decc.gov.uk (February 2012).

The key is that this calculator is applied to a known entity, the United Kingdom. Of course the United Kingdom is no closed state, like the former Soviet Union, or current North Korea, where only restricted flows of energy, people, and goods enter and leave. But since the United Kingdom is a political entity, the calculator simplifies the components of this known entity and shows how with these components, targets can be achieved. In principle, any entity with known energy behavior can be the subject of this game.

The definition of an energy landscape can be the geographic description of a territory in terms of energy behavior. The application of a calculator to this territory in order to change its energy behavior and develop new means of reducing carbon emissions turns this territory and geographical entity into an incubator. An incubator is, according the director of the Xue Xue Institute in Taipei, a cultural incubator, a "space for the creation of ideas." It also is the vehicle to develop economically viable businesses, even new business models. We aim to create urban incubators that, through their status as energy landscapes and cauldrons of emergent energy practices, become nurturing places for business models aimed at carbon reduction and the awareness of a society, that is, awareness of the need for the lifestyle changes necessary to meet national and global targets. An incubator forms the geographical space of a game, such as the one used to demonstrate the calculator. Its weakness is the need for a political or other governance entity equivalent to the territory of the incubator. Or that may, in the case of Chengdu's fifth ring and even more the Taiwan Strait, be its very innovative character: The Taiwan Strait as energy landscape and

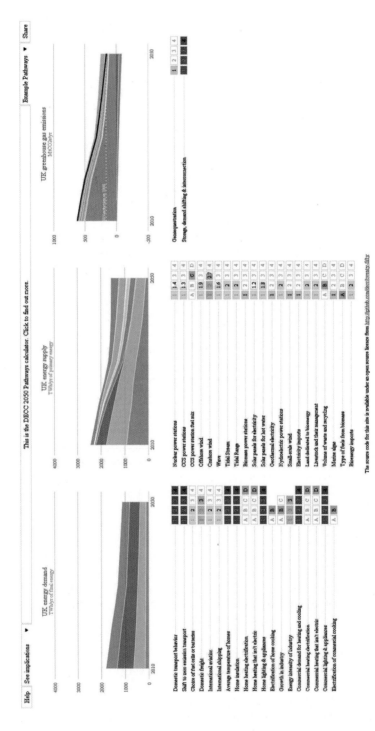

FIGURE 12.6
Screenshot of the DECC 2050 Pathways calculator.

designated incubator requires joint governing agencies and strong cultural and other cooperation. Since there exists a great desire for this, the Taiwan Strait could be the world's largest and most effective low-carbon incubator.

About CHORA and the Author

CHORA is a new kind of global practice linking research into new methods of urban planning with the aim to reduce carbon emissions to the implementation of these new methods in city planning and construction. CHORA has been involved in low-carbon studies and stakeholder game workshops for the cities of Xiamen, Taichung, and London and has acted as consultant to the London Development Agency on the planning of the Green Enterprise District, East London.

The author, Professor Raoul Bunschoten, is the professor of sustainable urban planning and urban design at the TU Berlin and is a senior lecturer at the London Metropolitan University. He has widely published and lectured on urban planning.

Acknowledgments

I would like to acknowledge Adam Atraktzi of CHORA London, who cooperated on research and composition. Smart City Chengdu, Chengdu Biennale for Art, Design, and Architecture had a range of contributors. a + p | CHORA (Shanghai) supported the event with materials, staff, and finance, as did TU Berlin and London Metropolitan University. Students from FHD Dusseldorf aided in preparatory research for the utopian plans, and Yuyang Liu provided curatorial support. Adam Atraktzi and CHORA London provided support for planning and logistics as well as for on-site management of the installation.

The Taiwan Strait Smart Region is the latest configuration of a long-standing project of CHORAs. This was supported by students and Professor Wang ShaoSen and Professor ShuennRen LiouTaiching at Xiamen University and TungHai University. Xiamen City and Taichung City local authorities have also supported the concept by commissioning studies for low-carbon plans for each city, which will be included in the book, while the Netherlands Architecture Funds provided some financial support. Over several years, a multitude of staff at CHORA London have assisted in the research and preparation of this ongoing project.

13

Conduit Urbanism: Rethinking Infrastructural Ecologies in the Great Lakes Megaregion, North America

Geoffrey Thün and Kathy Velikov

CONTENTS

> There is a growing sense that sustainability issues cannot be discussed in isolation. They must always be examined within their broader context. Every system is a component of another system, and is in itself, made up of systems.
>
> **James Kay, 2002**

> ... bio-fuelled traffic jams may be as plausible a prospect as solar boosted suburban sprawl ...
>
> **Peter Droege, 2006**

13.1 Introduction

It is almost 40 years since Horst Rittel and Melvin Webber, in their influential article "Dilemmas in a general theory of planning," identified that most problems planning and policy aim to tackle are "wicked problems," that is, problems so complex and difficult to define that they cannot be resolved through traditional analytic approaches, if at all (Rittel and Webber 1973). We posit that the design and planning of sustainable energy landscapes, like climate change, be identified as a "super wicked problem," since this engages challenges not originally identified by Rittel and Webber: the urgency of time, the authority vacuum within which the problem and possible solutions exist, the fact that those who try to solve this type of problem are also enmeshed in its creation, and the tendency for "hyperbolic discounting" that pushes immediately necessary action into the future (Levin et al. 2010). Given this, not only is it important for planners and designers to maintain a multiplicity of perspectives, definitions, methodologies, and operational strategies that are offered by this book, but also to recognize that these practices operate within specific times and places and cannot be considered solutions or universally applicable formulae.

The design-research project described in this chapter, *Conduit Urbanism*, is offered to reflect on questions of how the regional scales of planning and design might benefit from a synthetic approach that engages human and nonhuman agents within frameworks of urban ecologies. This work also raises the problem of multi-jurisdictional cooperation required in the development of sustainable energy landscapes and speculates on possible points of design leverage that might transform the role of regional mobility infrastructures when considered within the context of rethinking energetic, infrastructural, and land use systems relative to renewable energies, symbiotic relationships, complexity, and resilience (Stremke and Koh 2011; Stremke et al. 2011). The work is specifically intended to address these questions within the horizontally distributed urbanity of North America in an era characterized by transformations in energy supply, shifting population density, global economic competition, and competing land use pressures.

13.2 Operative Contexts and Methodology

Conduit Urbanism exists within discursive, epistemological, and geographic contexts, all of which inform its formulation and approach. First, we position the work within the contemporary discourses and theories of ecology, ranging from the complex system–based ecological method of James Kay (Kay et al. 1999; Kay 2002), to approaches in industrial ecology (Alberti 2008; Jacobsen 2006), to emerging ideas of infrastructure ecology (Xu et al. 2010), to concepts of integrated ecological and thermodynamic design put forth by the editors of this volume (Stremke and Koh 2011), to research in urban metabolism (Heynen et al. 2006; Kennedy et al. 2007), and to conceptions of the urban condition—which include the prefixes of post-, sub-, peri-, and ex-, and that define it as territory no longer distinct from the natural order but existing as a spatially unbounded constructed ecology, in complex and evolutionary interdependency with biophysical and geological systems, global economies, and politics (Belanger 2009).

At the regional scale, one of the greatest challenges in sustainable planning and design is the need to break down the jurisdictional and managerial boundaries placed on infrastructures:

> Given the magnitude of the sustainability challenge, new ways of conceptualizing cities as urban ecosystems, and thence designing and engineering urban infrastructure are needed. Traditional boundaries in urban infrastructure planning and management have considered transportation, water, energy, residential structures and industrial units to be compartmentalized sectors. Yet there are huge opportunities for synergy if a more holistic approach considers these subsystems (or sectors) to be inter-connected, addressing the combined impact of the total urban infrastructure system on air, water, land and biological resources (Ramaswami 2006, p. 414).

Epistemologically, the work situates itself within the contemporary context of massive information and data, much of it geographically, and increasingly temporally, encoded (Dangermond 2011), as well as the ubiquitous use of powerful digital tools and technologies, which enable statistical apprehension of large-scale terrestrial systems and processes with an unprecedented degree of rapidity and fine grain resolution of detailed data. Geo-design is a method of design and planning enabled through geographic information systems (GIS), which "tightly couples the creation of a design proposal with impact simulations informed by geographic context" (Flaxman 2010, p. 29). In many ways, this might be understood as a conflation of Ian McHarg's overlay method of environmental planning with concepts of real-time feedback on decisions and resources, as was introduced by Fuller in his "World Game" computer program (Fuller 1969). While the limits of this type of statistical knowledge system have been well argued (e.g., May 2008), the approach prioritizes a combinatory perspective that has the capacity to simultaneously consider ecologies (i.e., codependent relationships between things in their environment), economies (i.e., value systems organizing capital flows between things), energy infrastructures, and geographies as codependent agents in shaping landscapes and figuring future urbanization.

Geographically, *Conduit Urbanism* is situated within the context of emerging mega regions of North America, and specifically within the Great Lakes megaregion, which straddles the U.S.–Canada border and includes the cities of Chicago, Detroit, Toledo, Toronto, Buffalo, Pittsburgh, and Cincinnati and extends to the Atlantic port of Montreal, and the Midwest cities of Milwaukee, Columbus, Indianapolis, and St. Louis (Dewar and Epstein 2007). Jean Gottman, in his 1961 "Megalopolis," first identified this metropolitan formation of agglomerated networks of urban centers connected by sprawl development (Gottman 1961) that also define integrated labor markets, infrastructure, and land use systems and that share and organize complex and interdependent transportation networks, economies, ecologies, and cultures.* The global rise of megaregions has recently become the focus of several prominent land use and planning agencies such as the Regional Plan Association, Brookings Institute, Lincoln Land Institute, and the Rotman Prosperity Institute who predict that based on population and migration trends, over 90% of humanity will be living in the world's emerging megaregions by the year 2050. In the North American context, the geography of megaregional urban development is defined by the distributed network of highways and the diffuse low-density postwar urbanism that was the result of a variety of planned distribution policies ranging from national defense (Galison 2003) to the subdivision as an economic, political, and social instrument (Easterling 1999), all constructed in a context of seemingly unlimited fossil fuel supply (Droege 2006). The inheritance of this urban morphology poses significant challenges to imagining a truly sustainable future for a great deal of the North American urban landscape (Williams et al. 2000).

Our working methodology is one that moves through multiple scales of consideration in both analysis and design, and, within the previously described framework, combines perspectives of urban assemblage (de Landa 2006) and actor–network theory (Latour 2005; Farias and Bender 2010). The research begins with mapping-based analysis that draws out networks, intraregional industrial relations, exchanges, spatial, biotic, and material ecologies and flows as well as linkages between various regional actors. It also includes an analysis of the physical (constructed) nonhuman actors that produce and reproduce the relational, material, and spatial conditions under investigation, and the apparatus of policy or practice codes. Through this work, we look for patterns, tendencies, overlaps, existing synergies, and gaps that suggest locations where design interventions might leverage transformational, system-wide impacts. Interventions are developed as prototypical design propositions that are set within a context of plausible policy scenarios through explicative drawings and diagrams. They are evaluated relative to their perceived ability to produce positive feedbacks within the networks, ecologies, and morphologies mapped in the analysis as well as for their ability to develop symbiotic regional systems.

13.3 Sheds: Or Territorialized Material Ecologies

Working within large-scale regional territories, GIS-based research and analysis extend the principles of Ian McHarg's "ecological method" enabling a layered understanding of (1) place-based conditions as a result of interrelated processes with a prioritization of causality in understanding existing

---

* See, for example, Lang and Dhvale (2005) and Carbonell and Yaro (2005).

conditions (as well as underlying geologic, biologic, and constructed systems), enabling a designer to (2) interpret, proscribe, and predict compatible uses, and (3) make projections that consider demand, logistics, and economics (McHarg 1969). Our visualizations are predicated upon similar uses of layered indicators, capable of illustrating volumetric quantification and flow relations over time within a given system.

Ecological thinking, as previously noted, requires a breakdown of artificially created boundaries—conceptually, politically, and functionally. However, this opens up the problems of appropriate scales of consideration for design. The concept of the *shed* becomes a useful means for conceiving of systemic extent and interrelation as a geospatial boundary. Sheds define geospatial boundaries within which elements of a particular system retain a high degree of interconnectedness and interdependency yet can be recognized as being distinct from adjacent system characteristics. Interactions within a shed can be understood through the frameworks of self-organizing holarchic open (SOHO) systems: hierarchically organized, nested systems that operate through the self-organizational principles of complexity (Kay et al. 1999). The emergence of the system itself is dependent on the agents, infrastructures, conditions, and tendencies brought into proximity and interrelation within a given shed geography.

The Great Lakes megaregion can be understood as a shed based on statistics of population densities, commuter flows, and material exchanges whose boundaries are somewhat elastic depending on data considered (Dewar and Epstein 2007). Partially overlapping with the megaregional geography, the Great Lakes basin or watershed is defined by its geologically integrated hydrologic system. Its extents and governance are shared between two countries, eight states, and two provinces, and it constructs a complex and continually evolving regional politics of cooperation and conflict around urbanization and water-based infrastructures within and without its bounds (Annin 2006). *Conduit Urbanism* redefines this operative boundary by merging the human geography of the Great Lakes megaregion as defined by the Regional Plan Association (RPA 2005) with the hydrological geography Great Lakes watershed (Figure 13.1). This broad territory of consideration allows for holistic comprehension of interactions to be brought to bear on the work and on evaluation of questions such as sustainability, recognizing the relationships between underlying geography, geology, and linked biophysical systems with population and labor based descriptions of urbanization.*

Over the last several centuries, human activity within the Great Lakes region has resulted in a number of overlapping and intertwined shed geographies of

---

* Even in this extended context, the artificiality of creating ecological boundaries becomes evident: Minneapolis/St. Paul does not fall into the geographical boundary of the megaregion in terms of the criterion for continuous urbanity; however, it is closely networked in trade, labor, and other regional movements (Dewar and Epstein 2007).

FIGURE 13.1
The Great Lakes megaregion. Boundary assembled to capture hydrological geography of the Great Lakes. Watershed and the geography of the megaregion. (From RPA [Regional Plan Association], *America 2050: A Prospectus*, www.america2050.org/pdf/America2050prospectus. pdf, Retrieved March 16, 2008, 2005.)

networks and interactions. The *Conduit Urbanism* project begins with the mapping of four types of sheds—*mobility sheds, commodity sheds, resource sheds,* and *energy sheds* (Thün et al. 2011)—to examine the conditions of mobility networks within the context of energy transformation. Each shed geography describes a territory that specializes not only the material network of its actors and flows but also begins to reveal critical nodes and synergies within these networks. *Mobility sheds* map the geography of highway-based freight volumes (Figure 13.2) as well as air passenger and freight volumes (Figure 13.3), demonstrating the intensity of physical flows carried by regional mobility systems. When the *mobility sheds* are seen relative to the *commodity shed* (Figure 13.4), the high

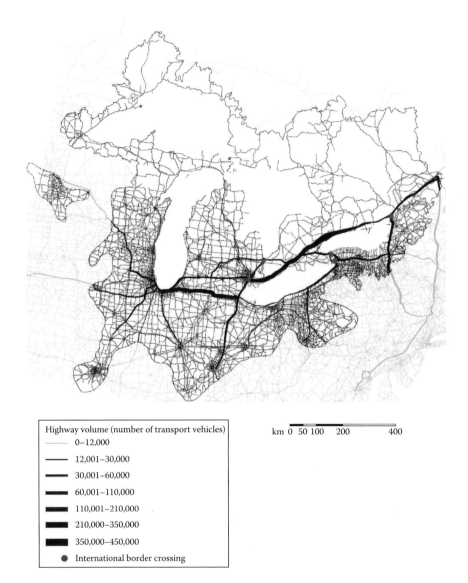

Highway volume (number of transport vehicles)

——— 0–12,000

——— 12,001–30,000

▬▬ 30,001–60,000

▬▬ 60,001–110,000

▬▬ 110,001–210,000

▬▬ 210,000–350,000

▬▬ 350,000–450,000

● International border crossing

km 0 50 100   200   400

FIGURE 13.2

Highway freight mobility shed: Annual transport vehicle volumes and border crossing locations within the Great Lakes megaregion. (From U.S. Department of Transportation, Estimated average daily Long-Haul truck traffic on the National Highway System: 2002, 2007, http://ops.fhwa. dot.gov/freight/freight_analysis/nat_freight_stats/docs/07factsfigures/fig3_4.htm, accessed June 2007; Ontario Ministry of Transportation, Provincial highways traffic volumes, 1988–2008, 2007, http://www.raqsa.mto.gov.on.ca/techpubs/TrafficVolumes.nsf/fa0278086478797 88525708a004b5df8/88c66a2279555c798525788d0048cca4/$FILE/Provincial%20Highways%20 traffic%20Volumes%201988-2008.pdf, accessed June 2007.)

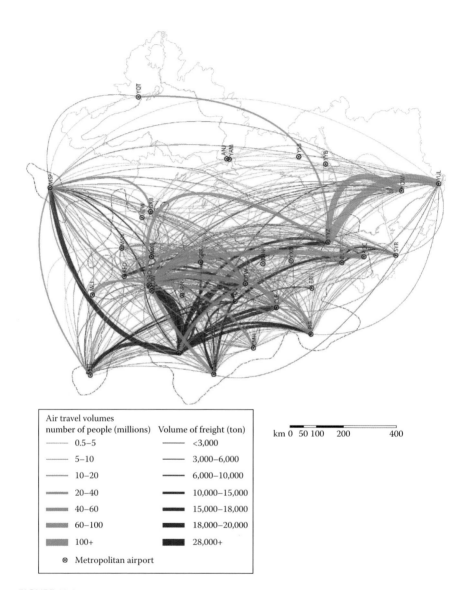

FIGURE 13.3
Air transport mobility shed: Annual passengers, freight tonnage, and annual airspace movements within the Great Lakes Region. (From Research and Innovative Technology Administration [RITA]; U.S. Department of Transportation Research and Innovative Technology Administration (RITA), Air carriers: T-100 domestic market, 2009a, http://www.transtats.bts. gov/Fields.asp?Table_ID=310, accessed July 2010; U.S. Department of Transportation, Research and Innovative Technology Administration (RITA), Air carriers: T-100 international market, 2009b, http://www.transtats.bts.gov/Fields.asp?Table_ID=260, accessed July 2010; Transport Canada, Transport Canada,Transportation in Canada Addendum, 2010, http://www.tc.gc.ca/ media/documents/policy/addendum2010.pdf, accessed July 2011.)

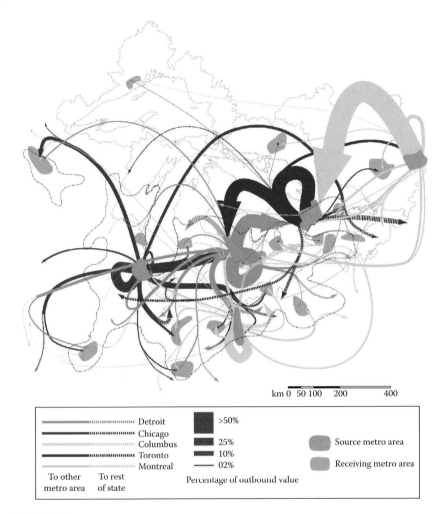

FIGURE 13.4
Commodity shed: Intraregional commodity exchange between Detroit, Chicago, Columbus, Toronto, and Montreal as a percentage of total annual metropolitan exports. (From U.S. Department of Transportation, Commodity flow survey 2007, 2008, http://factfinder2.census. gov/faces/nav/jsf/pages/searchresults.xhtml?refresh=t, accessed June 2008; Statistics Canada, Custom Marine and Railway Freight Tables by Commodity Origin and Destination, 2007-2009, Provided by request by Statistics Canada, accessed July 2011.)

level of infra-regional exchange as a percentage of individual civic economies becomes apparent relative to the vectors of movement. Daily, over $900 million worth of goods, or 25% of bilateral trade, crosses the Ontario–U.S. border via the highway system (RPA 2005).*

---

* As a statistical territory, the Great Lakes Megaregion constitutes the world's largest concentration of research universities and is home to 30% of North America's and 11% of the world's Forbes' 2000 international company headquarters (Affolter-Caine et al. 2007).

The *wastewater treatment shed* (Figure 13.5) spatializes the distributed sites of urban wastewater treatment facilities with the mining infrastructure of salt, lime, and electrolysis-based chlorine manufacture (all elements of the wastewater treatment process). Waterfowl and seabird migration has adapted to this condition, altering the trajectories of the Mississippi and Atlantic flyways to take advantage of the network of wastewater pools that remain unfrozen during winter migration.

The *current electrical power shed* (Figure 13.6) conflates the sites of current coal, hydro, and nuclear production with the geography of geologic seams, rivers, and water cooling necessary to each production process. While many of the shed mappings demonstrate a regional interconnectedness that seems to defy existing political boundaries, the *current electrical power shed* demonstrates clearly the differences in underlying geography linked to energy policy, practice, and implementation between the United States and Canada: U.S. electricity production is dominated by coal and its access to the coal seams in the south of the region, while Canadian electricity production is dominated by hydroelectric generation, harnessing the power of the many rivers that drain from the north into the basin. Both countries depend heavily on nuclear power. The heat sink capacity of the Great Lakes has also been the source of sustainable energy strategies, such as the deep lake water cooling system implemented by the city of Toronto in Canada (Kiel and Boudreau 2006).

The *potential renewable power shed* (Figure 13.7) spatializes the potential renewable source capacities of wind, solar, biomass, and untapped hydropower within the region. While the drawing of this shed does not outline a detailed analysis of energy and exergy-based sources and sinks (as described by Dobbelsteen et al. 2011), it does expose a number of broader patterns related to the future potential of the Great Lakes region to develop as a sustainable energy landscape.

The region territorializes significant amounts of renewable energy sources, with currently approximately 7.2 GW of hydroelectric energy potential, a copious potential for biomass energy provision, and an annual average of approximately 3.6–4.2 kWh/m^2/day of solar photovoltaic (PV) resource potential. However, the region's greatest energy potential contribution lies in estimated 320 GW of electricity that can be generated through offshore wind farms (Helimax 2008; Land Policy Institute 2008; U.S. DOE 2008). This shed underscores the primacy of the figure of the Great Lakes shoreline itself, expanding our awareness of this freshwater resource, and points to the necessity of more reciprocal and interregional agreements between national jurisdictions on energy policy and resource use (both energy and water related).

If fully exploited, offshore wind power in the Great Lakes is estimated to be enough to deliver 25% of the power needs of the United States and may also constitute a significant interregional export base. While current developments of offshore wind farms are being cautiously developed due to environmental and cultural assessment impacts, the province of Ontario has been actively moving forward with large-scale land-based renewable

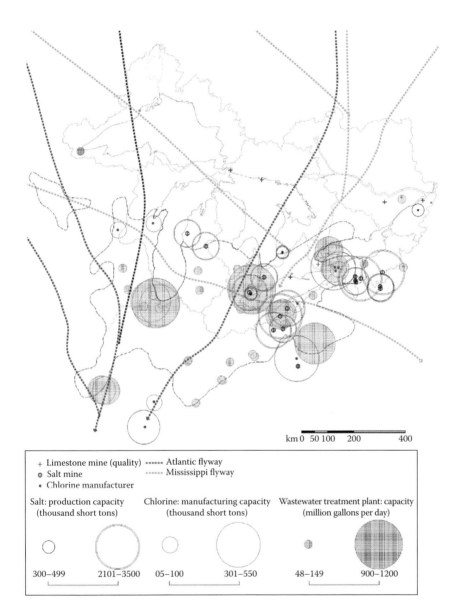

FIGURE 13.5

Wastewater treatment shed: Primary material production, wastewater treatment capacity, and avian migration routes. (From Natural Resources Canada, *Canadian Minerals Yearbook*, 2009, http://www.nrcan.gc.ca/minerals-metals/business-market/canadian-minerals-yearbook/2009-review/statistics/3214, accessed July 2011; Chlorine Institute, Regional team map, 2011, http://www.chlorineinstitute.org/Emergency/content.cfm?ItemNumber=3634&navItemNumber=3533, accessed April 2011; USGS Mineral Yearbook, 2008–2009; EPA Clean Watersheds Needs Survey, 2009, http://water.epa.gov/scitech/datait/databases/cwns/index.cfm, accessed April 2011; U.S. Fish and Wildlife Services, Migratory bird flyways, 2011, http://www.fws.gov/migratorybirds/Flyways.html, accessed April 2011.)

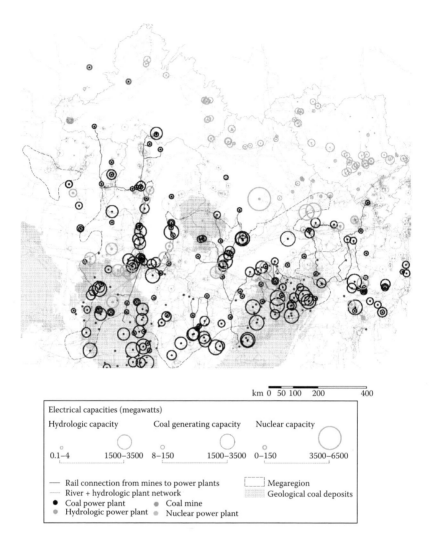

FIGURE 13.6
Current electrical power shed: Distribution and production volumes from regional hydroelectric, nuclear, and coal plants. (From National Energy Technology Laboratory Coal Power Plant Database, 2005, http://www.netl.doe.gov/energy-analyses/hold/technology.html, accessed June 2010; U.S. Information Administration (EIA) Existing Electrical Generating Units in the United States, 2008, http://www.eia.gov/electricity/capacity/, accessed June 2010; EIA Coal Transportation Trends, 1979–2001, http://www.eia.gov/cneaf/coal/page/trans/ratesntrends. html, accessed July 2011; Statistics Canada Electrical Power Generating Stations, 2000, http:// www5.statcan.gc.ca/bsolc/olccel/olc-cel?catno=57-206-X&lang=eng, accessed June 2010; Natural Resources Canada Atlas of Canada, Atlas of Canada base maps, 2007, http://geogratis.cgdi. gc.ca/geogratis/en/download/atlas.html, accessed June 2010; BTS National Transportation Atlas Database, 2009, http://www.bts.gov/publications/national_transportation_atlas_data-base/2010/, accessed June 2010; U.S. Geological Services [USGS] *Mineral Yearbook*, 2008-2009, http://minerals.usgs.gov/minerals/pubs/myb.html, accessed April 2011.)

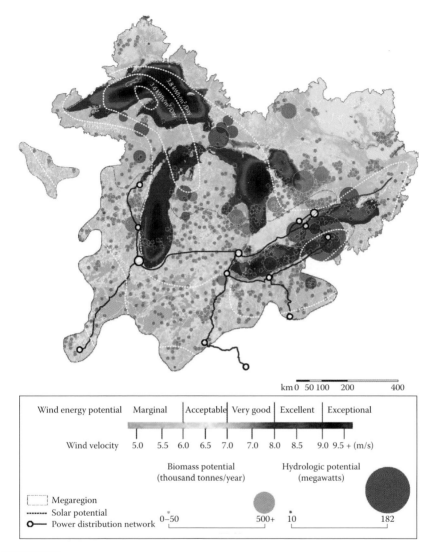

FIGURE 13.7
Potential renewable power shed: Spatialized intraregional sources of renewable energy provision: untapped hydroelectric, wind, solar, and biomass. (From U.S. DOE National Renewable Energy Laboratory [NREL], 2008; Natural Resources Canada [NRCan], 2008.)

energy installations throughout the province. This development has been partially due to the successful launch of the Green Energy Act in 2009, which includes an aggressive renewable energy feed-in tariff plan for both large- and small-scale producers (Bill 150 2009). At the time of writing, Ontario had installed 1413 MW of land-based wind farms, with an additional 719 MW in development (ISEO 2012) and 242 MW of photovoltaic installations approved (Ontario Ministry of the Environment 2012).

13.4 Leverage: Retooling Mobilities

Shed cartography sets the ground for the next phase of analysis, which delves deeper into policy frameworks, historical underpinnings, and potentials for intervention. The *Conduit Urbanism* project is specifically interested with the conditions of regional mobility systems—both as drivers of urban form (Thün and Velikov 2008) as well as significant consumers of energy (Lenzen et al. 2003).

The *highway freight mobility shed* (Figure 13.2) points to a condition of critical congestion along Southern Ontario's Highway 401, one of North America's busiest highways. At the city of Toronto, this highway has been expanded to 18 lanes and carries a daily average of 440,000 commuters and commercial vehicles (MTO 2008). As the primary conduit for both commuter and freight traffic that runs between Montreal and Windsor/Detroit, this artery is overloaded to the point of near terminal gridlock, and this condition will continue to be exacerbated by the provincial government's aggressive growth plan to locate a 30% population increase, or four million people, along the line by 2030 (Thün and Velikov 2008; Thün and Velikov 2009). This strategy is intended to foster higher density transit-oriented development with stronger community development potential while attempting to curb urban sprawl and protect natural resources, greenfields, and heritage landscapes (MOI 2006). On the U.S. side of the border, while rust belt cities and former manufacturing centers are still experiencing depopulation and job loss while struggling with urban restructuring relative to sprawl, there remains an overall projection for a population rise of 17%, or nine million people within the Great Lakes megaregion by 2050 (RPA 2005).

Mass urban migration to the globally emerging megaregions coincides with the "peak" of crude oil supply, which currently is the basis for 94% of transport (Gilbert and Pearl 2010). Often referred to as the post-carbon era, the demise of cheap and plentiful single-sourced hydrocarbon-based fuel is expected to shift energy sources away from fossil fuels (Lerch 2010) and will likely comprise a blended matrix of fuel types and production modes, with renewable energies expected to play a significant role in meeting the demand (Droege 2006).

Parallel to the unprecedented intensity of projected urbanization and crisis in fuel supply, planners, politicians, engineers, and industrial leaders foresee a future of increasing demand for mobility—especially infra-regional mobility, within the continuously populated megaregional corridors. Truck volumes are expected to double by 2035 (AASHTO 2007) and it is becoming clear that the rise of gasoline prices alone will not account for significant reductions in personal vehicle use (Austin 2008). This is also at a time when transport and energy infrastructures, largely built during the middle of the last century, are coming to the end of their service lives (ASCE 2009).

This "perfect storm" of diverse yet interrelated conditions of crisis that is being faced by the megaregion provides an opportunity for strategic design to find leverage within this system, and to imagine infrastructure-based interventions that could support future urbanization in the context of a complex and distributed system of maturing renewable energy sources within the region and in a condition of positive feedback and resilience between physical, biotic, and human ecologies. It is to be noted, however, that this proposition is set within a speculative context where the states bordering the Great Lakes follow Ontario's lead in developing robust renewable energy systems and also deploy sound sustainable decision making on the location of renewable energy technologies (RET).

Transport experts argue that the most effective and efficient technology for a mobility revolution resides in electrified high-speed rail directly tied to renewable energies, productively crossing mobility and energy distribution infrastructures (Gilbert and Pearl 2010; 2008). Maximum speed is coupled with minimal electrical conversion and distribution losses as vehicles receive a steady supply of electricity from the renewable energy distribution grid, resulting in a new type of hybrid infrastructure ecology that has the ability to move both people and energy between urban centers. The synergy is not only productive in the end result of the system: renewable energy systems not only implicate geographies—necessitating the territorial constructs of wind farms, dams, and solar farms—but the drastically increased electrical capacity demands a new infrastructural network of high-voltage transmission lines combined with strategies to store excess and fluctuating energy delivery (Delucci and Jacobson 2010). Similarly, electrified high-speed rail necessitates the construction of new rail lines, often elevated in order to achieve as horizontal a travel surface as possible for travel speeds of over 300 km/h while eliminating conflicts at grade.

Considering the highway's historical and strategic relationship to North American urban growth, industry, and mobility, an opportunity emerges in rethinking the distribution and mobility networks of people, goods, and energy. It is precisely the central and instrumental role that the highway has played in establishing the dominant patterns of distributed urbanism during the last century of development that now recommends it as a strategic site for intervention. *Conduit Urbanism* proposes a retooling of the highway cross-section as a conduit of bundled, parallel cooperative networks of transport, transit, and energy transmission infrastructures that could accommodate a variety of transportation and transit modes, including electrified high-speed or Mag-Lev rail, dedicated vehicle lanes at grade and high-voltage electrical transmission as well as high-speed data, freshwater supply, and waste movement (Figure 13.8).

These variable capacity vectors can be stacked and separated to maximize speed, safety, and accessibility thereby increasing conduit bandwidth, in addition to forming a resource umbilicus that can service and catalyze

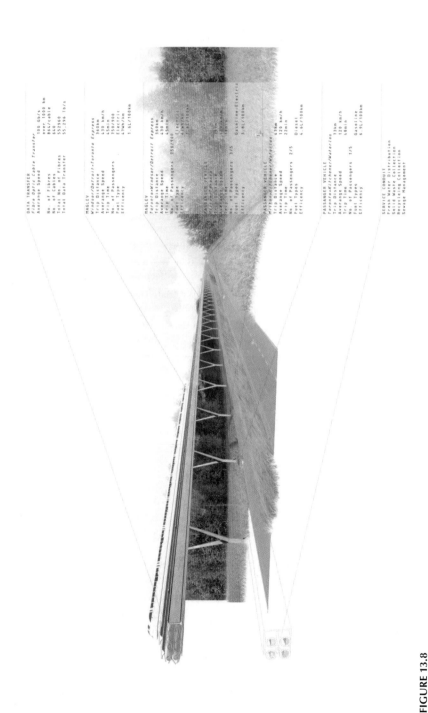

FIGURE 13.8
Bundled infrastructural systems: Elevated structures support renewable grid-tied Mag-Lev lines, high-voltage transmission, data distribution, and below grade conduits moving water resources and waste streams.

increasing densification and demand along the line. Flexibly dimensioned to operate within the right-of-way of existing major highways, this system facilitates ease of implementation by eliminating elaborate land acquisition and expropriation procedures, and recognizes the geography of the highway as the driver for future urban growth.

The introduction of this type of system has the capacity to link the urban centers from Chicago to Toronto in under 3 h, radically altering effective intraregional time and space (Thün and Velikov 2009). High-speed passenger rail could alleviate congested highway surfaces and could significantly decrease demand for short-haul flights, which encumber airports and consume disproportionate amounts of fuel, and might also play a *redistributative* role in future urban and economic development (Melibaeva et al. 2011), concentrating development in centers along existing urbanized corridors, now fueled by renewable energies, and protecting other landscapes from further sprawl development.

However, this proposal is not merely about high-speed rail development, which in North America has been a highly contentious issue with inconclusive life cycle assessment questions, especially when viewed from a context of return on investment (ROI)* based on ridership revenues alone or from the standpoint of the current energy costs in North America (Chester and Horovath 2010). If we take seriously the proposition put forth by the editors that sustainable energy landscapes are based on increasing urban system complexity toward system resilience, (Stremke et al. 2011, p. 159) then the diversity of bundled infrastructures carried along the new resource umbilicus could become a key part of a more mature and differentiated ecosystem within the region.

13.5 Nodes: Social Potential

The development of these new infrastructures has broader implications for the attendant territories proximate to the line, beyond the restructuring of development as mentioned earlier. Points of crossing and crossover, currently known through the ubiquitous and familiar typology of the highway interchange, now become charged with potential by virtue of their inherent availability for development and through the value they currently possess as points of access, transfer, and interface with dependent population concentrations. Surface lands orphaned by the interchange infrastructure—an average of 44.3 acres or 18 ha per interchange—become a potential site of

---

* We refer here to the tendency to evaluate large-scale infrastructural investments through narrow models of economic valuation versus a more holistic framework that includes consideration of the "true" costs of energy and other potential parallel synergies and benefits.

strategic design intervention along the line. A new megaregional typology of the *multimodal transfer interchange* is proposed as a hybrid development that might accommodate programs that include terminals and modal switch sites (from high-speed rail to local bus, light rail, vehicle, or airplane); facilities that can be used as renewable energy sinks* to further improve energy system reliability (such as greenhouses or parking lots with grid-tied electric vehicles); uses that capitalize on emerging industries (such as research hubs or multi-institutional academic clusters); social infrastructures accessible by multiple populations (such as regional health centers); or may even breed new synergistic opportunities for energy-conscious urban practices (such as regional farmer's markets that reduce farm-to-market transport while allowing access by diverse populations).

Not only can the hybrid programs housed within the nodes begin to take advantage of cascading energy relationships to reduce exergy destruction (Stremke et al. 2011) but wider development of public transportation systems can increase the availability of these social infrastructures to a wider constituency of participants than the traditional model of local or regional centers can currently facilitate, thus supporting sustainable community development within the periphery. In this way, *Conduit Urbanism* could enable the megaregion to become, as Sassen has suggested, a primary sociopolitical unit and a more positive and powerful enabler for diverse populations and economies than the "global city" (Sassen 2007).

A more systematic analysis of urban populations, economies, and the various resource and industry sheds begins to reveal patterns that implicate programs and strategic opportunities for various nodal sites. While nodal sites adjacent to major international airports and at border crossings have obvious programmatic pressures tied to intermodal transfer of passengers, less obvious influences also emerge. The agglomeration of technology-heavy postsecondary institutions, spin-off development agencies, and private research think tanks in the area of Kitchener–Waterloo–Cambridge, Ontario, suggest this site for the development of a technology research hub. Other contingent shed intensities suggest hybrid nodal developments in medical research, culture, tourism, and agriculture.

The design logic of these interchange structures begins with the spatial language of logistics, type, and prototype, developing contextual specificity as the systems and structures flexibly mediate the local geometry, program, and ecological pressures (Figure 13.9). Nodal interchanges might also become focal points for future development, as well as providing iconic figuration and physical identity[†] within a landscape currently defined by largely undifferentiated low-density sprawl.

---

* See Stremke and Koh (2011) for a definition of "energy sinks" and possible related strategies in the context of energy-conscious spatial planning and landscape design.
† As advocated, for example, by Lynch (1960) and subsequently discussed by D'Hooge (2011).

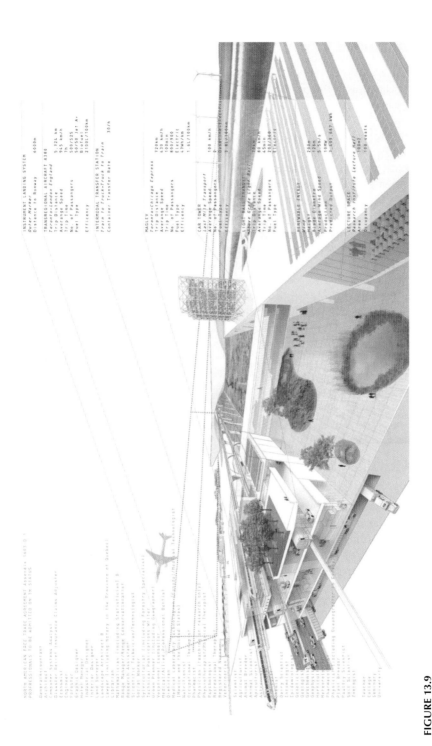

FIGURE 13.9

Visualization of a potential node: New architectural complexes at strategic points of exchange conflate access to mobility systems, proximity to resource distribution within combinatory architectural typologies as a new format of built infrastructure catalyzing surrounding urbanization.

13.6 Conclusion

By way of conclusion, we point to two areas of consideration. First, the challenges associated with the coordination and implementation of large-scale, synergistic regional projects such as *Conduit Urbanism* would require cooperation between a complex hierarchy of governmental agencies, utilities, transportation, transit, and freight authorities, as well as policy makers, planners, politicians, and civil engineers in two countries. In terms of the attempt to plan at the geographic scale, a valuable precedent can be found in the historic undertakings of the Tennessee Valley Authority in the 1930s and 1940s (Wolff 2007).

Second is the growing recognition that individual indicators, such as the production of compact urban form or increased use of renewable energy inputs, cannot be read as clear signifiers of sustainability (Guy and Marvin 2000). This is especially apparent when one looks at consumption-based models of urban metabolism that are not only related to energy of urban operational production, but also consider population wealth, which has a tendency to externalize consumption-related energy to other global regions (Rice 2007).

As the quotes at the opening of this chapter suggest, the design and planning of sustainable energy landscapes require not only new holistic models of system and ecology-based conceptualization with old boundaries undrawn and newly apprehended and synergistic interrelationships actively managed, but also deeper changes to human habits, behaviors, and expectations of our built environment.

Acknowledgments

The *Conduit Urbanism* project has received generous support from the Social Sciences and Humanities Research Council of Canada (SSHRC) and the Taubman College of Architecture and Urban Planning at the University of Michigan. All images reproduced in this text are courtesy of the authors, with recognition of the following individuals for their efforts in developing the work: Colin Ripley, Mary O'Malley, Zain Abseir, Matt Storus, Matt Peddie, Dan McTavish, Adam Smith, Lisa Sauvé, Julie Janiski, Sara Dean, Sonja Storey-Fleming, Maya Przybilski, and Mike Vortruba.

References

AASHTO (The American Association of State Highway and Transportation Officials). 2007. *A New Vision for the 21st Century*. www.transportation.org (retrieved March 17, 2008).

Affolter-Caine, B., J. Milway, and J.C. Austin. 2007. The Great Lakes: A world-leading bi-national economic region: Brookings Institute report. www.brookings.edu/speeches/2007/0328metropolitanpolicy_austin.aspx (retrieved February 20, 2009).

Alberti, M. 2008. *Advances in Urban Ecology.* New York: Springer.

Annin, P. 2006. *The Great Lakes Water Wars.* Washington, DC: Island Press.

ASCE (American Society of Civil Engineers). 2009. *Report Card for America's Infrastructure.* Reston, VA: ASCE.

Austin, D. 2008. *Effects of Gasoline Prices on Driving Behavior and Vehicle Markets.* Washington, DC: A CBO Study. Congress of the United States, Congressional Budget Office.

Belanger, P. 2009. Landscape as infrastructure. *Landscape Journal* 28(1): 79–95.

Bill 150 (Green Energy Act). 2009. www.ontla.on.ca/bills/bills-files/39_Parliament/Session1/b150ra.pdf (retrieved February 6, 2012).

BTS National Transportation Atlas Database. 2009. http://www.bts.gov/publications/national_transportation_atlas_database/2010/ (accessed June 2010).

Carbonell, J. and R.D. Yaro. 2005. American spatial development and the new megalopolis. *Land Lines* 17(2): 1–4.

Chester, M. and A. Horovath. 2010. Life-cycle assessment of high-speed rail: The case of California. *Environmental Research Letters* 5. http://iopscience.iop.org/1748-9326/5/1/014003 (retrieved February 6, 2012).

The Chlorine Institute. 2011. Regional team map. http://www.chlorineinstitute.org/Emergency/content.cfm?ItemNumber=3634&navItemNumber=3533 (accessed April 2011).

D'Hooge, A. 2011. The objectification of infrastructure: Elements of a different space & aesthetic for suburban America. *Projections 10: Designing for Growth and Change. MIT Journal of Planning* 10: 85–93.

Dangermond, J., interviewed by E.H. Jazairy. 2011. Geography by design. In *New Geographies 4: Scales of Earth*, E.H. Jazairy (Ed.). Boston, MA: Harvard University Press, pp. 152–155.

Delucci, M.A. and M.Z. Jacobson. 2010. Providing all global energy with wind, water, and solar power Part II: Reliability, system and transmission costs, and policies. *Energy Policy* 39: 1170–1190.

Dewar, M. and D. Epstein. 2007. Planning for "Megaregions" in the United States. *Journal of Planning Literature* 22: 108–124.

Dobbelsteen A. van den, S. Broersma, and S. Stremke. 2011. Energy potential mapping for energy-producing neighborhoods. *SUSB* 2: 170–176.

Droege, P. 2006. *Renewable City: A Comprehensive Guide to an Urban Revolution.* Chichester, U.K.: Wiley-Academy.

Easterling, K. 1999. *Organization Space: Landscapes, Highways and Houses in America.* Cambridge, U.K.: MIT Press.

EIA Coal Transportation Trends. 1979–2001. http://www.eia.gov/cneaf/coal/page/trans/ratesntrends.html (accessed July 2011).

EPA Clean Watersheds Needs Survey (2009). http://water.epa.gov/scitech/datait/databases/cwns/index.cfm (accessed April 2011).

Farias, I. and T. Bender. 2010. *Urban Assemblages: How Actor-Network Theory Changes Urban Studies.* New York: Routledge.

Flaxman, M. 2010. Fundamentals in Geodesign. In *Proceedings Digital Landscape Architecture 2010, Anhalt University of Applied Sciences*, E. Buhmann, M. Pietsch, and E. Kretzler (Eds.). Berlin/Offenbach, Germany: Wichmann, pp. 28–41.

Fuller, B. 1969. *Utopia or Oblivion: The Prospects for Humanity.* New York: Overlook Press, pp. 157–161.

Galison, P. 2003. War against the center. In *Architecture and the Sciences, Engineering Metaphors*, A. Picon and A. Ponte (Eds.). New York: Princeton Architectural Press, pp. 196–227.

Gilbert, R. and A. Pearl. 2008. *Transport Revolutions. Moving People and Freight without Oil.* London, U.K.: Earthscan.

Gilbert, R. and A. Pearl. 2010. Transportation in the post-carbon world. In *The Post-Carbon Reader*, R. Heinberg and D. Lerch (Eds.). Healdsburg, CA: Watershed Media, pp. 347–362.

Gottman, J. 1961. *Megalopolis: The Urbanized Northeastern Seaboard of the United States.* Cambridge, U.K.: MIT Press.

Guy, S. and S. Marvin. 2000. Models and pathways: The diversity of sustainable urban futures. In *Achieving Sustainable Urban Form*, K. Williams, E. Burton, and M. Jenks (Eds.). London, U.K.: Spon Press, pp. 9–18.

Helimax Energy Inc. 2008. *Analysis of Future Offshore Wind Farm Development in Ontario.* http://www.wrapwind.com/download/2008-11-helimax2006.pdf (retrieved March 17, 2008).

Heynen, N., M. Kaika, and E. Swngedouw. 2006. *In the Nature of Cities: Urban Political Ecology and the Politics of Urban Metabolism.* New York: Routledge.

ISEO (Independent Electricity System Operator of Ontario) www.ieso.ca/imoweb/siteshared/images/wind_generation_in_Ontario.pdf (accessed February 6, 2012).

Jacobsen, N.B. 2006. Industrial symbiosis in Kalundborg, Denmark: A quantitative assessment of economic and environmental aspects. *Journal of Industrial Ecology* 10: 239–255.

Kay, J.J. 2002. On complexity theory, energy, and industrial ecology. In *Construction Ecology. Nature as the Basis for Green Buildings*, C.J. Kilbert, J. Sendzimir, and G.B. Guy (Eds.). New York: Spoon Press, pp. 72–103.

Kay, J.J., H.A. Regier, M. Boyle, and G. Francis. 1999. An ecosystem approach for sustainability: Addressing the challenge of complexity. *Futures* 31: 721–742.

Kennedy, C., J. Cuddihy, and J. Engel-Yan. 2007. The changing metabolism of cities. *Journal of Industrial Ecology* 11(2): 42–59.

Kiel, R. and J.-A. Boudreau. 2006. Metropolitics and metabolics: Rolling out environmentalism in Toronto. In *In the Nature of Cities: Urban Political Ecology and the Politics of Urban Metabolism*, N. Heynen, M. Kaika, and E. Swyngedouw (Eds.). London, U.K.: Routledge, pp. 41–62.

Land Policy Institute. 2008. *Michigan's Offshore Wind Potential.* http://www.landpolicy.msu.edu/modules.php?name=News&op=viewlive&sp_id=71 (accessed March 17, 2008).

Landa, M. de. 2006. *New Philosophy of Society: Assemblage Theory and Social Complexity.* New York: Continuum. http://www.landpolicy.msu.edu/modules.php?name=News&op=viewlive&sp_id=71

Lang, R.E. and D. Dhavale. 2005. Beyond Megalopolis: Exploring America's New "Megapolitan" Geography. Metropolitan Institute Census Report Series. www.america2050.org/2006/01/beyond-megalopolis-exploring-a.html (retrieved April 10, 2009).

Latour, B. 2005. *Reassembling the Social: An Introduction to Actor-Network Theory.* Oxford, U.K.: Oxford University Press.

Lenzen, M., C. Dey, and C. Hamilton. 2003. Climate change. In *Handbook of Transport and the Environment*, D. Hensher and K. Button (Eds.). Oxford, U.K.: Elsevier, pp. 37–60.

Lerch, D. 2010. Preface. In *The Post-Carbon Reader*, R. Heinberg and D. Lerch (Eds.). Healdsburg, CA: Watershed Media, pp. xix-xxiv.

Levin, K., B. Cashore, S. Bernstein, and G. Auld. 2010. Playing It Forward: Path Dependency, Progressive Incrementalism, and the "Super Wicked" Problem of Global Climate Change. http://environment.research.yale.edu/documents/downloads/09/2010_super_wicked_levin_cashore_bernstein_auld.pdf (retrieved January 24, 2012).

Lynch, K. 1960. *The Image of the City*. Cambridge, U.K.: MIT Press.

May, J. 2008. *Statistical-Mechanical Geographic Vision. Perspecta 40—Monster*. Cambridge, U.K.: MIT Press, pp. 40–53.

McHarg, I. 1969. *Design with Nature*. New York: Doubleday/Natural History Press.

Melibaeva, S., J. Sussman, and T. Dunn. 2011. Comparative study of high-speed passenger rail deployment in megaregion corridors: Current experiences and future opportunities. ESD Working Paper Series. esd.mit.edu/wps (retrieved February 10, 2012).

MOI (Ontario Ministry Infrastructure). 2006. *Places to Grow: Growth Plan for the Greater Golden Horseshoe*. Ontario, Canada: MOI.

MTO (Ontario Ministry of Transportation). 2008. *Goods Movement in Ontario: Trends and Issues*. Ontario, Canada: MTO.

National Energy Technology Laboratory Coal Power Plant Database. 2005. http://www.netl.doe.gov/energy-analyses/hold/technology.html (accessed June 2010).

Natural Resources Canada Atlas of Canada. 2007. Atlas of Canada base maps. http://geogratis.cgdi.gc.ca/geogratis/en/download/atlas.html (accessed June 2010).

Natural Resources Canada. 2009. *Canadian Minerals Yearbook*. http://www.nrcan.gc.ca/minerals-metals/business-market/canadian-minerals-yearbook/2009-review/statistics/3214 (accessed July 2011).

Ontario Ministry of Transportation. 2007. Provincial highways traffic volumes 1988–2008. http://www.raqsa.mto.gov.on.ca/techpubs/TrafficVolumes.nsf/fa02780864787978852570a004b5df8/88c66a2279555c798525788d0048cca4/$FILE/Provincial%20Highways%20traffic%20Volumes%201988-2008.pdf (accessed June 2007).

Ontario Ministry of the Environment. www.ene.gov.on.ca/environment/en/subject/renewable_energy/projects/index.htm (accessed February 12, 2012).

Ramaswami, A. 2006. Engineering sustainable infrastructure. In *Sustainability Science and Engineering: Defining principles*, (Ed.). M. Abraham. Amsterdam, the Netherlands: Elsevier, pp. 411–434.

Rice, J. 2007. Ecological unequal exchange: Consumption, equity, and unsustainable structural relationships within the global economy. *International Journal of Comparative Sociology* 48(1): 43–72.

Rittel, H. and M. Webber. 1973. Dilemmas in a general theory of planning. *Policy Sciences* 4: 155–169.

RPA (Regional Plan Association). 2005. *America 2050: A Prospectus*. www.america2050.org/pdf/America2050prospectus.pdf (retrieved March 16, 2008).

Sassen, S. 2007. Megaregions: Benefits beyond sharing trains and parking lots. In *The Economic Geography of Megaregions*, K.S. Goldfield (Ed.). Princeton, NJ: The Policy Research Institute for the Region, Trustees of Princeton University, pp. 59–83.

Statistics Canada Electrical Power Generating Stations. 2000. http://www5.statcan.gc.ca/bsolc/olc-cel/olc-cel?catno=57-206-X&lang=eng (accessed June 2010).

Statistics Canada. 2007–2009. Custom Marine and Railway Freight Tables by Commodity Origin and Destination. Provided by request by Statistics Canada (accessed July 2011).

Stremke, S., Dobbelsteen, A. van den, and J. Koh 2011. Exergy landscapes: Exploration of Second-law thinking towards sustainable landscape design. *International Journal of Exergy* 8: 148–174.

Stremke, S. and J. Koh. 2011. Integration of ecological and thermodynamic concepts in the design of sustainable energy landscapes. *Landscape Journal* 30: 194–213.

Thün, G. and K. Velikov. 2008. Post-carbon highway. In *Fuel*, J. Knechtel (Ed.). Cambridge, U.K.: MIT/Alphabet City Press, pp. 164–211.

Thün, G. and K. Velikov. 2009. Conduit urbanism: Regional ecologies of energy and mobility. In *New Geographies 02: Landscapes of Energy*, R. Ghosen (Ed.). Boston, MA: Harvard University Press. pp. 83–96.

Thün, G., K. Velikov, and C. Ripley. 2011. Infra-/Eco-/Logi-/-Urbanism. *MONU 15 Post-Ideological Urbanism* 15: 101–112.

Transport Canada. (2010). Transportation in Canada Addendum. http://www.tc.gc.ca/media/documents/policy/addendum2010.pdf (accessed July 2011).

U.S. DOE (U.S. Department of Energy). 2008. *20% Wind Energy by 2030*. www.nrel.gov/docs/fy08osti/41869.pdf (retrieved March 17, 2008).

U.S. Department of Transportation. 2007. Estimated average daily Long-Haul truck traffic on the National Highway System: 2002. http://ops.fhwa.dot.gov/freight/freight_analysis/nat_freight_stats/docs/07factsfigures/fig3_4.htm (accessed June 2007).

U.S. Department of Transportation. 2008. Commodity flow survey 2007. http://factfinder2.census.gov/faces/nav/jsf/pages/searchresults.xhtml?refresh=t (accessed June 2008).

U.S. Department of Transportation. 2009a. Research and Innovative Technology Administration (RITA). Air carriers: T-100 domestic market, 2009. http://www.transtats.bts.gov/Fields.asp?Table_ID=310 (accessed July 2010).

U.S. Department of Transportation. 2009b. Research and Innovative Technology Administration (RITA). Air carriers: T-100 international market, 2009. http://www.transtats.bts.gov/Fields.asp?Table_ID=260 (accessed July 2010).

U.S. Fish and Wildlife Services. 2011. Migratory bird flyways. http://www.fws.gov/migratorybirds/Flyways.html (accessed April 2011).

U.S. Geological Services (USGS) *Mineral Yearbook*. 2008–2009. http://minerals.usgs.gov/minerals/pubs/myb.html (accessed April 2011).

U.S. Information Administration (EIA) Existing Electrical Generating Units in the United States. 2008. http://www.eia.gov/electricity/capacity/ (accessed June 2010).

Williams, K., E. Burton, and M. Jenks. 2000. *Achieving Sustainable Urban Form*. London, U.K.: Spon Press.

Wolff, J. 2007. Redefining landscape. In *The Tennessee Valley Authority: Design and Persuasion*, ed. T. Culvahouse. New York: Princeton Architectural Press.

Xu, M., J. Crittenden, Y. Chen, V. Thomas, D. Noonan, R. DesRoches, M. Brown, and S. French. 2010. Gigaton problems need gigaton solutions. *Environmental Science and Technology* 44(11): 4037–4041.

14

Bi-Productive Urban Landscapes: Urban Resilience through a Redevelopment of Postindustrial Space in the United Kingdom

Greg Keeffe

CONTENTS

14.1 Introduction

The aim of the chapter is not only to show that the modern U.K. city in its current form is particularly vulnerable to climatic and resource perturbations, but also that urban postindustrial space, which is currently problematic, holds the key to the future resilience of the city. This space, coined by the author as "the collapsoscape," is a landscape of collapse and fragmentation, but it is of a scale that meets the needs of the productive city and offers a new landscape that will increase sustainability and urban resilience by hosting new functions for the city that help to close material and energy cycles in an effective way, by improving the metabolism of the city.

The chapter first defines the dynamic that has made the current configuration and describes the issues with deindustrialized space and its communities and the issue of conflicting network scales that characterize the space. This is followed by an analysis of the urban issues surrounding sustainability and in particular resilience, particularly with regard to biomass and food production. These needs are large and the scale of the landscape needed is beyond the size of the neighborhood. Finally, the chapter describes, through a series of case studies of real and hypothetical projects—developed by the Bioclimatic Architecture Labs at Manchester School of Architecture and the Crash Test Labs at Leeds School of Architecture—how the insertion of these new functions is possible, not only increasing sustainability by closing cycles, but also improving legibility and coherence of the postindustrial space. The author shows that the scale of these interventions needs to be large and that their scale is often at odds with the city, and that with careful cross-programming it is possible to build a new resilient model for a productive and biospheric city, one that embodies closed cycles of energy crops (biomass) and material production and reuse.

14.2 Ecology of the Postindustrial City in the United Kingdom

The city is changing—no longer is it an aesthetic creation, nor purely an industrial powerhouse. It is becoming a living, breathing superorganism, with a myriad of multiple, competing functions enabling the city to dwell within its particular ecology. As a superorganism, the future city will be defined more by its metabolism than purely by its primary function or spatial form. These biospheric flows of biomass and materials will drive the new city and create new synergies for living.

The city must now be seen as *the* technology by which we live—not a landscape full of technologies. The city is like an iPhone; life without it is unimaginable, we need it in every aspect of life. The city as a whole needs to

FIGURE 14.1
Collapsoscape: West Gorton, Manchester, United Kingdom. (Photo by Gerald England.)

be effective, and its constituent parts—transport, industry, commerce, social functions, and leisure, for example—must fit within it and be seamless in use.

If the city is a body, an organism, can we really make one out of un-living, off-the-shelf parts? This would be like a Frankenstein city, undead but not alive. This analogy is clearer in postindustrial cities, where failing industrial sites sit adjacent to new housing, or old housing lies uncomfortably next to new transport infrastructure. This sprawling Frankenstein landscape has been coined "collapsoscape" by the author, a place of failing networks, where the network scale is unbodied and the resulting spaces are so large that any building designed into it is too small, so the building invariably collapses (Figure 14.1).

The present ecology of the city is complex and singular, yet at times generic and at others specific. This means that although cities are each unique places with their own history and spatial forms, they are often created out of a combination of standard components—motorways, city squares, workers housing, and suburban developments. This creates an opportunity, in that solutions for one might be seen as solutions for all, but this is often not the case. Many cities have been damaged by inappropriate development that worked in another place, but failed here. This, in bioethics, is known as the substitution problem, where adding a comparable ecological function to replace a lost one has dire unforeseen consequences.

Most cities in northern Europe and particularly the United Kingdom are in a postindustrial phase, where large-scale infrastructure developed in the past has become redundant and derelict. These new spaces, the collapsoscape, often seen as the blight of the city, are the resource that offers a new frontier to incorporate the new metabolisms of the city. These metabolisms are at present linear and global, but in the future they will be required to be more complex and less monocultural.

14.3 Collapsoscape

The collapsoscape is a new type of city form, which can be defined in many spatial and programmatic ways; however, it is really its scale, the programs that inhabit it, and its proximity to the center that defines it. Unlike the Drosscapes of the United States described by Berger (2007), it is particularly postindustrial and within walking distance of the historic center. Compared to the United States, the small scale of the United Kingdom means that the collapsoscape is characterized by multiple, bit-part conflicting functions in one oversized landscape.

The space itself is usually created by industrial collapse, and this collapse of the industry has left an abundance of unprogramed space, at a scale that impedes people-scale connection of walking and general socializing. This large-scale devastation of the city leads to the development of a mixture of new poor-quality housing in contained estates, with cul-de-sacs rather than streets, and large sheds, which create poor space syntax. A good example of this is the South Leeds Gateway in Yorkshire, United Kingdom (Figure 14.2).

The collapsoscape has elements of placelessness, where people are merely traveling through: rather like the description of the airport in Auge's book, 'Non Place' in which people do not travel to the collapsoscape to visit it, more to get through it (Auge 1995). There are many nonplaces in the landscape,

FIGURE 14.2
South Leeds Gateway. (Photo by Will Inglis Crash Test.)

but that is not its overall characteristic. Being next to the city center, this postindustrial space is currently used as a receptacle for new global functions that cannot because of their scale be housed within the heart of the city. In their book *Edgelands*, U.K. poets Farley and Roberts (2011) describe journeys through this space as if on safari, seeing it as a true wilderness, but this is an existential view of the space. The collapsoscape has two faces: one as a utility space for urbanists and another as a consecutive collection of nonplaces of edgelands. In particular, the collapsoscape can be characterized by four key elements that create a specific generic dynamic to the space: infrastructure, content, psychogeography of nostalgia, and policy.

14.3.1 Infrastructure

One of the defining elements in the space is new motorized infrastructure. This infrastructure is not tied in spatially to the existing spatial configuration. It consists mainly of expressways, motorways, and high-speed rail links; these are often high speed with limited or no connection to the neighborhood. This network grid is spatially massive and is not used to get around the neighborhood. The transport links are used to get in and out of the city, but as they do not engage with the in-between, they become hard urban edges that are difficult to permeate. Where they do engage, car parks on waste ground abound. Elsewhere there is a lot of unprogramed space, surrounded by fences. Fences are a major element of the area's infrastructure.

14.3.2 Content

Land values are low, and there is a high rate of urban perturbation—things do not last long. The collapsoscape consists of typically two sorts of buildings: housing in estates and industrialized sheds. The estates have poor space syntax and are badly connected to their surroundings, and they are usually for the poor; the sheds are usually either for retail and are massive or for distribution and are defensive. Vacant, unprogramed space abounds and retail sheds are the only landmarks, except for derelict Victorian church spires.

14.3.3 Psychogeography of Nostalgia

The collapsoscape always has an illustrious history; however, the neighborhoods that were the focus of this history no longer exist. At present there are just weird programmatic configurations of space, which no longer have a name (or the name is meaningless, such as South Leeds Gateway). People navigate by satellite navigation and postcode, as the space is illegible either in a car, or on foot, walking. Walking is generally inadvisable, due to the complex spatial organization, the abundance of fences, the high levels of crime, and lack of surveillance. In many spaces, for example, alongside expressways, walking is actually prohibited. Defensible space is at a minimum too, with large grass swathes around small houses.

14.3.4 Policy

The policies that are in place in these neighborhoods are of two types: They are of two extremes, either totally liberalized or totally constrained. In South Leeds Gateway, there is actually no development policy in some areas ("Leeds says Yes!"); however, in Birkenhead, Merseyside, the very opposite is true: here the development control framework has over 800 clauses, many protecting its illustrious heritage (http://www.wirral.gov.uk/my-services/environment-and-planning/planning/local-development-framework/unitary-development-plan). This illustrates the bipolar way the collapsoscape is viewed by city governance: one as an opportunity to engage with the unpalatable elements of the globalized city (motorways, large sheds, global retail), the other as a heritage landscape representing the genius loci of the city's raison d'être.

What is needed is a new engagement with this complex landscape. The problems of scale and spatial complexity need to be dealt with in a new way. A new program is needed for the landscape. These future requirements of urban resilience may allow a new connective landscape to be created that ties existing functions together and make them commensurate with the fragmented landscape.

14.4 New Function for the Collapsoscape

Resilience is the resistance to external forces. Today the main external forces on a city are global: pressure from a system that has many external agents. Globalization has continued apace during the early years of the century. The modern city is now powered by gas and oil from Alaska or Saudi Arabia or maybe eastern Russia or Kazakhstan; its water may be imported from 200 km or even abroad; materials for building may be from Japan, Australia, or Nigeria, yet processed in Germany; in its supermarkets there may be fresh asparagus from Peru, lamb from New Zealand, or tuna from the Pacific; clothing and consumer goods may be produced in China, Saipan, or Indonesia, yet grown in India. Although these global trades offer great opportunities, they also create complex webs of dependence that leave the city vulnerable, both functionally and economically, to perturbations. Increasing oil costs and global political strife could leave the global city in a very vulnerable position. This can be seen clearly in Nelson, Lancashire, where a study by Citylab (2006) showed 65% of the retail spend of the town (population of 28,000) is in one supermarket.

14.4.1 Strategies for Resilience

Strategically, resilience is based upon two key concepts, the first being nested scales of production, which build redundancy, and the second being effective use of resources by the upcycling of waste through the closing of material cycles. The forces that may be encountered could be wide and varied, but this chapter focuses on food and energy use and biomass production to illustrate how resilience might be achieved. According to City Limits (Best Foot Forward 2005), the ecological footprint of London food corresponds to nearly half its ecological footprint, and energy some 12%, and so these are key elements of the metabolism of the city.

14.4.2 Scales of Production

The author believes that urban resilience is based on nested scales of production, from the local—through neighborhood, city, and regional scales—to larger national and international ones. This nesting builds in a redundancy and duplicity that creates an overlapping stable system. The current scale of production creates a nonresilient system, which has no level of redundancy. This monoculture is clearly seen within food production. With the production of food, at present, very little is actually produced within the city, as most food is produced and sold on the global market. Thus, the city has little control over its major ecological and economic cycle. This problem is compounded by the just-in-time strategy of supermarkets such as TESCO, where there is little or no storage of food products on site, putting other pressures on the city. During the Eyjafjallajökull ash cloud of 2010, when flights were banned over the United Kingdom, there were empty food shelves in the supermarkets within just a few days. As fossil fuel supply reduces and other pressures come to bear, we may find that our globally induced habits might need to change, or we may be starved into submission.

Equally, reliance on global retailing of food creates other economic issues. Locally spent money is more valuable than globally spent money, as profits stay and are spent locally, versus global profits being money spent elsewhere. In his book "Local Money," Peter North (2010) talks of an almost factor-3 difference in the economic value of where the money is spent (£1 spent locally = £1.60 to the local economy; £1 spent globally = £0.60 to the local economy). The New Economics Foundation (NEF) in the United Kingdom claims it is even higher in their 2005 study of Northumbria, using their LM3 methodology: £1 spent locally was reckoned to be worth £1.76, yet only 36 pence if spent globally*.

There is also a disjuncture between the scale of the city and the scale of the agricultural (and retail) system to supply it. The relationship of the city neighborhood with these forces is rather like the way the body engages with new viruses. These forces cause the urban cell to engage in a process of connection. The scale of globalization is large (and getting larger) compared to the local scale. This can be seen clearly in the scale of the TESCO Extra supermarket in Walkden, North Manchester (floor area of 17,200 m²), versus my local vegetarian mini-supermarket Unicorn (800 m²) or my local Londis corner shop (84 m²) (Figure 14.3). It is also visible in the scale of farms that supply these shops: Unicorn, the vegetarian cooperative in Chorlton, Manchester, aims to obtain half its food from a 50 km radius of the shop. These local

---

* http://neweconomics.org/press-releases/buying-local-worth-400-cent-more.

FIGURE 14.3
U.K. retail scales. Clockwise from top: Londis, Beech Road, Chorlton; Unicorn Vegan Co-operative Supermarket, Chorlton; and TESCO Extra, Walkden, Manchester, United Kingdom. (Photo by author.)

producers have holdings that vary in size from 20 to 200 ha. These farms are small compared with those involved in wheat production on the American prairies (989 ha each, on average) (U.S. Wheat Associates 2011) or soya farms in Brazil (typically 8000 ha). The local neighborhood, with its fine grain and the global production facility, cannot easily share land with these larger spatial scales. This scale issue is a key challenge for urban designers, however, it may be possible to accommodate the smaller, local, food-producing farms.

These issues can be seen clearly in the project "Zero Food miles" by Crash Test's Andrew Goodwin (Keeffe and Fagan 2011). This hypothetical MacDonald's Drive-thru, which produces all its food within its own shop footprint in a vertical farm, is an incredible 30.6 km tall. The agricultural area needed to produce all the food for just one burger bar is around 400 ha. Beef production is the major component here; producing the burgers with goat meat has a dramatic shrinkage of the spatial requirement and producing soya-based burgers reduces the agricultural space by a factor of 30, to a manageable 14.5 ha (Figure 14.4).

Self-sufficient McDonalds: *Zero food miles*

Vertical McDonalds:

The average size of a UK McDonalds Restaurant is 381 m². Using this as a basic footprint guide, the diagram below indicates what a self-sufficient McDonalds would look like.

Chicken:
To house all the broilers for chicken meat would require 376 levels at approximately 1.127 km high

Pork:
To house all the pigs would require 2483 levels at a total of 7.448 km

Beef:
It would require approximately 6979 levels just for cows alone. This would account for 20.94 km of the overall structure

A passenger jet:
To give an example of how high the structure is, a standard passenger plane flies at around 10.5 km

Islington McDonalds:

The site in Islington is approximately 89,000 m². If we use that as the footprint for our self-sustained McDonalds restaurant we get different results. The new building would be 43 stories tall, reaching 129 m high. The new 'Mega-McDonalds' would dominate the cityscape of Liverpool.

Chicken:
To house all the broilers for chicken meat would require 1.6 floors at approximately 4.8 m high

Pork:
To house all the pigs would require 10.6 floors at a total of 31.8 m tall

Beef:
It would require approximately 29.9 floors just for cows alone. This would account for 89.7 m of the overall height of the structure

Liverpool metropolitan cathedral:
Possibly one of Liverpool's most famous historic landmarks. Standing at approximately 85 m tall, this building would be swamped by the new 'Mega-restaurant'.

30.39 km high!
Current meat-based farm

1.16 km high
Vegetarian burger farm

FIGURE 14.4
Zero food miles, Liverpool by Andrew Goodwin. (From G.P. Keeffe and D. Fagan, Crash test: Bio-spheric city. Internal report, Leeds School of Architecture, Leeds, U.K., 2011.)

However, meat production is not the only large-scale impact in our consumption habits. Even innocent-looking food retailers, such as a Starbucks coffee shop, have a huge ecological shadow. The Crash Test project "Starbucks" by Adam Leigh-Brown (Keeffe and Fagan 2011) shows the massive impact of drinking coffee. Coffee consumption for just one Starbucks is 57.6 kg of coffee per day (in addition to 429 L of milk) and would need a 21 ha estate (along with 9.2 ha of grazing for the cows). Even utilizing super high efficiency 24 h daylight hydroponic systems in a biospheric greenhouse, the most efficient way to grow the coffee beans, the farm would still be huge in comparison to the city scale (Figure 14.5).

Other food retailers have even worse footprints. The Australian-founded smoothie vendor Boost, for example, uses fruit from all around the world. One particular smoothie boasts a food miles total that reaches an incredible 32,000 miles (50,000 km). For this to be made sustainable, we would have to

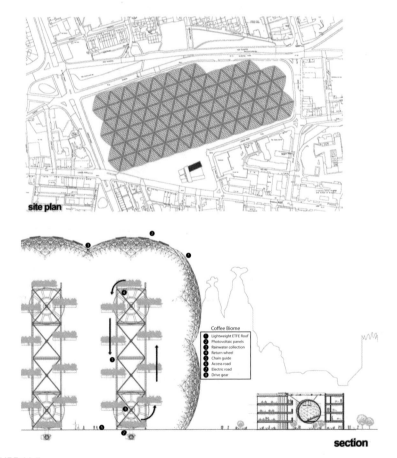

FIGURE 14.5
Sustainable Starbucks by Adam Leigh-Brown (crash test); note the difference in scale between biome and cafe.

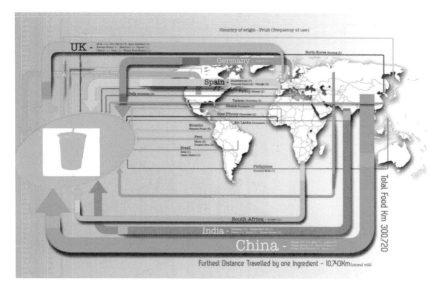

FIGURE 14.6
Food miles—Boost by Hall and Jenkins. (From G.P. Keeffe and D. Fagan, Crash test: Bio-spheric city. Internal report, Leeds School of Architecture, Leeds, U.K., 2011.)

change not only the linear process that produces the smoothie, but also its exotic ingredients, for something more local. The new Boost developed by Crash Test (Keeffe and Fagan 2011) utilizes local and seasonal fruit and vegetables produced within the city in a closed cycle way (Figure 14.6).

14.4.3 Climate Change

There are other global forces that will radically change our inhabitation of the city. One that is real and serious is climate change. Over the next century, this will put increased pressure on the city: climatic forecasts produced by the Prometheus Group at the University of Exeter, combined with urban heat island effects, show that Leeds in the United Kingdom (53°N) will have summer weather similar to Nice in the Mediterranean by 2050, and more like Casablanca in Morocco by 2080. This will make cities dangerous places to live in. My own research produced for the Technology Strategy Board "Design for Future Climate" (Keeffe 2011) showed that by 2080 homes produced to current U.K. Building Regulations or even PassivHaus standards might have internal temperatures above 28°C for over 28% of the year. The Chartered Institute of Building Services Engineers (CIBSE) in the United Kingdom sees 28°C as a high limit that should not be exceeded, and that presents a danger to health. The new infrastructures will have a role to play in helping the city mitigate and adapt to climate change. There are many studies, for example, the

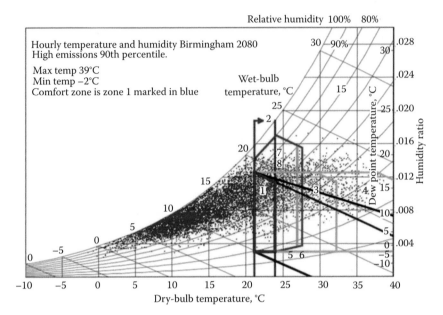

FIGURE 14.7
Birmingham U.K. Weather data for 2080, 1 in 10 year prediction, with high emissions. (From Keeffe, G.P., Facing forward: Strategies for future climate adaptation in elderly persons housing, In *Proceedings of the International Federation of Housing and Planning World Congress*, Tallinn, Estonia, 2011, Data from Prometheus.)

Rotterdam Climate Initiative (http://www.rotterdamclimateinitiative.nl), that show the urban heat island effect can be massively reduced by green space (Figure 14.7).

In addition to this change in summer temperatures, global warming creates other problems. The first of these is a threat to indigenous species and the need for migration. Cities presently create a severe barrier to land-based migration, and estimates by the University of York reckon that plants and animals may need to move at 20 m/h, 24 h/day, 365 days/year to keep pace with the changing climate. Ecological corridors will need to be created through our cities to allow flora and fauna to migrate without barriers (http://www.york.ac.uk/biology/research/featured/hot-stuff/).

The second problem is that of increased extreme weather events. Climate change in the United Kingdom will produce much wetter winters (UK Metrological Office estimate: 20% wetter by 2050), which puts strain on the current sewerage system, but combined with extreme events these can cause catastrophic problems. These can already be seen in the United Kingdom, where the Lake District in Cumbria in November 2008 had a 1 in 100 year flood, followed a year later by a 1 in 1000 year event! The future city will have to be far more permeable than at present to deal with these types of events, and it will need to accommodate green space that can mitigate the heat island effect and promote biodiversity.

14.5 Bi-Productive Cities

If we are to build resilience into our cities, it will have to be multilayered and multifunctional: the city is, as Richard Rogers so succinctly put, "the most complex thing man has designed" (Rogers and Power 1995).

The solutions will undoubtedly be complex, covering issues of economics, social structure, energy, climate, materials, water, transport, and food, to name just the main factors. How we engage with these complex systems will probably determine the shape of the future city.

The threats and opportunities for the modern city are clear: The need for a contraction of our demands is obvious, but also clear is the need for localized production. Nevertheless, the scale of these functions is so great that we will need a new land-use strategy: one that is bi-productive, where food and energy crops can be produced together in a livable landscape that incorporates leisure and biodiversity.

However, one thing is for sure—the city is not large enough to produce all it requires within its boundary in its present form. According to City Limits (Best Foot Forward 2005), London needs some 293 times its area to support itself. This approximates to a land mass the size of Spain. Presented with these numbers, urban resilience looks like an impossible dream: The United Kingdom itself needs 3.63 times its area to be self-sufficient, and even meeting this seems impossible. Reducing United Kingdom's consumption by a factor of four seems difficult, but seeing it as a halving of demand and a doubling of efficiency makes it seem more possible.

So how do we half the demand and double the efficiency, while still answering the issues described previously?

14.5.1 Biospheric City

The concept of the closed cycle city has been well documented by Girardet (2004), among others. In the book Cradle to Cradle, McDonough and Braungart (2002) developed the idea of a city without waste, where cyclical streams of material flow dominate the city. This new metabolism allows a radical reduction in material consumption and waste. This reduction in demand creates immediate improvements in resilience, as many of these processes can be accommodated within the city itself.

One key idea of the closed cycle city is upcycling, where waste products when recycled produce goods of at least the same value, rather than poorer contaminated projects. This idea was taken to the extreme in Crash Test's Silkworm City project by Natalie Hall (Keeffe and Fagan 2011). Inserted into the collapsoscape of Liverpool, this urban renewal project created high value, closed-cycle material streams that aided regeneration of the city. The project focused on the production of silk moths on mulberry trees in large urban greenhouses. The production of silk allows for a new business in

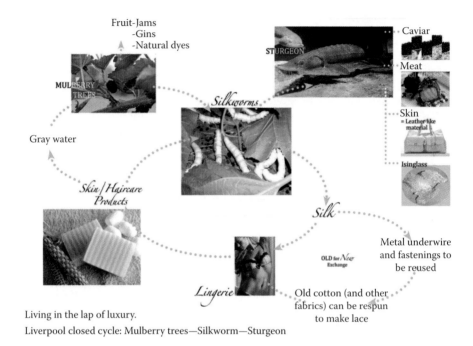

Fruit-Jams
-Gins
-Natural dyes

MULBERRY TREES

Gray water

STURGEON

Silkworms

Caviar

Meat

Skin
= Leather-like material

Isinglass

Skin/Haircare Products

Silk

OLD for *New* Exchange

Metal underwire and fastenings to be reused

Lingerie

Old cotton (and other fabrics) can be respun to make lace

Living in the lap of luxury.

Liverpool closed cycle: Mulberry trees—Silkworm—Sturgeon

FIGURE 14.8
Silkworm City closed cycle by Natalie Hall (crash test).

high-value lingerie to be developed, and the waste silk is used in the production of skincare products. Once the silk is removed from cocoons, the worms are used as food for the lake sturgeon, which produce expensive caviar and the fish themselves are used to produce leather foodstuff food and fertilizer for the trees, thus closing the cycle. Maximizing the value of the products produced show how upcycling can overcome issues of scale and economics with regard to urbanized food production (Figure 14.8).

14.5.2 Material Flows

The key biospheric flows of the city will be flows of materials and energy (biomass); however, other flows such as water and mobility will also be important. The bi-productive landscape will introduce not only solutions that aid resilience, but also develop new more livable spaces.

14.6 Urban Food Opportunities

The city has a voracious appetite for resources. These resources include energy, food, water, and materials, and their supply and use is the major constituent of the ecological footprint of the city. Food in particular is a real challenge, and although supplies needed are huge, there is an incredible waste inherent in the system. In Hungry City, Carolyn Steel (2009) states that in the United Kingdom over 30% of all food is wasted, worth over £10 billion per annum. This is mainly thrown away by retailers and consumers because we do not understand how to tell if food is going off and rely on randomly produced "sell by dates."

The scale of food production, particularly industrialized food production that maintains a meat-centric diet, is at odds to the scale of the city, even with the large unprogramed spaces of the collapsoscape, but it is possible to engage with the production through the development of nested scales of infrastructure.

These scales can be easily seen as

- *Pico*: localized home-grown garden produce
- *Nano*: neighborhood allotment production or microscale industrialized production, for example, greenhouses up to 5 ha
- *Mico*: small-scale commercial agriculture or hydroponics up to 30 ha
- *Meso*: medium-scale farming up to 500 ha—European arable or dairy farm
- *Macro*: up to 8000 ha—U.S. wheat or Brazilian soy or beef farms

14.6.1 Small-Scale Solutions

From this it can be seen that the three smaller scales of production can be accommodated easily in the postindustrial city. At the pico scale, reasonable amounts of food can be produced by engaging local people with production: in the project Back to Front, Alan Clavin and Emma Oldroyd worked closely with local inhabitants in Harehills Leeds to create food production at a very small scale (Clavin and Oldroyd 2011), in small front gardens, window boxes, and planters (Figure 14.9).

Taking this idea to the limit, my students, James West and Anthony Campbell, from the Bioclimatic Architecture Labs, developed the Edible Terrace, a refurbishment of a terraced house in Manchester that produced enough food for all the residents within the confines of the house and garden. The project not only created a carbon neutral house but also provided a meat- and fish-based diet as well as vegetables, for the occupants, by utilizing closed cycles, which fed fish (tilapia), pigs, and chickens on food waste, while vegetables (and algae for fuel) were grown with hydroponic systems with the facade of the dwelling (Figure 14.10).

14.6.2 Large-Scale Interventions

Connection and Production: Nutritionally complete city

Integrating at a larger scale can have profound effects on the city's landscape: the project Nutritious City, by Mike Kendrick in the Bioclimatic Architecture Labs, illustrates this point clearly. An analysis of the Ancoats area of Manchester showed that as an inner city neighborhood, it had most of

FIGURE 14.9
Back to Front, small-scale urban agriculture. (Photo by E. Oldroyd.)

the criteria to be a sustainable neighborhood, but it had poor connection with the economic heart of the city, had no localized food or biomass production, and had very little public space (Figure 14.11).

The city carpet scheme consisted of a new landscape inserted over the inner relief road. The aim of this was first to connect the neighborhood with the city center by covering the road, and additionally providing space for the new tram interchange, and second to provide an energy source for the neighborhood. The energy source in question was a public park created on the new landscape, which grows short coppice willow. This is very high in biodiversity and is robust enough to handle pedestrian movement. Furthermore, it is harvested and then burnt in a combined heat and power plant. This provides renewable heat and electricity for the neighborhood and the CO_2 produced is collected and sequestered in heliotropic hydroponic glasshouse arrays, which produce vegetables for the neighborhood (Figure 14.12).

14.7 Integrated Emergent Solutions

14.7.1 Bio-Port Liverpool

Liverpool, United Kingdom, is one of the fastest shrinking cities in Europe (Oswaltt 2005). Its future is uncertain; a poor geography, tied with endemic population decline, has left the city in seemingly terminal collapse. A new

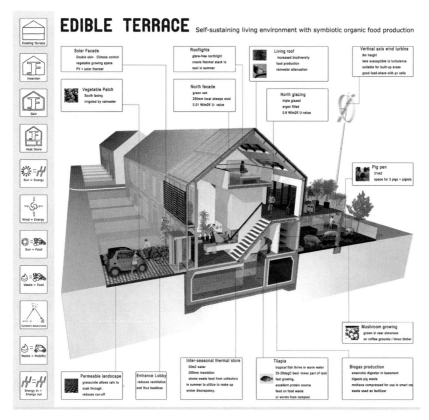

FIGURE 14.10
Edible Terrace (illustration by West and Campbell).

metabolism is needed to regenerate the city; one answer to this collapse is the proposal Bio-Port Free Energy City (Keeffe and Swietochowski 2009). Bio-port is an emergent, self-assembly, sustainable insertion of glass factories powered by biodiesel from algae arrays in the now redundant natural harbor of the Mersey Estuary, symbiotically combined with urban food production in the vacant urban space.

When farmed sustainably, biofuels offer a robust alternative to fossil fuel energy and they do not contribute to the rise in anthropogenic greenhouse gas emissions. However, they traditionally require a large area of land to produce a worthy amount of biofuel and compete with food crops for that land. Algae-derived aquatic biofuels offer energy yields over 100 times greater than traditional biocrops such as rapeseed and soy, and they do not compete for land with food crops. Scientific research shows that culti-vated algae could produce up to 150,000 L of biodiesel per hectare per year (Briggs 2004), given a suitable strain, but potentially this figure could reach 5,800,000 L/ha with further technological advancement. Could algae offer

FIGURE 14.11
Bi-productive landscape: nutritionally complete city.

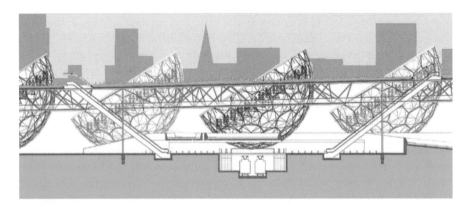

FIGURE 14.12
Nutritious city section, showing new biomass landscape and heliotropic hydroponic greenhouses.

not only a new sustainable and clean energy source, but also a new impetus to save the failing city?

When complete, Bio-port will produce some 4.4 TWh of electricity each year, provided by 6000 ha of algae matrix, photovoltaic arrays, and an off-shore wind farm. The large-scale algae matrix, by powering biodiesel-fueled cogeneration plant, produces over 3,400,000,000 kWh of electricity and more than 26,500,000 GJ of heat per year, enough energy to heat and power over a million homes.

Microalgae are an aquatic biofuel, exceptionally rich in bio-oil and natural hydrocarbons and capable of doubling their volume up to four times daily, a significant advantage over traditional biocrops that are usually harvested not more than once or twice a year. At a time of climatic crisis, one of the most relevant qualities of algae is its inextricable appetite for carbon dioxide which together with sunlight is used during photosynthesis. The synergy between carbon sequestration and natural algae-oil production is highly profitable: as more algae is cultivated, more oil is produced, but more importantly more carbon dioxide is removed from the atmosphere and converted to oxygen. Unlike terrestrial energy crops, algae have a continuous production cycle and sequester the carbon equivalent of twice their own weight—roughly 1 ton of marine algae biomass to 2 tons of carbon dioxide (GHGMP 2008).

There are said to be over 100,000 strains of algae, which vary immensely according to shape, size, color, and biological makeup. Seaweed is a typical form of algae, but so too is the thick green sludge, which can often be found resting on the surface waters of freshwater swamps and ponds. It is the strains found in these thick sludges or blooms that contain the most natural oil and hydrocarbons. A strain suitable for biofuel cultivation must not only have a high lipid oil content, but also have a fast growth rate to secure maximum energy yields. Two of the most productive strains are Spirulina and *Botryococcus braunii*, which produce natural oil up to 45% of their dry weight (Metzger and Largeau 2005).

It is possible to mimic the aquatic environments in which the algae grows in glass bioreactors, or "raceway ponds," which are effectively translucent cultivation tanks for growing algae. Once established, the thick green biomass is removed from the surface water and pressed to separate natural oil from waste pulp which itself has multiple applications from animal feed and fertilizer to soap and cosmetics. Once a crude bio-oil has been extracted, it can be processed into biodiesel and used to fuel reciprocating engines producing heat and electricity, or simply used as carbon-neutral transport fuel.

There are 6000 ha of bioreactors in Free Energy City, constructed on over 190 ha of existing derelict dockyards and further land reclaimed from the Mersey estuary and Liverpool bay. The energy is produced cleanly in vast quantities, superfluous to society and the city system, but wholly carbon-neutral provided by a constantly expanding algae matrix that not only produces enormous volumes of naturally cultivated energy but also acts as a colossal carbon sink that utilizes the waste gases of the city in its photosynthetic

FIGURE 14.13
Bio-port close up: raceway ponds, greenhouses, and algae-powered combined heat and power
plant.

growth cycles. The algae matrix is an organic energy factory which grows
symbiotically with urban growth, absorbing increasing volumes of carbon as
the urban realm and algae matrix mutually expand (Figure 14.13).

The collapsoscape abounds in Liverpool; in the waterfront alone, space of
over 1100 ha is considered derelict and unprogramed (Save Liverpool Docks
2008). The derelict space within each ward can be used to farm fresh produce
in greenhouses manufactured within the algae matrix. In the project, the
failing neighborhood of Anfield is the "tomato ward," where streets of derelict terraces are turned into linear greenhouses which imitate the physique
of terraced heritage, but with glass instead of crumbling brick. The greenhouses are filled with organic vines that in total produce over 12 million

FIGURE 14.14
Urban greenhouses in Anfield Liverpool.

tomatoes per year sold through farmers' markets locally and further afield. The process is organic and carbon neutral: high protein waste algae and recovered carbon dioxide is used to fertilize the vines (Figure 14.14).

Each greenhouse is heated during the winter by waste heat in an exergy cascade from Anfield's biodiesel combined heat and power plant, which is fueled by biodiesel produced in the algae matrix. Carbon dioxide is recovered from the flue and routed directly to the base of each vine, which not only stimulates growth but also minimizes carbon emissions from the energy manufacturing process. By turning vacant plots into organic vegetable factories, a proactive advance on regeneration occurs where not only the aesthetic is substantially improved, but also the nature of regeneration becomes purposeful and productive (Figure 14.15).

14.7.2 Paradigm Shift

Energy is a constantly fought over commodity, but Free Energy City demonstrates that by rationalizing a city's immediate hinterland, whether rooftops, rivers, estuaries or even derelict realm, and developing urban energy farms, such a prolific amount of electricity and heat can be developed that it becomes a clean and "free-to-use" asset. By running cities autonomously on decentralized independent networks, current alternatives like nuclear are completely redundant not only on efficiency and production grounds, but also on ecological terms. Farming algae to provide energy could not get more sustainable; algae are nature's answer to an enormous human race. The solution is symbiotic—the more algae we farm, the more carbon we absorb, but perhaps more important to the synthetic system is the more energy we produce (Figure 14.16).

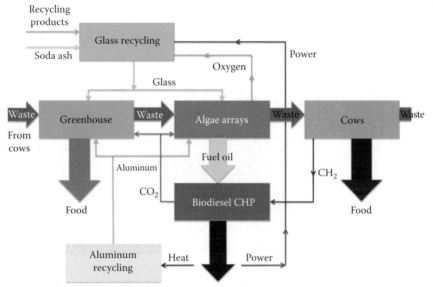

Bio-port closed cycles: Fuel, meat, dairy, vegetables, glass, aluminum

FIGURE 14.15
Bio-port closing cycles.

FIGURE 14.16
Bio-port. Finished master plan.

Synergy is ubiquitous in Free Energy City, and as a result the urban realm grows naturally without compromising its carbon-neutral physique. As each stakeholder within the synergy expands—whether biofuel production, land reclamation, glass manufacture, kelp farming, or even recovered heat and carbon dioxide—the holistic synergy expands with it. As a result, the algae matrix will expand naturally further into the Liverpool bay and the Irish Sea, cultivating more and more biomass as it grows. Urban growth is a direct result of farming algae biomass on the city's hinterland, which is used to attract new investment to the city. Uniquely however, the carbon sink of the algae matrix expands in parallel with urban growth and the carbon emissions associated a sprawling city. The scale of the insertion is large, but if seen as an ongoing 50 year process, it is smaller than the increase in size of Europort, Rotterdam, during an equivalent period in recent history.

In Free Energy City, energy is not associated with anthropogenic global warming. In fact it is quite the opposite: the model shows that specific renewable energy production can facilitate massive carbon sequestration and multiply biodiversity within urban space. Free Energy City represents a paradigm shift from urban consumption to urban production, where abundant volumes of carbon-neutral energy sources, organic produce, and industry have become symbiotic partners in a dynamic and sustainable urban model. It is the contemporary result of this mutual and symbiotic relationship that must become ubiquitous if humanity is to continue indefinitely within the self-regulating system that is "Gaia."

14.7.3 Total Integration

The production of these new multi-variant landscapes will be developed through new policies that create a new layered landscape that aids closing cycles and resilience. This will be achieved through a layering of programmatic function in three dimensions. The layering of function will allow the differing scales of network to work synergistically, connecting horizontally with itself and vertically at strategic nodes that will aid the transmission of resources to and from the landscapes. In addition, the insertions will change the city for the good. The urban agriculture will reenergize and reorder collapsing urban space and create jobs, and the large floating algae arrays will help prevent flooding due to storm surges in the Mersey basin.

14.8 Conclusion

Urban sustainability is multivariate in that it not only embodies ecological factors but also social and economic ones. The city after all is first and foremost an economic and social entity. Thus, the addition of ecological services alone into the city may not be enough to make the city resilient. In the postindustrial city, the collapse of social and economic services is the dominant trajectory and these need to be addressed primarily in urban redevelopment. The closing cycles of the silkworm in Silkworm City and those of algae in Bio-port show how the integration of productive landscapes in the city can produce economic benefit.

In this chapter, urban resilience is seen as the key service that makes the city sustainable and the chapter argues for a multi-scalar engagement with the productive cycles need by the metabolism of the city, in particular those of food and fuel. These will form a new nested form of sustainable energy landscape. Thus, urban resilience is more than the insertion of green space into the city fabric: It is a multi-variant design exercise, involving multilayered, multifactorial, and multi-scalar responses to perturbation. The new resilient city will be more complex than cities have been in the past, as it will use biological systems to provide a large range of services, from climate proofing to food and biomass production.

The resilient city will be nested; it will be productive at a range of scales, which starts with the building scale and ends globally. These nested scales will offer a redundancy that will protect the city against perturbation. The postindustrial city with all its inherent problems has within its form vacant space that at present is a problem. However, if reconfigured, this space can offer space for these new services to aid the city's resilience.

One of the key ideas of the bi-productive landscape is to use closing cycles as a new way of connecting resource production with the city's inhabitants. Many of the insertions described are dual-function, that is, not only do they provide "ecosystem services" but also connective space that reconnects the neighborhoods of the city. This new use for difficult space solves not only the issues of sustainable production of food and biomass but also makes a new city form that is more livable. This is clearly demonstrated in Nutritionally Complete City, where the creation of a new productive urban park generates electricity and heat and produces food. Nevertheless, it better connects an isolated urban neighborhood with the city center.

Sustainable energy landscapes at the urban scale need to be more than just productive spaces. They need to connect not only with the city, but also with its economy and its inhabitants. This new city will be better integrated: first ecologically, through the closing of resource cycles; second, economically, by the production of new industries using the products and bi-products of the landscapes; and finally, by socially filling the voids left by postindustrial collapse, which allows the reconnection of disparate urban communities. The solutions will thus be specific to the place, rather than a generic solution.

Acknowledgments

The author would like to acknowledge the work of Andrew Goodwin, Tilly Hall and Andrew Jenkins, James West and Anthony Campbell, Michael Kendrick, Simon Sweitochowski, and Adam Leigh Brown in the production of illustrations for this chapter.

References

Auge M. (1995). Non place: Introduction to an anthropology of supermodernity, trans. London, U.K.: John Howe, Verso.

Berger A. (2007). Drosscape: Wasting land in urban America. New York: Princeton Arch Press.

Best Foot Forward (2005). City limits: A resource flow and ecological footprint analysis of Greater London. www.citylimitslondon.com

Briggs M. (2004). Wide-scale Biodiesel production from Algae. University of New Hampshire Biodiesel Group. http://www.energybulletin.net/node/2364

Citylab (2006). Whitefield, Nelson Report to Pendle Council.

Clavin A. and Oldroyd E. (2011). *Back to Front Manual*. Leeds, U.K.: Infra Press.

Farley P. and Roberts M.S. (2011). *Edgelands. Journeys into England's True Wilderness*. London, U.K.: Jonathan Cape.

GHGMP (2008). Green house gas mitigation project: At Jacobs University. http://www.irccm.de/greenhouse

Girardet H. (2004). *Cities People Planet: Liveable Cities for a Sustainable World*. London, U.K.: Earthscan.

Keeffe G.P. (2011). Facing forward: Strategies for future climate adaptation in elderly persons housing. In *Proceedings of the International Federation of Housing and Planning World Congress*, Tallinn, Estonia.

Keeffe G.P. and Fagan D. (2011). Crash test: Bio-spheric city. Internal report, Leeds School of Architecture, Leeds, U.K.

Keeffe G.P. and Swietochowski S. (2009). Bio-port: Free energy city. In *SASBE09 Proceedings*, Delft, the Netherlands.

McDonough W. and Braungart M. (2002). *Cradle to Cradle—Remaking the Way We Make Things*. New York: North Point Press.

Metzger P. and Largeau C. (2005). *Botryococcus braunii*: A rich source for hydrocarbons and related ether lipids. *Applied Microbiology and Biotechnology*. February 66(5): 486–496.

North P. (2010). *Local Money*. London, U.K. Transition Books.

Oswaltt P. (Ed.) (2005). *Shrinking Cities*, Vol. 1. Ostfildern, Germany: Hatje Cantz.

Rogers R. and Power A. (1995). *Cities for a Small Planet*. London, U.K.: Faber and Faber. p. 9.

Save Liverpool Docks (2008). *Idle Land. Liverpool*. www.saveliverpooldocks.co.uk

Steel C. (2009). *Hungry City: How Food Shapes Our Lives*. London, U.K.: Vintage.

U.S. Wheat Associates (2011). US wheat: Quality, variety, dependability. www.uswheat.org

15

Spatial Modeling for Community Renewable Energy Planning: Case Studies in British Columbia, Canada

Olaf Schroth, Ellen Pond, Rory Tooke, David Flanders, and Stephen Sheppard

CONTENTS

15.1 Community Energy Planning in the Context of British Columbia

Community energy planning is defined here as comprehensive and integrated energy planning at the community scale, taking supply, transmission/ distribution, and demand into account. A community vision can be used to define a desired system that includes both sustainability and community perspectives and also makes the most efficient use of energy by matching energy supplies to energy services (Church and Ellis 2007; Salter 2008; Stremke et al. 2011). However, community energy planning in Canada has conventionally focused predominantly or wholly on demand management, with renewable energy less well integrated (St. Denis and Parker 2009). The work presented here focuses on the supply side in order to address this gap.

The province of British Columbia (BC), Canada, is unique in North America in terms of its regulatory framework. In 2007 and 2008, the provincial government passed two bills directly addressing climate change and greenhouse gas (GHG) mitigation (Province of British Columbia 2008, 2007). The Green Municipalities Statute Amendment of 2008 encourages municipal and regional government to deliver policies and actions to reduce GHG emissions, following provincial commitments to reduce GHG emissions 33% by 2020 and 80% by 2050. Since then, the province has set up the Community Energy and Emissions Inventory (CEEI) program that assembles energy consumption and GHG emissions data for transport, buildings, and waste for each community, as well as considering impacts of deforestation and agriculture at the regional district level. Based on the CEEI, municipalities have been encouraged to adopt community-wide the 80% GHG reduction targets set by the provincial government. While community planners are turning to higher building densities within communities in order to reduce energy demand, on the other side of the energy equation, local renewable supplies are touted as being able to meet at least some of the remaining required energy services. Planning instruments for addressing GHG mitigation and renewable energy planning include integrated community sustainability plans (ICSPs) and community energy emissions plans (CEEPs). CEEPs, in particular, fall under the broader rubric of community energy planning.

In BC, the CEEP process (Wilson 2008) is supported and funded in part by BC's Crown Corporation electrical utility, BC Hydro, which supplies the majority of BC's electrical power, 94% of which is hydroelectricity. There is, however, currently only one feasible (yet contentious) location for a new large-scale hydroplant remaining in the province. With increasing interest in low-GHG hydroelectricity for energy substitution due to the regulatory regime, BC Hydro supports the "Independent Power Producers" (IPPs) in finding alternate, nonfossil fuel energy supplies to meet community demand for space heating, hot water, and transportation, as well as the increasing demand for electricity. However, the historical availability of large-scale hydroelectricity has enabled low electricity prices compared to much of North America (much lower than in Europe), reducing the current economic competitiveness of other forms of renewable energy, particularly for producing electricity, in the province. In addition, the energy supply for much space heating, hot water, and other commercial/industrial uses, as well as transportation, continues to rely heavily on fossil fuels although that might change due to increasing fossil fuel prices and the BC carbon tax.

This chapter describes a set of BC case studies that have tested spatial modeling tools in order to assess local potential for the sustainable exploitation of solar, biomass, wind, and microhydro resources. Working at several scales— from regional forested areas to a small city downtown core—the spatial modeling of renewable energy supplies has taken into consideration factors ranging from fish habitat and forest soil carbon levels to sun angles and shading.

Understanding of scales is crucial for community energy planning but currently available models usually address larger scales. For an assessment of the many different options toward sustainable energy system, more modeling of "in-between" neighborhood and community scales is required (Kellett et al. 2008). In addition to the scale of the overall considered area, there is also the scale of different energy systems, that is, the overall energy system in a built area including distant and/or localized, transmission, and demand, as they relate to the regions' transportation and buildings. Such energy systems range from regionally or even provincially operated wind farms (discussed in Section 15.3.1) down to individual solar photovoltaics at the building scale (discussed in Section 15.3.3.2). Some renewable energy technologies (RETs) require individual building-owner decision making, while others need to be considered in the context of other land uses, such as biomass harvesting and managing scenery values for tourism. Localized renewable energy planning involves community members and requires political decisions about trade-offs, land-use conflicts, and social acceptability. Engaging with citizens and communicating with decision makers about renewable options is thus a critical part of sustainable community energy planning.

The research presented here is part of the wider field of planning sustainable energy landscapes; emerging work is exploring ways to integrate renewable energy into regional, urban, and landscape planning. Stremke (2010: 1) defines "sustainable energy landscapes" as "landscapes that are well adapted to renewable energy sources without compromising other landscape services, landscape quality or biodiversity." Both demand and supply* vary across space and time, offering various opportunities and limitations for local generation (Smil 2003). In addition, the quality (or usefulness) of energy varies and can be expressed as *exergy* (a measure of work capacity).†

---

* Stremke and Koh (2010) conceptualize areas with large energy demand as "energy sinks." Areas where energy provision exceeds consumption are considered "energy sources."
† See Stremke et al. (2011) as well as Dincer and Rosen (2007) for more information on exergy.

Electricity, for example, has a higher exergy than heat because electricity can be converted into other forms of energy including heat.

Internationally, considerable research has estimated renewable energy potentials on larger scales, from global assessments (Salameh 2003; Jacobson and Delucchi 2011) to national and state scales (Nielsen et al. 2002; Sawyer et al. 2012). Potentials are quantified for different energy sources and Nielsen et al. and Sawyer et al. use interactive online atlas systems, that is, interactive collections of web maps, for the visualization. MacKay (2009) allocates space on a per capita basis but, in his energy potential calculation, he does not use spatial landscape analysis and geographic data.

Angelis-Dimakis et al. (2011) provide a review of methods and tools to assess solar, wind, wave, biomass, and geothermal energy from small to large scales, which consider site-specific environmental characteristics. They also propose recommendations on how to bridge different scales and evaluate design options, although most approaches fall short of incorporating local social and economic constraints. In the Netherlands, based on the Grounds for Change study, Dobbelsteen et al. (2011, 2009) have developed a comprehensive energy potential mapping (EPM) method that is quantifiable and spatial, works across scales, and considers energy demand and supply (also see Chapter 4). For Switzerland, Wissen Hayek and Grêt-Regamey (2009, 2012) propose a method building on ecosystem services and visualization (also see Chapter 6).

Common characteristics of the investigated methods are that they are spatial, quantifiable studies on a regional or even smaller scale. They do not only consider the potentials but also constraints to renewable energy provision—especially with regard to sustainability. Furthermore, most approaches include some kind of stakeholder participation and visualization although the visualization methods vary largely (also see Chapter 7). Most methods that work on local to regional scales, and that include some form of multi-criteria analysis, are from Europe, for example, the EPM method developed in the Netherlands. Certainly, there is a lot to learn from EPM and similar approaches in Switzerland. However, area size, topography, political context, and population density differ in northern America in general and particularly in BC, so that it will be interesting in the future to discuss how far the European approaches can actually be transferred.

15.2 Approach to Spatial Modeling and Energy Potential Mapping in British Columbia

As shown in Figure 15.1, assessing supply is part of community energy planning. Flanders et al. (2009) had proposed a comprehensive framework for mapping energy potentials, similar to Dobbelsteen et al. (2011, 2009). The case studies presented in this chapter, however, target more specific

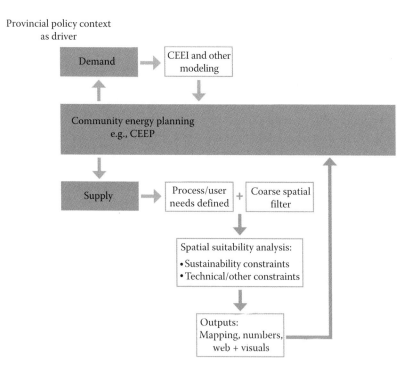

FIGURE 15.1
Application of the framework for assessing renewable energy systems (From Flanders, D. et al., *A Preliminary Assessment of Renewable Energy Capacity in Prince George, BC*. Prepared for NRCan. Collaborative for Advanced Landscape Planning, UBC, Vancouver, British Columbia, Canada, 2009: 14) as part of wider community energy planning process.

user- defined energy types and specific goals as applications in various contexts and with different motivations. In that respect, Figure 15.1 is a practical application of the overall framework proposed by Flanders et al. (2009). Since the demand side is already rather well elaborated in BC, we focus on the supply side. In contrast to Dobbelsteen et al. (2011, 2009), who cover all renewable resources, we start with the energy resources that local stakeholders identified as most feasible for their needs. Then, a coarse spatial filter is applied to identify opportunities with regard to local resources, geography, climate, and natural and built environment.* Viable local renewable energy resources can be identified for a given area, as several alternative energy sources are more general in their requirements, with fewer local specific restrictions.

Geothermal energy (Evans 2008), solar thermal heat (for heating water), photovoltaics, passive solar or solar air heat (for space heating), and biomass are

---

* Wind resources, for example, have been modeled nationwide in the Canadian Wind Energy Atlas. Solar potentials have been mapped on a provincial scale, and biomass potential can be derived from the BC Vegetation Resources Inventory (VRI). Although local assays are more accurate, these maps can provide a general indication of whether or not it is viable for a community to investigate their wind generation potential.

examples of RET that have less restrictive local constraints and may be found in many communities (Flanders et al. 2009). Other alternative energy sources have more strict local constraints: wind, microhydro, tidal, sewage, residual heat, and landfill gas are much more site specific. Microhydro (either in the form of small hydro installations, or run-of-river approaches) requires specific characteristics from any proposed water system, including slope or head potential and characteristics of the catchment area. Spatial suitability analysis plays a critical role in assessing how specific local renewable energy resources may be implemented in a community. BC is data rich; readily available or easily generated spatial data include topography, existing land uses, zoning, soil, and forest cover by type. The work presented here uses spatial suitability analysis methods (McHarg 1969; Steinitz et al. 2003; Malczewski 2004) and, in the case of biomass assessment and solar roof potential, constraints models (for details, see Sections 15.4.2. and 15.4.3) to assess renewable energy potentials. Nonspatial issues including technical considerations such as efficiencies, costs, or other modeled data such as forest productivity inform the analysis.

Results are visualized in maps, or web interfaces, or produced in the form of energy supply numbers for use in a planning process. In contrast to Dobbelsteen et al. (2011, 2009), we use the number of average households provided with energy as a benchmark measure, rather than GJ/ha. Working with average households is more useful for local residents in northern America, although it complicates international comparison. The community energy planning process potentially tested in these case studies inform CEEPs. The CEEP is covering stakeholder participation, using visualization and other facilitation techniques and includes demand management and other issues that are not considered in this chapter.

15.3 British Columbia Case Studies in Wind, Solar, Microhydro, and Biomass

The following five case studies cover various scales of the renewable resources in BC. The selection of case studies is not spatially inclusive and comprehensive but the goal was to cover the renewable energy sources that are most likely to become important for BC in the future (according to Evans 2008), and to cover case studies from the highly populated Metro Vancouver area, which has the highest energy demand, as well as case studies from the north of BC, where previous studies located the highest potentials for wind energy and biomass (Environment Canada 2011).

15.3.1 Regional Wind Energy Mapping for Northern British Columbia

According to BC Hydro (2002), BC has an energy potential of approximately 730 MW (1600 GWh/year) and capacity factors between 18% and 38% for wind energy. The intermittent resource wind would be complementary to the existing hydropower facilities that could store energy. However, most wind energy potential lies in the north of BC and would require major infrastructure projects to become useable. Other BC-specific factors have to be considered: topographic (accessibility, slope), environmental (old-growth forest, wildlife areas, wetlands, important bird areas), legal (e.g., First Nations land title and treaties), and social and visual (BC visual landscape inventory) constraints to wind energy. Public perception of wind farms raises important issues of social acceptance, as shown by wind farms in the province of Ontario and some European countries. To address these issues, a research project was funded through the Pacific Institute for Climate Solutions to

1. Identify typical and feasible energy landscape options for BC
2. Visualize these energy landscape options
3. Assess the social acceptance of the various options against each other

The case study discussed in this section refers to the first step of the research project, which aimed at identifying suitable locations for wind farms at different scales as the basis for a later three-dimensional (3D) visualization and a survey on social acceptability.

Until now, wind energy potential mapping for BC has been based on conventional methods such as mean annual wind speed (Environment Canada 2011), and a range of net capacity factors (Garrad-Hassan 2008; Angelis-Dimakis et al. 2011). Our project included an extended geographic information system (GIS)-based suitability analysis in order to add important bird areas, nature parks, and other protected land-use designations as constraints, informed by recommendations from countries that started investigating wind energy earlier, such as the Swiss Federal Office of Energy (Bundesamt fuer Energie BFE 2004). The resulting energy potential map was compared with existing wind energy potential maps (BC Hydro 2010) and showed a subset of land where wind energy could be used in a sustainable way. In the next step of the research project, the identified potential wind energy locations and representative locations for solar, hydro, and biomass production are illustrated as 3D landscape visualizations and tested in an online survey with regard to public landscape preferences. This helps to assess the visual impact of different locations (Figure 15.2).

15.3.2 Energy Potential Mapping in the North Shore Region of Metro Vancouver, British Columbia

This case study assessed the potential of microhydro and biomass resources for the North Shore region of Metro Vancouver, a region of 28,500 ha that includes the districts of West Vancouver and North Vancouver and the city of North Vancouver, with a combined population of approximately 175,000 persons in 2011.

FIGURE 15.2
GIS-based suitability analysis considering opportunities (wind energy potential, proximity to existing power lines) and constraints (proximity to settlements, national park areas, important bird areas, slope, scenic areas) on a regional scale. Suitability analysis of local potentials and constraints. (From BC Hydro, Resource options mapping (ROMAP) report, Figure 2-14, 2010, http://www.bchydro.com/etc/medialib/internet/documents/planning_regulatory/iep_ltap/2012q1/2010_resource_options19.Par.0001.File.2010ResourceOptionsReport_Appendix5-A14.pdf, accessed on May 21, 2012.)

15.3.2.1 Microhydro at the North Shore

The North Shore appears to be a favorable location for microhydro due to its steep slopes, significant precipitation, large number of creeks, and larger catchment areas (mountain valleys) than are found in most other Metro Vancouver municipalities (Figure 15.3). For this study, potential microhydro electrical generation was assessed for North Shore streams using a spatial, quantitative technique, drawing on several source studies. Many factors, such as the capacity rating of the installed generation facility, the number of days of the year the turbine is generating energy, and balancing ecological values (such as fish habitat) with energy priorities, will ultimately impact annual generation. Ecological factors have only recently been implemented; they are necessary to ensure the development of truly sustainable energy landscapes (see, e.g., Stremke et al. 2011).

The preliminary analysis conducted in our study can be used to help determine streams that may warrant further investigation, to give a preliminary idea of the estimated total capacity, and to better understand trade-offs inherent within the North Shore's energy planning options. The highest flows that are best for diversion for energy generation occur in the spring after a gradual increase through the winter. Generation may have to be shut off during the summer months in order to preserve base-line stream flows during this time. For this study, potential microhydro facilities were proposed to be located either above known fish habitat and fish sightings in the streams, or in streams that are within watersheds with multiple branches to avoid serious impacts to fish populations (Flanders and Sheppard, 2009). In summary, the formula

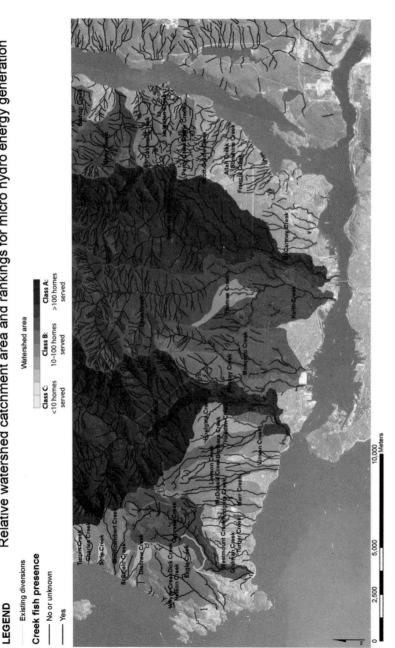

FIGURE 15.3
Watersheds included in estimates of run-of-river and reservoir-based hydroelectric generation in this sustainable hydropower energy landscapes at the North Shore in BC, Canada.

distinguished low flow (October–January), high flow (February–May), and potential shut down (June–September) months and considered streamflow data (if available) for the calculation of annual "realizable" energy generation. Here, "realizable" refers to a realistic generation estimate based on a multi-criteria assessment that includes a balance between energy generation and, for example, ecosystem conservation. Converting potential energy supply numbers to number of homes supplied, based on average household energy demand, was used as a way of expressing potential energy in understandable (lay) terms. Assuming an electrical use of 9.6 MWh/year/dwelling unit (NRCan CEUDT, 2009), 11,085 homes could potentially be served by realizable energy generation from microhydro at the North Shore.

15.3.2.2 Biomass at the North Shore

Highly productive and fast growing, the North Shore forests have the potential to supply biomass for low-carbon energy to local municipalities. However, prior to this study, quantitative, spatial production estimates have never been generated for forested landscapes near Metro Vancouver with its large energy demand. Furthermore, the ecosystem and land-use implications of biomass extraction on the landscape have not previously been considered. Note that biomass harvesting at non-sustainable rates could have adverse public reaction, and degrade local ecosystems and yields over time.* For the study presented in this section, the local biomass resources have been mapped and quantified by models while considering how the harvests may be managed to ensure long-term biomass supply while maintaining forest stand health and ecosystem productivity.

15.3.2.3 Biomass Assessment Methods

GIS maps of vegetation cover were used to feed into the FORECAST Forest Ecosystem Model, developed at the University of British Columbia. FORECAST estimates total biomass produced by forest stand based on the species composition and other biophysical parameters. It also allows for the exploration of various management/harvesting strategies and their effects on the sustainability of the resulting ecosystem. This, in turn, allows researchers to address the public and scientific community's concern of over-extraction. Thirty meter wide riparian zones for all creeks, protected areas, and protected watersheds were excluded from the forested lands available for biomass production. The FORECAST model was run to simulate both soil and stand organic matter over 150 years of simulated harvesting. Harvest rotation times and percent thinning parameters were calibrated for stand species composition using FORECAST to ensure that soil nutrients in the forest ecosystem remained consistent while allowing for continuous harvests

---

* Use of industrial fertilizers, produced from natural gas, would add GHG emissions to the forest resource.

without the need for fertilization or loss of forest productivity. Higher, more intensive yields from required regular harvesting of biomass throughout numerous BC Hydro transmission line corridors (to maintain access and line clearance) were calculated outside of the FORECAST analysis.

15.3.2.4 Biomass Energy Generation Results

Figure 15.4 maps the biomass yield by forest stand within the North Shore based on sustainable management practices, shown here as a Google Earth coverage. Biomass yields from the FORECAST model varied according to stand type, with a range of <1 to 7 bone dry ton (BDT)/ha/year. Averaged across all stand types throughout the North Shore, the sustainable harvesting rate suggested by FORECAST simulations would be 3 BDT/ha/year. This is almost three times higher than the sustainable yield calculated for the provincially interior and northern Prince George Community Forest using a similar methodology, indicating the higher productivity on North Shore stands. Site conditions vary between stands, as does modeled management strategy: percent thinning rates ranged from 25% to 50% across the 10 stand types identified.* This is a novel and ecological approach to the

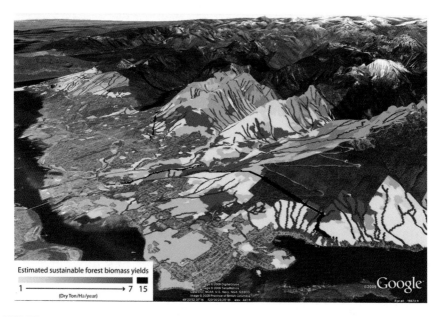

FIGURE 15.4
3D image of estimated sustainable forest biomass yields at the North Shore. Visualized in Google Earth (Flanders et al. 2009; © 2009 Google, Image © 2009 DigitalGlobe, Image © 2009 TerraMetrics, Data SIO, NOAA, U.S. Navy, NGA, GEBCO, Image © 2009 Province of British Columbia).

---

* Note that yield is not associated to a specific tree species, rather to site-specific stand conditions, ensuring that trees are not overharvested in less productive areas of the landscape.

planning of sustainable biomass energy landscapes. Intensive harvesting areas in the transmission right-of-ways (the red linear features in Figure 15.4) were assumed to yield 15 BDT/ha/year, based on recent research findings examining hybrid poplar and willow plantations with 3–5 rotations in the Quesnel, BC area (Green et al. 1999; Powell 2009).

For the following calculation, we assume that a combined heat power (CHP) plant would per BDT of biomass generate 1.398 MWh of electricity (Stennes and McBeath 2006) at an efficiency of 20% (Bolhàr-Nordenkampf 2003; Carmichael 2009) and 18 GJ of heat at an efficiency of 70% (Carmichael, 2009). The average North Shore household electricity consumption is 9.6 MWh/year (34.56 GJ) and heat consumption is 86 GJ/year per (NRCan CEUDT 2009). Considering the above figures on average household electricity and heat consumption, 9278 or 14% of all North Shore dwelling units could be supplied with space heating and hot water from biomass heat energy generation, assuming they could be connected to the CHP. If part of a district heating system, or having a very high efficiency furnace, an average home* requires almost 2 ha of sustainability managed forest to meet its annual space and hot water heat needs. For intensively managed woodlot with high yield species (e.g., hydropower transmission right-of-way), less than half a hectare would suffice.

15.3.3 Solar Energy for the City of Prince George and the District of North Vancouver

Solar energy received at the surface of a city, including building rooftops, trees, and bare terrain, is highly variable due to the complexity of urban 3D form and land cover. Solar energy determines not only the potential supply for renewable solar technologies, such as photovoltaic panels and domestic hot water systems, but is also critical for assessing surface climates (Voogt and Oke 1997), building electrical and thermal energy demand (Ratti et al. 2005), and human thermal comforts (Lindberg and Grimmond 2011). Therefore, understanding and assessing the local variation in solar energy is fundamental to the planning of urban energy landscapes in general and for community energy planning in particular.[†]

Two case studies assessing solar energy potential are presented: both use spatial modeling to assess incoming solar radiation at detailed, local scales in response to municipal planning needs. However, they differ in method due to data availability, address different areas, and have different scalability and replicability. They also hold different variables constant or dynamic and have different applications in terms of intended use and target groups.

---

* An average home in Metro Vancouver requires 86 GJ/year in space and hot water heating and has 2.5 inhabitants (NRCan CEUDT, 2009; Statistics Canada, 2006).
† In Canada, solar thermal (hot water) generation is developing into a renewable energy success story, with demonstrated effectiveness and a relatively short payback period (CANMET 2004; NRCan 2003).

In Section 15.3.3.1, the total rooftop solar thermal heat potential was modeled within a small urban core in Prince George including 255 office buildings and minimal vegetation. In Section 15.3.3.2, residential roof areas for solar thermal installation were modeled and visualized for a municipality of 26,000 homes in North Vancouver, with extensive forest and tree cover. While both studies used spatial (GIS) data, they used different 3D modeling tools, as the first case study had access to a 3D city model in Sketchup, while the second case study had access to detailed light detection and ranging (LiDAR) data (high-resolution terrain data).

15.3.3.1 Municipal Solar Thermal Potentials in Prince George

The total solar thermal potential from rooftops was modeled for the downtown area of the northern city of Prince George, as part of a Smart Growth planning exercise.* The suitability mapping method developed in this case study is easily replicable and can be applied to urban areas with minimal vegetation and multiple flat-roofed buildings, in order to assess the amount of solar thermal energy that could be generated using rooftop collectors, provided that GIS data and a 3D city model are available. The method, however, would be difficult to scale up, because it was not automated, a key difference from the LiDAR study presented later. The Prince George case study was intended to provide a single thermal energy supply number to the larger planning process that was independently investigating demand.

The essential characteristic required to compute solar thermal potential is the amount of solar insolation hitting the collector panel, which requires the solar insolation values for the site in question, and an approximation of the collector area. Monthly solar insolation at the solar collectors was calculated for March through September using data from Natural Resources Canada (2009). Potential energy generation for the solar panels, before system losses, was calculated by multiplying monthly insolation (KWh/m^2) by conservatively estimated system efficiency of 40%. For this case study, the challenge was to assess collector area, and a Sketchup and GIS-based suitability analysis was developed to approximate the solar hot water collector area at a neighborhood scale. With 255 buildings in the study area, detailed individual building analysis was not practical. Therefore, a suitability analysis for building roofs based on inter-building shading and the percentage of available (non-obstructed) roof area was used to determine the roof area on which solar collectors could be located.

First, a 3D SketchUp model enabled evaluation of inter-building shading for the vernal equinox, summer solstice, and autumnal equinox, for the hours of 9 am–5 pm; each building was assigned a weight based on a visual examination of daily shading, from no significant shading, to moderate shading[†] to sig-

---

* See Smart Growth BC (2009) and Flanders et al. (2009) for a detailed description of the methodology.
[†] Less than 50% shading for less than 2 h.

nificant shading.* Next, a photo-interpretation analysis was conducted for all buildings to estimate the available roof area for solar thermal hot water panel installation. Sloped roofs (not facing southeast to southwest), heating, ventilation and air conditioning (HVAC) equipment, stacks, sectioned roofs, and unidentified objects were all considered obstacles to panel installation; weightings were based on 0%, 25%, 50%, 75%, and 100% roof availability. A composite building suitability map for solar thermal panel installation was created by adding the weights for building shading and available roof area (Figure 15.5); buildings with no significant or moderate shading and minimally 25% of available roof space were considered suitable for solar thermal installations.[†]

Next, the amount of roof space that could be used by solar thermal collectors was estimated by determining the necessary distance between rows of panels to prevent them from shading each other, using Sketchup. The potential installed solar thermal collector area was then calculated to be 19% of the total roof area in the study area. This, combined with the suitability analysis provided earlier, allowed us to calculate the total area of fixed, roof-mounted, south-facing collectors.

The total amount of solar thermal energy generated for March through September was then calculated at 414,556 kWh (1,492 GJ). The results suggest that solar thermal is a viable source of on-site renewable energy, able to augment hot water supply during the months of March through September; solar thermal would need to be incorporated with other systems, such as heat–cold storage or biomass-based heating to meet year-round demand. This case study was part of a larger planning process, including multi-stakeholder design charrettes. Given that hot water demand for the area was not provided to us, we instead compared the downtown supply to single-family home demand in order to facilitate nonenergy expert understanding of the potential energy supply. The total supply is roughly equivalent to providing hot water for approximately 41 single-family dwellings in the Prince George area for 1 year.[‡]

This case study assessed flat roofs with the capacity to install multiple racks per roof with optimal southern orientation, with the goal of evaluating the total maximum thermal energy that could be generated for a compact downtown area; the method would require adaptation in order to evaluate single-family dwellings with peaked roofs and various orientations to south. Such a residential analysis would provide further data about the capacity and feasibility for meeting municipal demand for hot water using solar thermal, which could reasonably be expected to supply a portion of residential hot water demand outside the downtown core. The next case study continues where this one leaves off, using LiDAR data to scale up assessment, improve accuracy, and automate some of the analysis.

---

* Greater than 50% shading for greater than 2 h.
[†] See Flanders et al. (2009) for the detailed weighting matrix.
[‡] Based on an estimate of 10,000 kWh of energy to provide domestic hot water to a single-family dwelling for a year (Flanders et al. 2009).

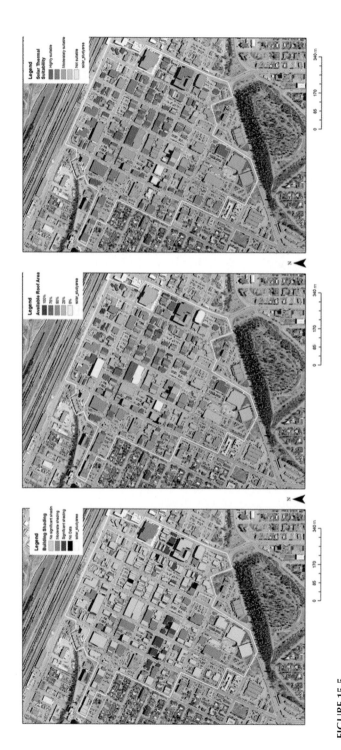

FIGURE 15.5

Cumulative suitability (from left to right: shading, available roof area, and cumulative weighting) for each study area building.

15.3.3.2 *Mapping Solar Energy Potentials from LiDAR for North Vancouver*

The case study presented in this section used LiDAR in order to assess the availability of residential roof area for solar thermal installation. LiDAR technology emits and receives laser pulses typically from an aerial platform and can be used to generate highly accurate 3D terrain models often at submeter spatial resolutions. LiDAR enabled us to assess 26,000 buildings, although its replicability is limited to locations with LiDAR data.* In terms of intent, this case study embedded the solar thermal assessment within an online interactive tool to help homeowners make informed decisions.

By representing the vertical relief of a landscape, LiDAR datasets facilitate the extraction of key parameters for solar energy models, including surface elevation, orientation, slope, and shading features. Additionally, the detailed elevation data can be used to classify objects within the urban landscape including individual buildings and trees (Goodwin et al. 2009). Together, the extraction of solar energy modeling parameters and urban feature classification leads to a highly automated analytical framework that can be applied to assess solar resource availability across entire regions where LiDAR datasets have been acquired. Furthermore, the spatial scale at which the LiDAR data are captured allows information to be generated on a building-by-building basis, offering a unique opportunity to engage with individual citizens and homeowners. The spatial nature of LiDAR data lends itself to visualization through maps or spatial 3D visualizations. By facilitating a process to automate solar energy assessment over large areas, LiDAR can be used to highlight those locations within a city that are more or less suitable for installing solar technologies.

The interactive nature of a web tool can help increase awareness and public engagement with local issues directly related to sustainability and energy. One example of a LiDAR-generated interactive solar map is the "Solar" application, part of the district of North Vancouver's (DNV) GEOweb GIS website. This application integrates monthly estimates of incoming solar radiation with their interactive web-mapping platform. An interactive interface was designed to allow users to enter building occupancy, demonstrating offsets to utility energy cost and carbon emission reductions from installing a standard solar domestic solar hot water system on the sunniest area of a roof (Figure 15.6). Results showed substantial solar energy potential within the municipality due to the abundance of south-facing roofs and the low-density residential character of the region. Furthermore, the application allows individuals to accurately assess important criteria for their decision-making process; users are provided with economic information including annual savings in addition to the potential CO_2 reductions (Figure 15.7).

---

* A detailed description of the methodology can be found in Dean et al. (2009).

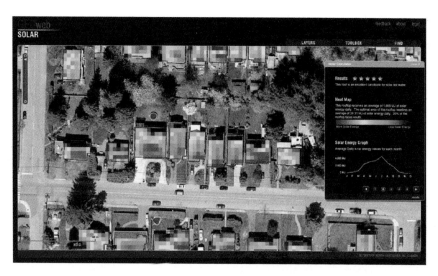

FIGURE 15.6
Solar energy map generated from LiDAR data, as part of the district of North Vancouver's web-mapping platform. (http://www.geoweb.dnv.org/applications/solarapp/)

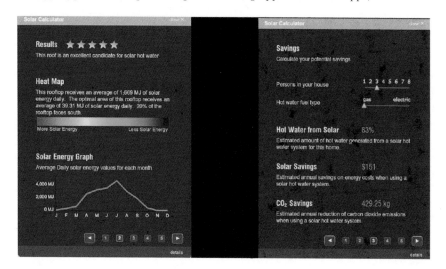

FIGURE 15.7
Example of the district of North Vancouver's "Solar" application interface, demonstrating location-specific carbon reductions and cost savings from installing a rooftop solar domestic hot water system.

Compared to the aforementioned case studies, the North Vancouver solar energy potential project (this section) did not compute a total solar thermal potential for a centralized planning process. Instead, it enables decision making at the level of the individual household, helping to inform citizens of their potential contribution to developing sustainable urban energy landscapes.

15.4 Discussion

Comparing all five case studies, it can be seen that they all share the basic criteria described in Section 15.2, that is, they are spatial, quantifiable, and take spatial criteria as well as local potentials and constraints into account (Table 15.1). All the case studies benefited from a data-rich environment.[*] Where available, LiDAR data are another promising data source, for example, allowing 3D shadow analysis for solar potential.

Comparing the case methods with the literature reviewed earlier, the overall methodology is not yet as comprehensive as Dobbelsteen et al. (2011, 2009). Instead, at this point, individual energy supply case studies are being tested that will enable more comprehensive sustainable energy assessments in the

TABLE 15.1

Comparative Overview of BC Case Studies

	3.1 Regional Wind in the North	3.2.1 North Shore Microhydro	3.2.2 North Shore Biomass	3.3.1 Prince George Solar	3.3.2 District North Vanc. Solar
Users					
Researchers	x	x	x	x	
Planners		x	x	x	x
Public	*Proposed*		*Discussed*	x	x
Data					
Topography	x	x		x	x
Vegetation	x		x		x
LiDAR					x
Building footprints				x	x
Tools					
Numeric modeling		Spreadsheet	FORECAST	x	
GIS suitability analysis	x	x	x	x	
3D visualization	GIS map 3D visualization proposed	Google Earth	Google Earth	Sketchup	
Web tools					x
Approach					
Spatial	x	x	x	x	x
Quantifiable	x	x	x	x	x

---

[*] Due the historic importance of forestry, BC has rich provincial geodata on vegetation and a visual inventory.

future, currently unaccounted for in the provincial Community Energy and Emissions' Inventories. This BC supply-side method can be linked to other planning tools (e.g., CEEI providing GHG inventories) and processes, particularly municipal CEEPs that assess demand and demand management options, and may include a local participatory component and a multi-criteria analysis. All these are part of the overall mapping framework described by other authors but, in BC, may continue to be separate yet linked projects.

The charts in Figure 15.8 help to highlight the limitation of locally feasible RET in offsetting current demand trends for the North Shore region of BC (Miller et al. 2011). As a result, approaches to sustainable energy planning must consider demand side management strategies. The provincial government has set an ambitious target to acquire 50% of the state-owned electricity company's incremental resource needs to be met by energy efficiency

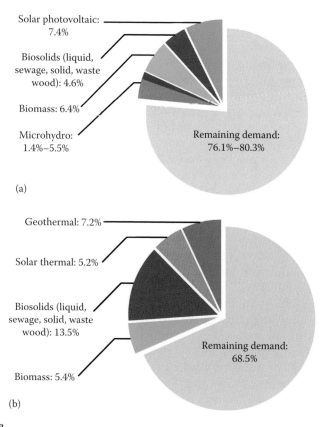

Solar photovoltaic: 7.4%

Biosolids (liquid, sewage, solid, waste wood): 4.6%

Biomass: 6.4%

Microhydro: 1.4%–5.5%

Remaining demand: 76.1%–80.3%

(a)

Geothermal: 7.2%

Solar thermal: 5.2%

Biosolids (liquid, sewage, solid, waste wood): 13.5%

Biomass: 5.4%

Remaining demand: 68.5%

(b)

FIGURE 15.8
Potential share of local renewable energy sources in a community energy mix for the North Shore. (a) North Shore low-carbon electricity supply mix (% of total demand). (b) North Shore low-carbon heat energy supply mix (% of total demand). (From Flanders, D.N. and Sheppard, S.R.J., *A Preliminary Assessment of Renewable Energy Capacity in the Northshore, BC—Draft Summary Report*, Vancouver, British Columbia, Canada, 2009.)

measures by 2020 (Ministry of Energy, Mines and Petroleum Resources 2007). In addition, the carbon tax was introduced to discourage fossil fuel based transportation and industrial GHG emissions. A distinct feature of the BC approach is the importance of GHG emissions as a main driver and indicator due to provincial legislation.* It has to be seen whether that will have any positive or even negative impact on the sustainability of the proposed energy-conscious interventions.

The BC case studies presented here are helping to address larger challenges, including how much local renewable supply may contribute to the overall goal of local community GHG reductions, what demand reductions and system efficiencies may also be required, and whether remaining demand can or should be met with renewable production elsewhere. Temporal availability is another issue to consider. Though BC has large wind energy potentials—an intermittent energy source—some researchers argue that wind energy in BC could be used more effectively on a provincial or even national scale if wind farms were linked to the large hydroplants by means of a smart power grid (see, e.g., Evans 2008).

This answers to the question of how much energy supply can be relocalized, or whether growing urban demand should be met by "industrial" scale renewable energy in the vast rural areas of the province. Benefits from a shift toward localized provision include increased community resiliency and efficiency by making use of residual heat, enhancing scalability (increasing or decreasing system size and capacity), facilitating fuel switch in the future, keeping energy expenditures within the community, reducing vulnerability to production and transmission disruptions elsewhere, creating green job opportunities (Globe Foundation 2010; Parfitt 2011), and improving compatibility with mixed-use urban forms and complete communities.

Nevertheless, additional generation in upstate BC may be inevitable. However, an integration of a diverse suite of small-scale interventions within local jurisdictions could satisfy interim goals and GHG targets as stepping stones to or elements of larger, long-term solutions. Together, the case studies presented here illustrate the difficulties in decentralizing renewable energy supply in BC and suggest that a more complex mix of energy supplies and geographies will be required in the future.

The cases also show that finding an appropriate mix of technologies will depend upon local resources and constraints. In this context, energy mapping can provide a valuable tool for communities to assess their current energy supply, identify areas for future consideration and investment, and plan for development and infrastructure. Spatial and visual modeling tools help communities to understand their local sustainable energy capacity, with the goal of contributing to improved decision making about how to meet energy demand with local to regional energy.

---

* This is not supported in or even contradicted by federal legislation.

15.5 Conclusion

The transition from fossil fuels to renewable sources means that more land area will be required for energy provision, to match the demand of urban areas in BC. In the process, land-use conflicts may occur between energy provision and other uses. This goes beyond inventory mapping and must assess trade-offs between, for example, biomass production and the traditional forest products industry, wind farms and tourism, microhydro generation and habitat conservation, or corporate investment and local control over decentralized energy plants.

A key political decision will be how much of the energy mix is provided locally, possibly by community-owned RETs, and how much is imported from large-scale facilities in remote areas. These decisions will have major impacts for the landscape in BC, and public participation in such decision making is strongly recommended. Key issues include not only community resilience but also social acceptability and whether communities will "learn to love the landscapes of carbon-neutrality" (Selman 2010). Communities in BC have multiple options, and finding a suitable trade-off between demand and supply management and a suitable energy "breadbasket" for each community are major challenges.

The case studies described here are a first step to push this discussion and to provide tools for sustainable community energy planning and dialogue in the province of BC, Canada. Although most of the policy and economic context as well as the spatial characteristics of the discussed case studies are unique, some of the methods and perhaps some of the key issues to solve may apply to communities outside of Canada as well.

Acknowledgments

Special thanks to Jon Salter for his contribution and advice on the case study that was considered for this chapter. We would also like to thank the Pacific Institute for Climate Solutions (PICS), Metro Vancouver, the City of Kimberley, and the GEOIDE network for their support.

References

Angelis-Dimakis, A., Biberacher, M., Dominguez, J., Fiorese, G., Gadocha, S., Gnansounou, E., Guariso, G. et al. 2011. Methods and tools to evaluate the availability of renewable energy sources. *Renewable and Sustainable Energy Reviews*, 15(2), 1182–1200.
BC Hydro. 2002. *Green Energy Study for British Columbia Phase 2: Mainland. October.* Vancouver, British Columbia, Canada.
BC Hydro (2010). Resource options mapping (ROMAP) report. Figure 2-14. http://www.bchydro.com/etc/medialib/internet/documents/

planning_regulatory/iep_ltap/2012q1/2010_resource_options19.Par.0001.File.
2010ResourceOptionsReport_Appendix5-A14.pdf (accessed on May 21, 2012).

Bundesamt fuer Energie BFE, Bundesamt fuer Umwelt, W., und L. B., and Bundesamt fuer Raumentwicklung ARE. 2004. *Konzept Windenergie Schweiz. Grundlagen fuer die Standortwahl von Windparks.* Bern, Switzerland.

CANMET. 2004. *Clean Energy Project Analysis: RETScreen Engineering and Cases Textbook.* Natural Resources Canada, Ottawa, ON.

Carmichael, J. 2009. Personal communication. Metro Vancouver, Policy and Planning.

Church, K. and Ellis, D. 2007. *Community Energy Planning: A Guide for Communities. Volume 2—The Community Energy Plan.* Ottawa, ON, Canada.

Dean, J., Kandt, A., Burman, K., Lisell, L., and C. Helm. 2009. Analysis of web-based solar photovoltaic mapping tools. *Proceedings of the 3rd International Conference on Energy Sustainability ES2009.* July 19–23, 2009, San Francisco, CA.

Delucchi, M. and M. Z. Jacobson. 2011. Providing all global energy with wind, water, and solar power, Part II: Reliability, system and transmission costs, and policies. *Energy Policy,* 39(3), 1170-1190.

Dincer, I. and M. A. Rosen. 2007. *Exergy: Energy, Environment and Sustainable Development.* Oxford, U.K.: Elsevier.

Dobbelsteen A. van den, Sergeev, V., Sarukhanyan, S., Mallon, W., and R. Roggema. 2009. Energy Potentials. In: R. Roggema (Ed.), Adaptation to Climate Change: A Spatial Challenge, pp. 253–288. Dordrecht/Heidelberg/London/New York: Springer Netherlands.

Dobbelsteen A. van den, Broersma, S., and S. Stremke. 2011. Energy potential mapping for energy-producing neighborhoods. *SUSB Journal (Sustainable Building Technology and Urban Development),* 2(2), 170–176.

Environment Canada. 2011. Canadian Wind Energy Atlas. Retrieved from http://www.windatlas.ca/en/index.php (accessed on February 10, 2011).

Evans, R. L. 2008. *Alternative Energy Technologies for BC.* Victoria, BC, Canada: Pacific Institute for Climate Solutions.

Flanders, D., Salter, J., Tatebe K., Pond, E., and S. Sheppard. 2009. *A Preliminary Assessment of Renewable Energy Capacity in Prince George, BC.* Prepared for NRCan. Collaborative for Advanced Landscape Planning, UBC, Vancouver, British Columbia, Canada.

Flanders, D. N. and S. R. J. Sheppard. 2009. *A Preliminary Assessment of Renewable Energy Capacity in the Northshore, BC—Draft Summary Report.* Vancouver, British Columbia, Canada.

Garrard-Hassan Canada Inc. 2008. *Assessment of the Energy Potential and Estimated Costs of Wind Energy in British-Columbia.* Report for BC Hydro.

Globe Foundation. 2010. *British Columbia' s Green Economy. Building a Strong Low-Carbon Future.* Vancouver, British Columbia, Canada.

Goodwin, N. R., Coops, N. C., Tooke, T. R., Christen, A., and Voogt, J. A. 2009. Characterizing urban surface cover and structure with airborne Lidar technology. *Canadian Journal of Remote Sensing,* 35(3), 297–309.

Green, D. W., Winandy, J. E., and D. E. Kretschmann. 1999. Mechanical properties of wood. Chapter 4 In: *Wood Handbook: Wood as an Engineering Material. General Technical Report 113.* US Department of Agriculture, Forest Service, Forest Products Laboratory, Madison, WI, p. 463.

Jacobson, M. Z. and M. Delucchi. 2011. Providing all global energy with wind, water, and solar power, Part I: Technologies, energy resources, quantities and areas of infrastructure, and materials. *Energy Policy,* 39(3), 1154–1169.

Kellett, R., Cavens, C., Miller, N., and J. Salter. 2008. Climate change and urban design: Science, policy, education and best practice. Paper presented at the *Third International Council for European Urbanism Congress*, Oslo, Norway, September 14–16.

Lindberg, F. and C. S. B. Grimmond. 2011. The influence of vegetation and building morphology on shadow patterns and mean radiant temperatures in urban areas: Model development and evaluation. *Theoretical and Applied Climatology*, 105(3–4), 311–323.

Mackay, D. J. C. 2009. Sustainable energy—Without the hot air. In *Energy* (p. 383). UIT Cambridge Ltd, Cambridge, U.K. Retrieved from http://www.withouthotair. com/, May 21, 2012.

Malczewski, J. 2004. GIS-based land-use suitability analysis: A critical overview. *Progress in Planning*, 62(1), 3–65.

McHarg, I. L. 1969. *Design with Nature*. New York: Wiley.

Miller, E. J., Sheppard, S. S., Grond, K., Kellett, R., Miller, N., Pond, E., Schroth, O. et al. 2011. *Visualizing Urban Futures: Geomatics Decision Support for Canadian Cities. Review Literature and Arts of the Americas* p. 195.

Ministry of Energy, Mines and Petroleum Resources. 2007. *The BC Energy Plan. A Vision for Clean Energy Leadership*. Victoria, British Columbia, Canada.

Nielsen, J., Innis, S., Pollock, L. K., Rhoads-Weaver, H., and A. Shutak. 2002. *Renewable Energy Atlas of the West Renewable Energy Atlas of the West: A Guide to the Region's Resource Potential*. Boulder, CO.

NRCan CEUDT. 2009. Natural Resources Canada Comprehensive Energy Use Database Tables. Retrieved from http://oee.rncan.gc.ca/corporate/statistics/ neud/dpa/trends_res_bc.cfm (accessed on September 30, 2009).

NRCan (Natural Resources Canada). 2003. *Solar Water Heating Systems: A Buyer's Guide*. Natural Resources Canada, Ottawa, Ontario, Canada.

NRCan (Natural Resources Canada) 2009. Solar atlas of Canada. Last accessed 2012, http://atlas.nrcan.gc.ca/site/english/maps/archives/5thedition/environment/ climate/mcr4076

Parfitt, B. 2011. *Making the Case for a Carbon Focus and Green Jobs in BC's Forest Industry*. Report for the Canadian Centre for Policy Alternatives. Vancouver, British Columbia, Canada.

Powell, G.W. 2009. Woody Biomass Crops: Biophysical Suitability and Environmental Assessments for the Quesnel Area. Prepared for Quesnel Community and Economic Development Corporation.

Province of British Columbia. 2007. *Bill 44—Greenhouse Gas Reduction Targets*. Retrieved from http://www.leg.bc.ca/38th3rd/3rd_read/gov44-3.htm (accessed on January 20, 2011).

Province of British Columbia. 2008. *Bill 27: Local Government (Green Communities) Statuses Amendment Act*. Retrieved from http://www.leg.bc.ca/38th4th/3rd_ read/gov27-3.htm (accessed on January 20, 2011).

Ratti, C., Baker, N., and K. Steemers. 2005. Energy consumption and urban texture. *Energy and Buildings*, 37(7), 762–776.

Salameh, M. G. 2003. Can renewable and unconventional energy sources bridge the global energy gap in the 21st century? *Applied Energy*, 75(1–2), 33–42.

Salter, J. 2008. *An Exploratory Look at Community Energy Planning: Participation and Information*. Unpublished comprehensive exam manuscript, University of British Columbia.

Sawyer, S., Brouillette, M., McAlenney, P., Haberle, M., and R. Wright. 2012. *Renewable Energy Atlas of Vermont*. Retrieved January 25, 2012, from www.vtenergyatlas.com

Selman, P. 2010. Learning to Love the Landscapes of Carbon-Neutrality. *Landscape Research*, 35(2), 157–171.

Smart Growth B. C. 2009. *Downtown Prince George Concept Plan*. Report prepared for the City of Prince George.

Smil, V. 2003. *Energy at the Crossroads: Global Perspectives and Uncertainties* (p. 452). MIT Press, Cambridge, MA.

St. Denis, G. and P. Parker. 2009. Community energy planning in Canada: The role of renewable energy. *Renewable and Sustainable Energy Reviews*, 13(8), 2088–2095.

Statistics Canada. 2006. Household Size by Census Metropolitan Area. Retrieved 2009, from http://www40.statcan.gc.ca/l01/cst01/famil53c-eng.htm (accessed on September 30, 2009).

Steinitz, C., Arias, H., Bassett, S., Flaxman, M., Goode, T., Maddock III, T., Mouat, D., Peiser, R., and Shearer, A. 2003. *Alternative Futures for Changing Landscapes: The Upper San Pedro River Basin in Arizona and Sonora*. Washington, D.C.: Island Press.

Stennes, B. and A. McBeath. 2006. *Bioenergy Options for Woody Feedstock: Are Trees Killed by Mountain Pine Beetle in British Columbia a Viable Bioenergy Resource?* Canadian Forest Service, Pacific Forestry Centre, Victoria, British Columbia, Canada.

Stremke, S. 2010. *Designing Sustainable Energy Landscapes. Concepts, Principles and Procedures*. Wageningen University, the Netherlands.

Stremke, S. and J. Koh. 2010. Ecological concepts and strategies with relevance to energy-conscious spatial planning and design. *Environment and Planning B: Planning and Design*, 37(3), 518–532.

Stremke, S. and J. Koh. 2011. Integration of ecological and thermodynamic concepts in the design of sustainable energy landscapes. *Landscape Journal*, 30(2), 194–213.

Stremke, S., van den Dobbelsteen, A., and J. Koh. 2011. Exergy landscapes: Exploration of second-law thinking towards sustainable landscape design. *International Journal of Exergy*, 8(2), 148–174.

Stremke, S., van Etteger, R., de Waal, R.M., de Haan, H.J., Basta, C., and M. Andela (Eds.). 2011. *Beyond Fossils*. Envisioning desired futures for two sustainable energy islands in the Dutch delta region, Wageningen, the Netherlands: Wageningen University.

Tooke, T. R., Coops, N. C., and Voogt, J. A. 2009. Assessment of urban tree shade using fused LIDAR and high spatial resolution imagery. *2009 Joint Urban Remote Sensing Event*. IEEE.

Voogt, J. and T. Oke. 1997. Complete urban surface temperatures. *Journal of Applied Meteorology*, 36(9), 1117–1132.

Wilson, M. 2008. *Community Energy & Emissions Planning: A Guide for B.C. Local Governments*. Vancouver, British Columbia, Canada.

Wissen Hayek and Grêt-Regamey. 2009. Advanced analysis of spatial multi-functionality to determine regional potentials for renewable energies. Presentation at *REAL CORP 2009*. Sitges.

Wissen Hayek and Grêt-Regamey. 2012. Identifying the regional potential for renewable energy systems using ecosystem services and landscape visualizations Renewable energy systems—A new pressure on landscape development. Presentation at *IALE 2012*. Beijing.

16

Initiating and Analyzing Renewable Energy Transitions in Germany: The District, Village, and Farm Scale

Peter Schmuck, Marianne Karpenstein-Machan, and André Wüste

CONTENTS

16.1 Introduction

The transition from fossil and nuclear fuels to renewable energy sources implies many questions such as "What is the appropriate role of bioenergy in the renewable energy mix?" and "How can renewable energies be made compatible with sustainability requirements?" Our main hypothesis is that the implementation of the *participation principle* simultaneously at three scales (i.e., the district, the village, and the farm scale) is a necessary precondition to enable and support sustainable energy transition. To answer these questions and to prove the hypothesis, it seems to be necessary to collaborate in both interdisciplinary and transdisciplinary teams.

This chapter will focus on an approach developed at the University of Göttingen since 1998. An approach that is combining interdisciplinary cooperation between scientists and close collaboration with practitioners outside university (i.e., transdisciplinary) to foster sustainable energy transitions in villages and districts and to overcome the fossil/nuclear primacy of today's energy generation with its detrimental consequences for society and nature.

We will start with proposing sustainability criteria of bioenergy projects as a framework of orientation, because not all renewable energy projects are in line with sustainability requirements. Then we will describe the essence of our approach and, as a historical background, the results of the first years of research prior to the year 2009—when the project focused on in this chapter started. The main part of this chapter will describe the actual project "Sustainable use of bioenergy: Bridging climate protection, nature conservation and society." We will report how we initiated the cooperation with partner districts and model farms, and how planning workshops were performed between 2009 and 2011. In Section 16.3 we will give an impression regarding the concrete changes in the model farms, the village of Immenrode and in the district of Wolfenbüttel. We will conclude with some recommendations regarding the active role of scientists in the transformation toward a sustainable energy future in general, and energy landscapes designed according to sustainability criteria in particular.

16.1.1 Principles of Sustainable Development

Why should one reflect about sustainable development principles before starting renewable energy projects? We think because renewable energy projects may defy the norm of sustainability. To give an example, bioenergy can positively contribute to climate goals and rural development; however, if not implemented carefully with regard to ecological criteria and not in cooperation with the people concerned (for instance, mega bioenergy generation sites associated with monoculture and long transport distances), it may exacerbate the degradation of soil, water bodies, and ecosystems and even increase greenhouse gas (GHG) emissions. Therefore, in our communal bioenergy projects, we followed a holistic approach based on ecological, economic, technical, and social guidelines, which we are trying to consider simultaneously.

The concept of "sustainable development," directed at intra- and inter-generational justice in sharing life chances, serves as the base of our work.* We formulated six principles, which, in our opinion, are required to reach the general goal of sustainable development[†]:

1. The *respect principle* assigns all forms of life the same right to live (Schweitzer 1991; Gorke 1999). This principle should be considered in energy crop cultivation concepts by combining the different utilization options of landscapes to produce food, fodder, and energy with requirements to support wild life (Karpenstein-Machan 2011). Integrated cultivation concepts are created with the goal to overcome the antagonism between the agricultural utilization of farmland and the protection of landscape. They are designed to harmonize utilization/production and protection of landscapes. This may happen if the production of energy crops is explicitly avoiding the loss of biodiversity, the degradation of land, and the pollution of ground water.

2. The *precautionary principle* is aiming at avoiding irreversible human caused changes in our biospheric balance. According to Komiyama and Takeuchi, "The primary objective is ... to achieve, as soon as possible, substantial improvements in ... the interaction between the sciences and decision-making, using the precautionary approach." (2006, p. 5). With respect to energy crops, the use of genetically modified crops should be refused due to environmental, social, and economic risks coming along with genetic engineering technology. Using such technologies in our opinion is clearly injuring the precautionary principle.

3. The *justice* or *sufficiency* principle. If available resources shall be distributed fairly, humankind needs new life patterns based on substantially lower consumption of resources, compared to the consumption level in today's industrialized countries. Instead of high-level material consumption, nonmaterial potentials for a fulfilling human life have to be unfolded; for example, creativity, social networking, or collective art activities. In many of our energy village projects, inhabitants try to become self-sufficient based on their own renewable sources and unfold social activities contributing to an inner fulfillment and satisfaction.[‡]

4. The *principle of participation* is ensuring a broad public population to take part in searching for, evaluating, and implementing sustainable ways of life. Many chapters in the Agenda 21 emphasize this principle (UNO 1992, Chapters 14.22, 31.1, 35.6, 36.9). In our

---

* Some limitations of the concept are discussed in Schmuck and Schultz (2002).
† The principles can also be found here: www.izne.uni-goettingen.de
‡ For the theoretical base see Schmuck and Sheldon (2001), for data Eigner-Thiel and Schmuck (2010).

projects, citizens are involved in the development of renewable energy solutions. We, the scientists, invite villagers to participate in the planning, investment, and implementation of renewable energy plants in order to gain a high rate of acceptance and commitment to these new installations among the rural population.

5. The *efficiency principle* is directed at avoiding the wasting of limited resources. In our projects, we recommend to make use of locally available animal manure and other organic residuals as energy sources in order to reduce land-use competition and to curb GHG emissions. We recommend cultivating energy crops only in the surroundings of the study area to avoid transports violating the efficiency principle. Further, in the case of biogas, we only recommend combined heat and power systems utilizing both products and, by doing so, displacing efficiently the necessity of fossil fuels.

6. The *consistency principle* is directed at replacing finite resources (actually the main base of our economy) by renewable sources and minimizing waste production, being in line with naturally occurring biospheric cycles. In the energy sector, renewable sources have to replace fossil fuels and nuclear power to safeguard supply as well as to reduce GHG emissions. Digestates and ashes left from energy generation have to be used as fertilizer in the fields to recycle nutrients and to reduce the use of mineral fertilizer.

16.1.2 Göttingen Approach of Sustainability Science

Our approach of sustainability science, based on several international political agreements regarding sustainable development since Rio de Janeiro (1992) and on proposals from the scientific community calling for a new type of scientific activities, consists of seven elements comprising the specific tasks scientists have to fulfill during the whole research cycle.* The first task is being defined as the traditional scientists' role: performing research to produce scientific knowledge. The other six tasks are different problem-solving activities, based on the application of scientific knowledge, and should be engaged by inter- and transdisciplinary teams. The research activities are distributed over the whole cycle, whereas the problem-solving activities are modeled as consecutive. Let us start with the detailed description of the latter.

16.1.2.1 Problem-Solving Activities

(1) The *selection of a critical global problem* constitutes the first activity. Looking in this early phase at the global level ensures that the really

---

* For more details, see Schmuck et al. (in press).

serious problems get priority.* (2) For the second step of problem solving— *formulation of possible solutions*—the district level seems to be appropriate because scientists usually do not have the power or the mission to change world politics in a direct way. Therefore, the creative process has to be started in a particular area, where an active group of scientists lives and works. (3) *Search for political and financial support.* The great majority of scientists are specialists in particular subjects of science and do not have the explicit assignment nor the financial means to perform inter- and transdisciplinary sustainability science projects. Nevertheless, it is critical to get both political and financial support for such projects. To get this support, it is helpful to refer to international and (where applicable) to national political agreements regarding the promotion of sustainable development. Agenda 21 may serve as an example: "Governments ... with the cooperation of ... non-governmental organizations, should ... promote the research, development, transfer and use of technologies and practices for environmentally sound energy systems, including new and renewable energy systems" (UNO 1992, Chapter 9.12). There are two ways for scientists to become active in sustainability science: One is to wait until governments or funding agencies create funding programs for sustainability research. The other possibility is for scientists to take the first step: to propose ideas for sustainability research to the political authorities. This is what happened in our project. Details will be described in the next section. (4) *Search for practice partners.* The next step comprises the motivation of practice partners—decision makers, stakeholders, and citizens outside the research community—to collaborate in the sustainability project. (5) *Perform pilot project on local level.* Scientists focuses on providing scientifically based advice to the practitioners during the implementation of the project. "The transcendent challenge is to help promote the relatively 'local' (place- or enterprise-based) dialogues from which meaningful priorities can emerge, and to put in place the local support systems that will allow those priorities to be implemented" (Clark and Dickson 2003, p. 8059). (6) *Transfer toward the district, national, and global level.* After the successful accomplishment of a/multiple pilot project(s), the task is to actively support the transfer of the knowledge gained in the pilot project(s) to other districts (Ruppert et al. 2010) and where applicable to other countries.

16.1.2.2 Research Activities

Results of traditional research are, as far as available, the base of all problem-solving activities. The selection of critical problems, for example, requires a

---

* Looking at the urgency of global problems such as climate change, one may conclude that the world's scientists should bundle their energy on the most pressing problems, if they want to contribute to the prevention of further real and predictable catastrophes like the 2010 oil disaster in the gulf of Mexico, the 2011 nuclear disaster in Fukushima, or the ongoing melting of glaciers.

close consideration. To enable corresponding evaluations, the actual scientific knowledge has to be checked, for instance regarding the potential problem fields of water, energy, health, agriculture, and biodiversity.* Later on, during the problem-solving process, scientific and technological knowledge should guide decision making. Before first alternative demonstration models are realized, hypotheses regarding consequences, for instance, within longitudinal research, should be designed and may be tested after the proposals are implemented. This also means, new scientific knowledge is produced based on alternative demonstration models.

16.1.3 Results until 2009

Following the aforementioned framework, our team initiated and facilitated the development of the first German bioenergy village between 2000 and 2005: the village Jühnde near Göttingen (Schmuck et al. 2007). The positive experiences in this successful pilot project have been communicated via public relations activities[†] and reached many interested members of Germany's rural population, especially farmers, village, and small town majors as well as district administrators.

As a consequence, several further projects were initiated. From 2006 until 2009, four more villages in the district of Göttingen, namely, Reiffenhausen, Wollbrandshausen, Krebeck, and Barlissen, again supported by the team of scientists, followed Jühnde and implemented similar communally organized bioenergy installations. Meanwhile, at the time of writing, there are more than 50 villages in Germany that changed their electricity and heat supply to make use of locally available bioenergy sources, many of them inspired by the example of Jühnde.[‡]

In parallel to this "success story," a couple of scientific studies analyzed the changes in pre/postexperimental designs from different perspectives (see, e.g., Eigner-Thiel and Schmuck 2010; Karpenstein-Machan and Schmuck 2010; Schmuck et al. in press).

16.2 Lower Saxony Sustainable Bioenergy Districts Project (2009–2012)

In the main part of this chapter, we report about the first steps of an ongoing project, which is focused no longer on the village level but on the district level instead. The ministry for science and culture of the federal state Lower

---

* They represent the so-called *WEHAB* priority targets as defined on the Johannesburg Summit and described by Clark and Dickson (2003, p. 8060).
† Such as mass media, scientific publications, practical guides formulating generalized principles for the conversion process.
‡ Also see: www.wege-zum-bioenergiedorf.de

Saxony invited our group to continue our work on the district level. Our basic idea is, similar to the bioenergy villages, to support a few selected and well-suited districts in their sustainable energy transition with the goal that other districts, later on, will follow the approach of the pilot districts. It is of paramount importance to identify districts with high realization chances, because a failing project (a "project ruin" instead of a pilot model) might have detrimental effects on the societal position toward sustainable energy transition.

16.2.1 Selection of Partner Districts

The challenge of finding partners willing to cooperate within the project was solved as follows: Based on our experiences on the level of villages, we started a competition between the 38 districts in the federal state Lower Saxony. The price for three winning districts was to get support from us in the further process of transition toward a sustainable energy system. This has the advantage that the activity and responsibility for starting the cooperation are balanced between the practitioners on the one side and the scientists on the other, ensuring that districts with the best motivated political decision makers are identified. In their application, the participating districts had to report on the status quo of renewable energy development, to indicate the political will to change toward renewable energy, and to list the strengths and weaknesses on that way in their own district. Twelve districts applied for cooperation.

In 10 districts, we organized meetings with leading representatives of agriculture and forestry, local politics, administration, nature protection, electric power companies, bioenergy networks, and with planners and operators of biogas plants as well as heads of relevant research projects. To compare the interested districts systematically and to select three partner districts, we asked (in these meetings) for many details and evaluated input by means of a systematic list of five suitability criteria that were developed by the team of scientists. The criteria and corresponding indicators are as follows (also see Table 16.1):

1. *Political will* to foster renewable energy transition including bioenergy and to perform climate protection activities: Indicators were (a) political resolutions supported or initiated by the highest political level of the district, (b) development of instruments directed at measuring environmental data (landscape framework plan), and (c) active support of sustainability projects by the administration.
2. *Motivation of main actors* to cooperate with a team of scientists: As indicators served (a) the variability of stakeholder groups present in the first meeting, (b) the interest in the topics of our research project, (c) the verbally expressed readiness for discussion, and (d) the cooperation during the initial meeting.

TABLE 16.1

Scores of the Five Main Criteria for the 10 Districts (A–K), Overall Sum, and Ranking[a]

	(1)	(2)	(3)	(4)	(5)		
District	Political Will	Motivation of Main Actors	Fitting with Contact Persons	Farmers' Readiness for Cooperation	Social Cohesion	Overall Sum	Ranking
A	−2	2	3	4	2	9	8
B	7	7	5	4	3	26	1
C	−1	−3	−2	−2	1	−7	10
D	3	7	6	3	3	22	2
E	5.5	1	1	2	0	9.5	7
F	7.5	0	7	1	3	18.5	4
G	6	7	3	1	3	20	3
H	2	2	2	−2	2	6	6
I	5.5	2	4	−2	2	11.5	5
K	0	0	2	−2	−1	−1	9

[a] Criteria with sums of zero or lower are marked with gray color, indicating that the criterion is not or at best partially fulfilled. Bold numbers indicate the highest overall sums and best ranking values.

3. *Interpersonal fitting* and readiness for cooperation/reliability of the contact persons: As indicators, we used (a) the quality of organization of the first meeting and (b) the completeness and timeliness of application documents. Note that the interpersonal fitting was indicated by our subjective feeling of sympathy with our contact persons.

4. *Farmers' readiness for cooperation* and reliability: Indicators were (a) the willingness of farmers to share data regarding agricultural practices and (b) readiness of single farmers to cooperate with our team as a "model farm for bioenergy."

5. *Social cohesion* of different stakeholder groups in the district: Indicators here were (a) the presence of relevant actors in the first meeting, (b) their engagement during the meeting, and (c) the presence of positive and constructive discussions in the first meeting.

Table 16.1 shows the comparison between the districts regarding the aforementioned primary criteria. The scores in the table show to which extent the criteria were fulfilled: negative scores indicate that the criteria are not fulfilled (averaged over the single indicators). The higher the positive scores, the more the single indicators were fulfilled, respectively. Where normative evaluations from our side were necessary, the three authors, who were present at all 10 meetings, made the corresponding ratings individually and independently from each other. The criteria and the results

FIGURE 16.1
Partner districts Wolfenbüttel, Goslar, and Hannover in Southern Lower Saxony, Germany.

of evaluations were presented and discussed in several internal project meetings, contributing to a final consensus within the team on which districts were best suited.*

Table 16.1 shows that there are three districts (B, D, and G), which show in all criteria positive sum scores. These three districts also have the highest overall score thus forming the top three in the ranking. The districts selected for further cooperation are Wolfenbüttel, Goslar, and Hannover (Figure 16.1).

16.2.2 Selection of Model Farms

As mentioned earlier, we were looking for interested farmers to cooperate and study how to improve conventional energy crop systems toward more ecological sound systems. In each of the three districts, we started the cooperation with one interested farmer, respectively. Note that all cooperating farmers are owners or joint partners of a biogas plant. The aim of the concept "model farms" is to introduce new energy cultivations concepts and to develop jointly with the farmers sustainable cultivation concepts for bioenergy, food, and feed crops. The long-term goal is that these model farms act as "trailblazer" to give other farmers new impulses to change their cultivation systems toward more sustainability and higher productivity.

---

* For details of the procedure see Schmuck et al. (in press).

16.2.3 Performing of Planning Workshops

Between March 2010 and November 2011, a total of 15 planning workshops took place in the three districts. The general goals of these workshops were (1) to formulate goals and visions for renewable energy transition, (2) to share and discuss experiences with initiators of already existing successful and sustainable projects, and (3) to initiate model projects relevant for the district level. Each workshop lasted about 4 h and had between 15 and 75 participants. They were representatives of the different stakeholder groups like local politicians, district administration, farmers' associations, and citizens' initiatives. The general format of the workshops consisted of *short information input elements** and *future workshop elements.*[†] Two workshops focused on the forming of visions, five workshops on sustainable agricultural practices, six workshops on how to initiate renewable energy communities, and two more workshops were directed at specific topics.[‡] Note that all workshops will be evaluated in the final phase of our project by analyzing the active participants' opinions regarding the adequacy and effectiveness of the elements we used in the workshops.

16.3 Results on Three Scales: The District, the Village, and the Individual Farm

In this section, we describe results of our research starting with a district (i.e., Wolfenbüttel), followed by a village (i.e., Immenrode), and some findings on the work at the scale of the individual farms.

16.3.1 Initiating the Bioenergy District Wolfenbüttel

One of the goals formulated and prioritized by the participants of "vision workshop" in the district of Wolfenbüttel was to initiate several bioenergy villages following the Jühnde model. In one of the next workshops, an expert who had facilitated the transformation of two villages gave a motivating presentation on the chances of communal use of bioenergy. Two months later, public servants and local politicians were invited on a bus trip to visit several bioenergy villages near Göttingen. Mayors and

---

* Across the 15 workshops, 28 input elements were delivered by scientists and 24 by experts from different practical fields.
† Participatory arrangements included circle of chairs, metaplan technique, visualized priority indications for goals and realization options. For details, see Jungk and Müllert (1991).
‡ Namely, using contaminated soils for renewable energy solutions and re-municipalization/ de-privatization of electricity grids.

project managers presented the history and the actual state of renewable energy use in their villages. Thus, the intention to start a similar initiative in their own district was consolidated by the decision makers who were impressed by the living examples. Few weeks later, the districts' authorities decided to give financial support for technical and social feasibility studies for bioenergy villages in the district Wolfenbüttel. In 2011, a starting meeting for interested villagers and farmers took place, giving them information on how to apply for support during the transition. A further workshop served to clarify specific questions regarding chances and possible problems related to bioenergy. The feasibility studies for suited villages in the district will be conducted in 2012 by an engineering office and our team.

16.3.2 Planning the Renewable Energy Village Immenrode

The village Immenrode (district Goslar) is located in the north of the Harz Mountains, in the eastern part of Lower Saxony. Immenrode has about 1690 inhabitants and belongs to the municipality of Vienenburg.* Currently, there exists a biogas plant, a number of wind turbines, and some solar plants (photovoltaic [PV]), all privately owned.

During the first workshop in the district Goslar—at the larger scale—the farmer who runs the biogas plant and the mayor of Immenrode formulated their vision of a self-sustaining village Immenrode with a complete supply of electricity and heat based on locally available renewable energy sources. Henceforward, this vision was a main topic in all of the moderated workshops in the district Goslar because the districts' stakeholders participating in the workshops agreed that a "lighthouse project Immenrode" could be a powerful example for the entire district.

The intention of the actors in Immenrode is to get their own, stable energy supply by directing all available renewable energy flows (bioenergy, wind, sun, and water) into one combined power plant. Furthermore, they want to buy the local electric power grid and operate it, to be fully independent from the four big energy companies in Germany.

For the second workshop, our team invited an expert from the village of Feldheim in the federal state of Brandenburg. Feldheim is the first village in Germany with a community-owned electricity grid that is exclusively powered by locally available renewable energies (i.e., bioenergy, wind, and solar energy). In addition, a biogas plant combined with a local heat network delivers heat to households and public buildings. After the presentation, possible steps for the implementation of the goals in Immenrode were identified and discussed amongst the members of the planning workshop. Another idea that emerged during the second workshop was the foundation of a participative

---

* Also see www.vienenburg.eu/ortschaften/immenrode.html

operating company involving the inhabitants of Immenrode. Two important numbers that would need to be studied was the amount of electricity generated by the biogas plant and the four wind turbines as well as the total electricity demand for the village.

In the third workshop, the steps toward energy transition became more concrete. The amount of electricity generated by renewable sources in Immenrode, we had found out, is already twice as high as the total demand of the 850 households in the village.

The fourth and (for now) last workshop in the district Goslar took place in Immenrode and started with a presentation about combined power plants by an expert from the Fraunhofer Institute for Wind Energy and Energy System Technology in Kassel.* Then, a second expert shared his experiences with the wind park Dardesheim in the German federal state of Saxony-Anhalt.[†] Shortly after the German reunification, in 1993, some pioneers had started renewable energy generation, first by installing wind power turbines. Later, large PV roof panels, solar boiler heating systems, and a biogas plant were installed. In Dardesheim, the project initiators have achieved a broad acceptance and high project identification by an intensive information and participation process. Another presentation was given by an engineer living in Immenrode. He presented his computer-based simulation tool that can help to identify the potentials of renewable energy sources, in order to become 100% self-sufficient.[‡]

Inspired by these presentations, among others, the participants of the planning workshop discussed and stated further steps to be taken in Immenrode. They decided to involve the neighboring village Weddingen in the transition process. The mayors of the villages Immenrode and Weddingen decided to organize a village meeting to inform inhabitants and to engage more citizens in the project. The owner of the wind turbines in Immenrode intents to "repower" his installations[§] and expressed the willingness to offer opportunities for financial participation of the inhabitants. Finally, our university team and a representative of the association *Verein Goslar mit Energie* ensured the continuous support for the projects in Immenrode and Weddingen (Figure 16.2).

Few months later, the mayors of Immenrode and Weddingen organized a village meeting (now without assistance of our team) to inform and recruit more engaged citizens. Here, a local engineer presented the vision for the "renewable energy self-sustaining villages Immenrode and Weddingen." In his presentation, he described the status quo of renewable energy generation in these two villages and showed further potentials for renewable energy necessary to reach this goal. During a very positive and constructive

---

* For details, see www.kombikraftwerk.de and Rohrig (2010).
[†] Dardesheim is located only a few kilometers away from Immenrode but behind the former border between West Germany and the GDR.
[‡] Personal communication Hans-Heinrich Schmidt-Kanefendt (2011).
[§] That means to replace old turbines by stronger ones.

FIGURE 16.2
Planning workshop in Immenrode.

discussion, the plenum agreed to support further activities in Immenrode and Weddingen. At the end of the village meeting, a few eager citizens founded a working group to create new ideas and to start the implementation of the renewable energy self-sustaining villages Immenrode and Weddingen.

16.3.3 Activities on the Level of the Individual Farm

16.3.3.1 Analyses of Main Ecological Challenges

The three farms produce market, fodder, and energy crops for biogas use under heterogeneous soil and climate conditions. Soil types range from sand, loam, lossial loam to organic soils. The heterogeneity in soils is reflected in the soil fertility code, which ranges from 30 to 100. Currently, the most frequently cultivated crops are maize, winter wheat, and sugar beet. Due to the one-sided crop rotation with only a few crops (mainly maize, sugar beet, and wheat), many problems arise.

With maize and sugar beet, two humus wasting crops are in one rotation. Due to the relatively low competitive power of young maize plants against weeds, weed management with herbicides or mechanical weed removal is necessary. The uncovered soil in the beginning and the long vegetation period of maize until autumn leads to humus degradation. Furthermore, the amounts of residuals that remain on the field after harvest are very low. Especially on organic soils, maize and sugar beet cultivation lead to strong humus degradation and enormous GHG emissions. Further problems are connected with "tight" rotations including maize and sugar beet. Crop-specific pests and diseases are supported by rotations with poor diversity or monoculture. The loam and organic soils are endangered for soil compaction, and danger of nitrate leaching is very high on karst soils and sandy soils.[*]

16.3.3.2 Solutions for More Ecologically Sound Crop Rotations

New energy crops,[†] cultivation systems,[‡] and perennials were proposed and proofed in the model farms in order to improve landscape quality, biodiversity, food production, and other life-supporting ecosystem services (see Figures 16.3 and 16.4). Based on these farm trials, models for sustainable crop rotations were developed for three "biogas farms."

FIGURE 16.3
Red fescue (*Festuca rubra*) undersown in maize.

---

[*] For more information, please see Karpenstein-Machan (1997, 2001, 2004), Finckh and Karpenstein-Machan (2002), and Karpenstein-Machan and Weber (2010).
[†] For example, winter triticale, winter rye, sunflower, and sorghum.
[‡] For example, double cropping, mixed cropping, undersown crops, and flowering bands.

FIGURE 16.4
Herbicide-free flowering bands in maize.

At first, crop rotations were agronomically optimized to increase biodiversity, to stabilize humus content, to reduce pesticide input and nitrate leaching, and to stabilize biomass and crop yield against biotic and abiotic stress. Figure 16.5 shows an example for a mixed crop rotation with food and energy crops, optimized for humus stability and crop diversity. With winter cereals combined with undersown field grass (harvested for biomass use) and catch crops over wintertime, the humus content can be stabilized, and the crop diversity can be increased. Together with maize (fodder or biomass for biogas use) and market crops (winter wheat and sugar beet), a sustainable crop rotation can be created. On ecological sensitive soils, perennial crops (*Silphium perfoliatum* L.) and wild herbs for biogas use were planted (see Figure 16.6).

Optimization of the crop rotations leads to higher crop diversity and better humus balance. Calculating the marginal income of the farms before and after the reorganization of the crop rotations shows that, taking into consideration the results of the field trials and positive rotation effects, the new crop rotations do not reduce farm income (Karpenstein-Machan 2011). In the long run, these crop rotations result in better protection against biotic and abiotic stress and stabilize biomass and crop yield (Karpenstein-Machan and Finckh 2002; Gross et al. 2011). In addition, energy farms can provide other

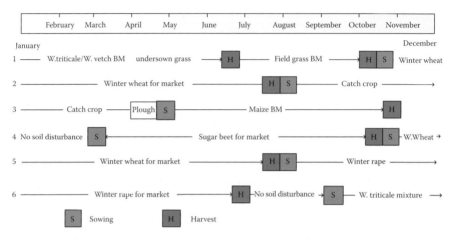

FIGURE 16.5
Example for mixed crop rotation with energy and food/fodder crops for market (BM, biomass crops).

FIGURE 16.6
Perennial crop Silphie (*Silphium perfoliatum* L.) grows in second year.

"environmental services" such as herbicide-free buffer strips and flowering bands (Figure 16.4) that help to improve habitat quality for many wildlife species in the open farm lands.

16.3.3.3 Implementation Chances for Ecologically Sound Crop Production

Proposals for new crop rotations were developed together with the farmers of the model farms. With increasing experience, farmers plan to reorganize their farms step by step to more sustainable concepts.

To increase implementation chances on the model farms and to enable the transfer of experience to other farmers, visits to the model farms were organized. Members of various nature protection organizations, politicians, civil servants, journalists, and villagers took part in the guided tours. These meetings of people with different backgrounds enriched the discussion on a number of topics, also beyond crop cultivation, and established a better understanding for different positions. Furthermore, if farmers get a good press, the motivation increases to break new ground toward more sustainable cultivation concepts.

On the district scale, "energy farmers" and landscape planners should work together to facilitate sustainable landscape development, in general, and a sustainable implementation of renewable energy technologies, in particular (e.g., biogas plants, wood-chip fired plants, and ethanol plants). This way, different societal groups and disciplines can influence the process and support the development of sustainable energy farms, bioenergy villages, and self-sufficient districts.

16.4 Conclusion and Outlook

Across the projects outlined in this chapter, the authors, in addition to the traditional role of scientists as objective analyzers, have been active as initiators of transitions toward sustainable energy systems. Our experiences demonstrate that it is possible for scientists to support (1) the envisioning processes with regard to renewable energy goals, (2) the broad participation in the implementation of new modes of energy generation, and (3) the identification and furtherance of sustainability criteria for renewable energies. The double role of scientists working in the field of sustainability research may be considered as an effective approach to contribute to act adequately regarding the challenges of the global ecological and social crisis.*

The results of the case studies described in this chapter show how scientists from different disciplines cooperate with different societal groups to promote the transition toward renewable energies at the scale of the individ-

---

* Also see Sheldon et al. (2000) and Schmuck et al. (in press).

ual farm, villages, and districts. Our contribution demonstrates how action research, a tool with a long tradition in social science,* may be embedded in the broader framework of sustainability science and thus contribute to the emergence of well-accepted (rural) sustainable energy landscapes.

Future effort has to be directed toward controversial strategic questions with high relevance for the integration of renewable energy technologies in future landscapes. One decision that has to be made is to choose between primarily centralized large-scale energy technologies such as the DESERTEC project and a more decentralized energy landscape with many medium- to small-scale energy technologies like biogas plants and CHPs.

For rural landscapes, we favor decentralized renewable energy technologies because we are convinced that only these *require and allow* broad participation—one of the key sustainability criteria we outlined at the outset of this chapter. The more people are discussing the advantages and problems of future options, the bigger is the chance to find optimized combinations of renewable energy technologies, which are then, by the way, much more accepted in the population as compared to the highly controversial "super-grids" (or plans for "carbon capture and storage"). The latter are associated with centralized decision making that, in many parts of Germany, resulted in acrimonious resistance. Supporting the development of scenarios for sustainable energy landscapes, in that way, also presents a contribution to peaceful, socially fair, and ecologically sound life patterns in the world of tomorrow.

References

Brydon-Miller, M., Greenwood, D., and Maguire, P. (2003). Why action research? *Action Research*, 1, 9–28.

Clark, W. and Dickson, N. (2003). Sustainability science: The emerging research program. *Proceedings of the National Academy of Sciences USA*, 100, 8059–8061.

Eigner-Thiel, S. and Schmuck, P. (2010). Gemeinschaftliches Engagement für das Bioenergiedorf Jühnde—Ergebnisse einer Längsschnittstudie zu psychologischen Auswirkungen auf die Dorfbevölkerung. *Zeitschrift für Umweltpsychologie*, 14, 98–120.

Finckh, M.R. and Karpenstein-Machan, M. (2002). Intercropping for pest management. In D. Pimentel (Ed.), *Dekker Encyclopedia of Pest Management* (pp. 423–425). New York: Encyclopedia Entry.

Gorke, M. (1999). *Artensterben. Von der ökologischen Theorie zum Eigenwert der Natur.* Stuttgart, Germany: Klett-Cotta.

Gross, G., Krause, A., Lenssen, C., Müller, U., von Buttlar, C., Karpenstein-Machan, M., Bauböck, R. et al. (2011). *Klimafolgenmanagement in der Metropolregion.* Hannover, Germany: GeoBerichte, p. 18.

---

* Also see Lewin (1946), Whyte (1991), and Brydon-Miller et al. (2003).

Jungk, R. and Müllert, N. (1991). *Zukunftswerkstätten. Mit Phantasie gegen Routine und Resignation. (Future Workshops. How to Create Desirable Futures).* München, Germany: Heyne.

Karpenstein-Machan, M. (1997). *Konzepte für den Energiepflanzenbau.* Frankfurt, Germany: DLG-Verlags-GmbH.

Karpenstein-Machan, M. (2001). Sustainable cultivation concepts for domestic energy production from biomass. *Critical Reviews of Plant Science*, Special Issue on *Bioenergy*, 20(1), 1–14.

Karpenstein-Machan, M. (2004). Neue Perspektiven für den Naturschutz durch einen ökologisch ausgerichteten Energiepflanzenbau. *Naturschutz und Landschaftsplanung*, 36, 2.

Karpenstein-Machan, M. (2011). Implementation of integrative energy crop cultivation concepts on biogas farms. In M. Savolainen (Ed.), *Proceedings of the International Nordic Bioenergy Conference 2011* (pp. 127–133). Jyvaskyla, Finland: Finbio Publication 51. ISBN 978-952-5135-51-0.

Karpenstein-Machan, M. and Finckh, M. R. (2002). Crop diversity for pest management. In D. Pimentel (Ed.), *Dekker Encyclopedia of Pest Management* (pp. 162–165). New York: Encyclopedia Entry.

Karpenstein-Machan, M. and Schmuck, P. (2010). The bioenergy village in Germany: A lighthouse project for sustainable energy production in rural areas. In M. Osaki, A. Braimoh, and K. Nakagami (Eds.), *Local Perspectives on Bioproduction, Ecosystems and Humanity* (pp. 184–194). Tokyo, Japan: United Nations University Press.

Karpenstein-Machan, M. and Weber, C. (2010). Energiepflanzenanbau für Biogasanlagen: Veränderung der Fruchtfolgen und der Bewirtschaftung von Ackerflächen in Niedersachsen. *Naturschutz und Landschaftsplanung*, 42(10), 313–320.

Komiyama, H. and Takeushi, K. (2006). Sustainability science: Building a new discipline. *Sustainability Science*, 1, 1–6.

Lewin, K. (1946). Action research and minority problems. *Journal of Social Issues*, 2, 34–46.

Rohrig, K. (2010). Kombikraftwerk 2.0. In Agentur für Erneuerbare Energien (Ed.), *Kraftwerke für jedermann. Sammelband Dezentralität* (pp. 76–79). Berlin, Germany: Agentur für erneuerbare Energie.

Ruppert, H., Eigner-Thiel, S., Girschner, W., Karpenstein-Machan, M., Roland, F., Ruwisch, V., Sauer, B., and Schmuck, P. (2010). *Wege zum Bioenergiedorf. Leitfaden.* Gülzow, Germany: Fachagentur für Nachwachsende Rohstoffe.

Schmuck, P., Eigner-Thiel, S., Karpenstein-Machan, M., Ruwisch, V., Sauer, B., Girschner, W., Roland, V. et al., (2007). Das Bioenergiedorf Jühnde (The bioenergy village Jühnde). In G. Altner, H. Leitschuh-Fecht, G. Michelsen, U. Simonis, and E.U. von Weizsäcker (Hrsg.) (Eds.), *Jahrbuch Ökologie 2007* (pp. 104–112). München, Germany: C.H. Beck Verlag.

Schmuck, P., Eigner-Thiel, S., Karpenstein-Machan, M., Sauer, B., Roland, F., and Ruppert, H. (in press). Bioenergy villages in Germany: The history of promoting sustainable bioenergy projects within the "Göttingen Approach of Sustainability Science". In M. Kappas and H. Ruppert (Eds.), *Sustainable Bioenergy Production: An Integrated Approach.* Heidelberg, Germany: Springer.

Schmuck, P. and Schultz, W. (2002). Sustainable development as a challenge for psychology. In P. Schmuck and W. Schultz (Eds.), *Psychology of Sustainable Development* (pp. 3–19). Boston, MA: Kluwer Academic Publishers.

Schmuck, P. and Sheldon, K. (2001). Life goals and well-being: To the frontiers of life goal research. In P. Schmuck and K. Sheldon (Eds.), *Life Goals and Well-Being. Towards a Positive Psychology of Human Striving* (pp. 1–18). Seattle, WA: Hogrefe & Huber.

Schmuck, P., Wüste, A., and Karpenstein-Machan, M. (in press). Applying sustainability science principles of the Göttingen approach on initiating renewable energy solutions in three German regions. In M. Kappas and H. Ruppert (Eds.), *Sustainable Bioenergy Production: An Integrated Approach*. Heidelberg, Germany: Springer.

Schweitzer, A. (1991). *Menschlichkeit und Friede. Kleine philosophisch-ethische Texte.* Berlin, Germany: Verlags-Anstalt Union.

Sheldon, K., Schmuck, P., and Kasser, T. (2000). Is value-free science possible? *American Psychologist*, 55, 1152–1153.

United Nations Organization (UNO) (1992). Agenda 21. The Rio Declaration on Environment and Development. http://www.un.org/esa/dsd/agenda21/ (Retrieved May 3, 2011).

Whyte, W. (Ed.) (1991). *Participatory Action Research*. Thousand Oaks, CA: Sage Publications.

17

Energy-Conscious Planning Practice in Austria: Strategic Planning for Energy-Optimized Urban Structures

Gernot Stoeglehner and Michael Narodoslawsky

CONTENTS

17.1 Introduction

Energy supply is one of the key infrastructures of society. As already pointed out in Chapter 10, this system has to undergo substantial changes in order to increase the overall sustainability of society and economy. In the following sections, we show that including spatial dimensions in the analysis, visioning, and action planning for the restructuring of energy supplies and recognizing the effects spatial structures have on energy efficiency, energy saving, and renewable energy supplies in an integrated spatial and energy planning approach do not only offer new opportunities for the design of energy-optimized spatial structures but also support communities on the local and regional level in the participatory planning of energy landscapes.

In order to fulfill these goals, methods and tools have to be designed and implemented in practice that are considerably different from usually applied energy balances, energy flow analysis, etc. of energy planning. The tools shall add the spatial dimension to plan energy demand and energy supply, and translate energy issues in the language and spheres of knowledge of planners and stakeholders involved in spatial planning. Such planning methods have to model complex systems, consider a broad science base, and prepare an information base for participatory discussion and decision making, which also means that nonexperts can understand the results. These methods have to support sustainable decisions and, in order to achieve this, allow for social learning on the level of values and the level of facts. Earlier, Stoeglehner 2010a has referred to such methods as "strategic planning and assessment methods" (Stoeglehner 2010a).

The aim of this chapter is to define a normative framework for strategic planning and assessment methods based on our previous work and to give two examples from Austrian planning practice and discuss them in the light of the normative framework. We introduce two planning methods that were developed in the interdisciplinary research projects, *PlanVision* and *ELAS—Energetic Long-Term Analysis of Settlement Structures*, and designed as role models to implement this vision of strategic planning and assessment methods.

The chapter is structured in the following way: In Section 17.2, we explain our understanding of strategic planning and assessment methods and draw a normative framework. Section 17.3 introduces *Energy Zone Mapping*—a Geographic Information System (GIS) based method to determine the energy demand of settlement structures for room heating and hot water production as well as examining area-based potentials for district heating. Section 17.4 introduces the *ELAS-calculator*—a model for holistic determination and assessment of energy demand and supply of residential areas. Section 17.5 concludes the chapter.

17.2 Strategic Planning and Assessment Methods: A Normative Framework

Our understanding of strategic planning and assessment methods developed from work on decision making with ecological footprints (see Krotscheck and Narodoslawsky 1996, Narodoslawsky and Stoeglehner 2010, Stoglehner 2003, Stoeglehner and Narodoslawsky 2008, 2009) and on strategic environmental assessment in spatial planning (see Stoeglehner 2010a,b, Stoeglehner et al. 2009) and sustainability assessment (see Stoeglehner and Neugebauer in press). The framework we offer is normative and determines features decision-making methods should have, being aware that hardly any decision-making method will fulfill all criteria. We propose that strategic planning and assessment methods have five features:

1. *Be strategic*: Being strategic means to focus on the general picture, to be proactive, oriented on visions and objectives and future states, to consider alternatives (Noble 2000), and to aim for reaching positive effects instead of mitigating negative impacts of proposed plans

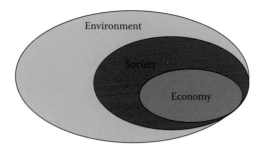

FIGURE 17.1
The "egg of sustainability": Need to integrate environmental, social and economic issues.

(Hacking and Guthrie 2008). Stoeglehner (2010a,b) proposes that especially system alternatives (addressing questions of demand, kinds of measures, technological options) offer the potential of being strategic, although site alternatives (location of proposed measures) and technical alternatives (technical design of a proposed measure on a specific location) have decreasing degrees of strategicness.* According to this definition, a system change as needed in the design of energy landscapes is strategic in its nature, and, therefore, the methods applied to design this change have to be as well.

2. *Be integrated*: In order to support sustainable development, decision-support methods should be embedded in a system of integrating environmental, social, and economic issues of sustainable development. We propose that, from a systems perspective, capacity limits have to be determined and considered in any decision for sustainable development as depicted in the "egg of sustainability" (Figure 17.1): environment is the overall system, society is a subsystem of the environment, and economy a subsystem of society (and environment). If one subsystem goes beyond the limitations of the surrounding system(s), the whole system is endangered and might collapse in the long run (see, for example, Birkmann 2000, Busch Lüty et al. 1990, Meadows et al. 1992). From this model, which can be associated with "strong sustainability" (see, e.g., Mayer 2008), we generated the model of the "indicator pyramid" (Stoeglehner and Narodoslawsky 2008; Figure 17.2) that has to be read top-down and shows that during a decision-making process the information load increases. We propose that on the level of preassessment, alternatives should be excluded that can be identified as unsustainable by applying general indicators. Unsustainability according to the egg of sustainability means that they do not comply with environmental, social, or economic

---

* For example, technological options (system alternatives) can only be implemented if suitable sites are available or if site-specific technical problems (e.g., protection of neighborhoods) can be solved.

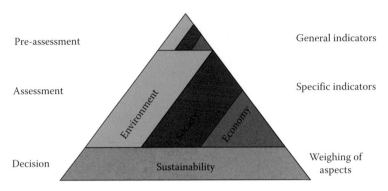

FIGURE 17.2
Model of the "indicator pyramid."

capacity limits. This screening of alternatives decreases the information load on decision makers. For the environmental pillar of sustainability, we suggest that modified or alternative calculation methods of the ecological footprint are suitable, for example, the *sustainable process index* (Krotscheck and Narodoslawsky 1996) or the *energy footprint* (Stoeglehner 2003). Only alternatives that pass the "unsustainability test" should be elaborated and assessed in detail. In this way, planning and assessment become effective, and the information load on decision makers can be considerably decreased.

3. *Support social learning*: Any planning and assessment process can be perceived as a social learning experience among the persons and institutions involved in such a process. Reflecting visions, actions, and anticipated consequences, social learning can happen on the level of facts and the level of values. Depending on the scope of the debate, two forms of learning can be distinguished: *single-loop learning* and *double-loop learning* (Argyris 1993, Innes and Booher 2000). Single-loop learning means learning about facts and is positioned between proposed actions and anticipated consequences. Without questioning the vision underlying and general outline of the planning process, actions will be modified or compensation measures proposed. These are typical outcomes for (environmental) assessment processes (see, for example, Wallington et al. 2007, Jha-Thakur et al. 2009, Stoeglehner 2010b). Double-loop learning goes deeper, especially when single-loop optimization does not provide for acceptable impacts, and addresses the underlying vision and values. Planning strategies are questioned starting from the demand for a planning proposal, so that completely different system alternatives might be proposed, different visions and objectives elaborated. Double-loop learning is needed if system changes as in the energy sector shall be successfully designed and implemented. From this model of learning, it also becomes clear that we do not separate planning and assessment methods, but that in our

understanding assessment methods directly support planning outcomes and should facilitate the designing of alternatives.

4. *Support rational–communicative planning*: Two dominating paradigms of planning theory are *rational planning* and *communicative planning*.* While in rational planning, decision making is an expert-driven, science-based exercise, in communicative planning dialogue between planners, decision makers, stakeholders, and the general public is put in the center. Critics of rational planning raise the objections that the science-based approaches fail because of lack of knowledge, incompleteness of information, and insufficient discussion of values present in a planning process. Against communicative planning is brought forward that interests not present in the planning dialogue are not heard (e.g., future generations), that the dialogue is dominated by the powerful, and that planning problems where consensus is hard to reach cannot be solved in communicative debate. Stoeglehner (2010b) proposed a rational–communicative planning paradigm, where the two planning paradigms are combined in order to utilize the benefits of both of them as well as overcoming at least some of the pitfalls, acknowledging that a planning process needs a sound science base and a deliberation of values. Applying a model of decision making, where a decision means to combine a level of facts and a level of values using certain aggregation rules, we outlined the following tasks: the communicative part should provide for an agreement about the system of values as well as the aggregation rules to be applied, whereas the rational part provided the facts and the application of the aggregation rules. Therefore, we propose that strategic planning and assessment methods allow for a deliberation of the value base, suggest and explain a decision-support model that has to be agreed for the planning process, and provide for a science-based collection and processing of information according to the decision-support model agreed before.

5. *Provide consistent guidance*: Decision making for sustainable development takes place on each level of decision making, from the global to the individual. In order to implement sustainable development, all decisions have to point in the same direction, as can be seen from the "decision-making pyramid" (Stoeglehner and Narodoslawsky 2008; Figure 17.3). Implementing sustainability strategies is not only a matter of frameworks and regulations, but also a matter of awareness and information for individual decisions. Each decision has an environmental, social, and economic dimension, which means that, for example, environmental decisions affect social and economic goals or economic

---

* The aim of this chapter is not to give a literature review about planning theory. A comprehensive review about planning theory in relation to environmental assessment is included, for example, in Lawrence (2000) and Stoeglehner (2010a,b).

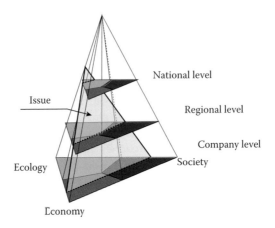

FIGURE 17.3
Model of the "decision-making pyramid."

decisions affect environmental and social targets. There is no pure environmental or social or economic decision, which raises complexity and the need for careful planning and assessment of alternatives on all levels of decision making. For shaping of sustainable energy landscapes at the regional and local level, decision-support methods should help to draw visions and action plans, assess potential consequences, comply with overall (global, supranational, or national) strategies, create awareness and guide decisions at the individual level toward fulfilling the agreed and desired vision, and implement the actions. This is especially needed, for example, for creating energy-optimized spatial structures, as individuals have to take decisions and invest private money and resources to implement public planning strategies. Even if regulations and subsidies are available, it still needs awareness of individual decision makers and the possibility to recognize the own contribution to an overall strategy to take action.

In the following two sections, we describe two methods for the design of sustainable (urban) energy landscapes. First, we outline the functioning of the method. Then, we describe an application. Finally, we discuss the method in the light of the normative framework laid out earlier.

17.3 Energy Zone Mapping

Energy Zone Mapping is a tool developed in the PlanVision project (Stoeglehner et al. 2011b) in order to determine the energy demand, energy-saving potentials, and the potential to supply urban neighborhoods with biomass-based district heating.* As laid out in Chapter 10, district heating is an important, environmentally friendly, and efficient possibility to supply room heating and warm water (or even cooling) and a key element to use waste heat in energy cascades.† Therefore, the motivation behind *Energy Zone Mapping* is to supply all areas where it is economically feasible and resource efficient with district heating and search for individual solutions only outside the district heating zones. Furthermore, we determine the economic feasibility according to the Austrian subsidy criteria for biomass district heating systems. *Energy Zone Mapping* was developed in a co-research process between the project team and the town of Freistadt—a rural small town with about 7500 inhabitants located in Upper Austria. The "community of practice" (see, e.g., Greenwood and Levin 1998) comprised the Committee for Spatial Planning and Energy Supply of the Municipal Council of Freistadt and the contracted spatial planners for the revision of the spatial development strategy and the local land use plan. All maps and plans in this chapter are taken from this Freistadt planning process. The results of *Energy Zone Mapping* also influenced the planning process as will be explained after the tool that has three steps:

1. *Estimation of the energy demand:* The energy demand for room heating and warm water is determined for neighborhoods, the so-called energy zones. The zones divide the urban fabric into areas that can be supplied by a heating grid. The energy survey includes residential, public, and commercial buildings. It is either based on primary surveys and interviews with the users or based on statistical data that are available in Austria and quite precisely informs about the energy demand per square meter floor space and is differentiated in different kinds of buildings and building periods. In GIS, we process the floor space of each building, multiply it with the energy demand per square meter (which includes warm water), and add the results for each building to the respective energy zone. The result is a map showing the energy demand of urban areas (Figure 17.4).

2. *Estimation of energy-saving potentials in buildings:* In the second step, the energy-saving potential is determined in scenarios. Figure 17.5 shows a scenario where all energy-saving potentials are exhausted, which would bring a 50% reduction of the energy demand for room heating and warm water production, including all residential, public, and commercial land uses. The saving potential can be easily calculated by multiplying the floor space of buildings with the target values differentiated in building periods and building types and calculating the delta between the actual and the targeted energy demand.

---

* The creation of the GIS tool, the underlying assumptions, and method of energy zone mapping are documented in the PlanVision project report so that GIS users like urban and regional planners can rebuild the tool by themselves or amend it to their specific situation.
† For more information on energy cascades, see Chapter 12 and Stremke et al. (2011).

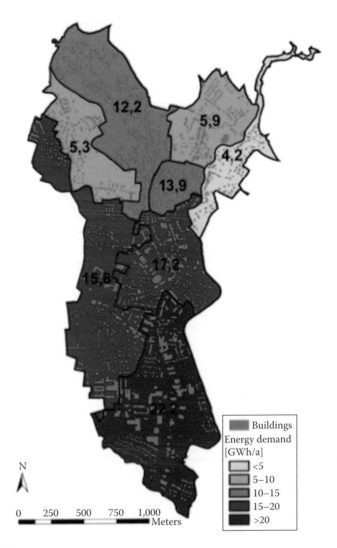

FIGURE 17.4
Map presenting the urban energy zones in Freistadt and their respective demand.

All values used are state of the art. This estimate is important to judge feasibility of solutions in the light of the prevailing trend to increase building efficiency.

3. *Determination of potential biomass district heating supply areas*: To judge the energy efficiency of biomass-based district heating supply, we use the subsidy guidelines for such systems. If subsidies can be acquired, the system is economically feasible, given the underlying parameters for the granting of subsidies. Basically, two criteria are important. The first one is the amount of energy that can be sold per meter grid

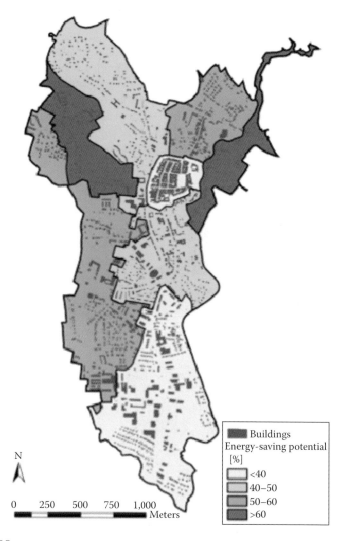

FIGURE 17.5
Map presenting scenario with maximum energy savings resulting in 50% reduction of energy demand.

length, and the measure is called *energy density*. At the moment, the energy density has to be over 900 kWh/m (including the main grid and the house connections) and is easy to calculate in GIS by dividing the energy demand of buildings by the length of the respective (potential) grid sections, which can be determined for areas like the energy zones shown previously, for certain streets, and might include all buildings existing and planned. It becomes obvious that the delineation of the areas determines the result. Optimization can take place in scenarios from an economic perspective, which means only the most feasible areas are chosen, or an environmental perspective, where the potentially largest economically feasible area is determined. The second criterion is the *energy loss* in the grid, which can also be calculated by dividing an approximated amount of energy loss per meter pipe length and the energy demand. In actual guidelines, the loss in the overall system has to be less than 20%. The result is shown in Figure 17.6, where potential supply areas are mapped in Freistadt based on the actual energy demand. What can

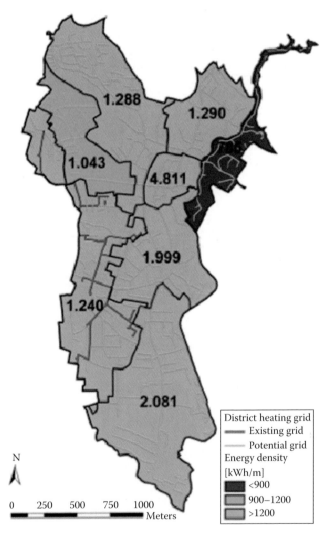

FIGURE 17.6
Map of supply areas in Freistadt based on the actual energy demand.

be seen is that taking current demand into account, practically the whole town can be supplied with district heating as the settlement structures are quite dense and multifunctional, so that big energy consumers* that are important for the grid design are present in large numbers, and small buildings can be supplied by connecting the bigger consumers.

What happened with the results in the Freistadt planning process? First, the town of Freistadt includes biomass district heating supply areas for the existing grid and biomass district heating development areas in the spatial development strategy and lays out that the further building development will take place in the district heating supply and development areas. Second, the decision makers realized the enormous potentials for climate and environmental protection and strengthening the regional economy. They started a survey of how many land owners would connect to a second district heating system in the north of Freistadt and committed themselves to connect all public buildings to the new district heating system.†

With the result they approached us and using the *Energy Zone Mapping tool*, we could tell them within a few minutes of processing time that with the number of already interested persons, a new district heating project can be started, leaving room in the future for other house owners to connect later. As a result, a site for a second biomass district heating plant in Freistadt was zoned in the land use plan, the detailed planning for the project started, with the start of the new system (grid and heat plant) being scheduled for 2012.

Our criteria for strategic planning and assessment methods are met in the following way:

1. The method works on the level of system alternatives, supporting the choice of technological options not only by testing the feasibility of district heating, but also by elaborating scenarios about the future energy demand. In that way, the result of measures to enforce certain energy standards of buildings can be seen.

2. The method is designed to integrate the following sustainability issues: climate protection by low-carbon energy supply from forest biomass, environmental friendliness of the proposed district heating systems (e.g., no emissions from combustion from individual heating systems), the social aspect of secure, convenient, and readily available regional energy, the macroeconomic aspect of regional resource use and income generation, and the microeconomic aspect of the feasibility of the district heating grid. With this information assessed, an optimal regional or local energy efficiency policy

---

* Such as residential multistory buildings, public buildings, and commercial buildings.
† Many buildings such as schools and the indoor pool are already connected to the existing biomass district heating system.

concerted with supply potentials for biomass district heating can be drafted, fine-tuned by taking individual opinions into account, decided, and implemented as in the Freistadt case.

3. The method supports social learning about the reduction of energy demand, the increase of energy efficiency, and the supply potentials for district heating. As the method is interactive, by connecting single objects of a street to the grid, it can be determined in real time, for example, in stakeholder and participatory public dialogues, if a feasible district heating grid can be established. This might also change the attitudes toward technologies, so that single- and double-loop learning can be achieved, for example, the focus of further settlement developments in the district heating supply areas.

4. The rational–communicative planning paradigm can be supported, as values and attitudes can be negotiated and influence the aggregation rules (e.g., the targeted energy density, the targeted energy efficiency standards), and the design and adoption of the decision model are science-based considering the system interrelations of spatial planning, energy efficiency policies, and district heating technology.

5. Finally, the method gives advice consistent with climate protection policies, energy policy concerning efficiency, and renewable energy supply considering subsidy schemes. By applying the model, the respective community and individuals can recognize their energy-saving potential and their options for renewable energy supplies.

17.4 ELAS-Calculator

Other than the *Energy Zone Mapping tool* that addresses the specific questions of energy-saving potentials for room heating and hot water production as well as the supply with district heating, the *ELAS-calculator* offers a holistic approach for estimating the energy demand of settlements including embodied energy and determines environmental as well as socioeconomic impacts of the respective energy provisions. The *ELAS-calculator* is designed to (1) estimate the energy demand of residential settlements, (2) determine environmental pressures as ecological footprint using the *sustainable process index method* as well as CO_2-*life-cycle-emissions*, and (3) consider socioeconomic effects of the project as macroeconomic turnover, regional revenue, and regional jobs. *The ELAS-calculator* is freely available on the Internet* and has a German and an English language version. As the socioeconomic analysis

---

* www.elas-calculator.eu

based on regional input–output analysis takes only the economic situation in the nine Austrian provinces into account, for sites outside Austria only the energy calculation and the environmental assessment are available.

The *ELAS-calculator* has two modes: a "municipality mode" for professionals, decision makers, and stakeholders working in comprehensive spatial planning, and a "private mode" for individuals assessing their own existing or planned home. The *ELAS-calculator* comprises a model that takes construction, operation, and maintenance of buildings as well as public infrastructure and mobility of the inhabitants into account. This includes electricity demand, demand for room heating and hot water of the residents, all activities related to public infrastructure (e.g., maintenance of streets, sewage systems), mobility of the residents depending on the supply structure of the village/town, centrality of the municipality, and distances to towns/cities with high centrality as well as the age structure of the residents.

The calculations can be done by entering data about the settlement concerning the location, distances to local and regional centers, kind of houses, number, and age structure of inhabitants, energy standards, energy supply, etc., which are the basis for the energy demand calculation. The database for the *ELAS-calculator* comes partly from existing studies, statistics, and also from a proprietary database based on a survey of the energy demand structure of about 600 households and about 1100 persons.

The model is complex as, for example, 75 different modal splits depending on centrality and age structure of the inhabitants are used by the *ELAS-calculator*. The *ELAS-calculator* calculates the life-cycle energy demand of buildings and infrastructures and includes embodied energy of mobility. Furthermore, the *ELAS-calculator* has a survey mode for the analysis of existing settlements and a planning mode that allows for the assessment of (1) changes of existing settlements and (2) new settlements. Different alternatives for residential areas, for example, the choice of different sites, changes of the densities, types of buildings, thermal insulation, change of energy sources (both fossil and renewable), change of mobility patterns, or fuels for means of transport, can be modeled and the effects on energy demand, environmental pressures, and socioeconomic effects evaluated. We point out that socioeconomic effects should be interpreted carefully: there is a strong correlation that environment-friendly settlement structures create less economic turnover. Yet, this is positive as private households get more disposable income for regional economic activities and improve their quality of life. Furthermore, the revenue of energy supplies might be smaller, but is regionally bound because of renewable energy use, while most of the income based on fossil fuels benefits other countries.

The *ELAS-calculator* offers highly differentiated results. We do not present a result of a calculation here, because this makes only sense in relation to the input data and a general description of the assessed settlement, but we invite the readers to visit the German or English version of the homepage and find out how the calculator works. In the German version also, the full documentation and project report of the ELAS-project are available. The results are astonishing, as, for example, in newly built low-density single-family-house areas, public infrastructure consumes more energy than all private activities together but the ratio totally changes in medium dense settlement areas. In general, the embodied energy of residential buildings is small compared to other parameters of the energy demand.

The *ELAS-calculator* is a role model for our understanding of strategic planning and assessment methods and meets our criteria as follows:

1. The *ELAS-calculator* combines system alternatives (e.g., settlement densities, greenfield vs. brownfield development), site alternatives (evaluating the effects of different sites to the same project), and technical alternatives (e.g., change of insulation, energy sources, design of technical infrastructure like streets, sewage systems). The *ELAS-calculator* allows designing energy-optimized (urban) energy landscapes, producing and instantly assessing a multitude of scenarios and alternatives to iteratively learn about the effects of certain measures compared to the overall situation of the settlement. As the *ELAS-calculator* allows for cross-cutting evaluations of all types of alternatives, it can be used for planning and development as well as design processes.

2. The *ELAS-calculator* is designed according to the "indicator pyramid" and provides measures for the preassessment, so that the "unsustainability test" of planning alternatives for residential settlements can be achieved by using the calculator (at least for Austria).

3. The method supports individual and social learning concerning both single-loop and double-loop learning. Values, actions, and consequences can be assessed using general indicators at a low level of detail concerning the elaboration of a residential project. Therefore, the model can work on early planning stages so that little inherent necessities are created, and the scope for change is large.

4. Concerning the rational–communicative planning paradigm, the *ELAS-calculator* provides a model where the effects of different values can be made visible, for example, living in a single family house, size of parcels, using cars extensively, etc., and reflected (see double-loop learning). The model itself is solidly science based and highly complex, which is necessary to be trusted by the users. As many values as possible are visible and changeable by the users, who can refer to the whole documentation and report on the ELAS-webpage.

5. Finally, the method gives consistent advice from municipal action to private action. The choice between "municipal" and "private" mode was implemented to this end. The *ELAS-calculator* makes visible how settlement projects comply with overall policies like energy strategies and climate protection strategies.

17.5 Conclusions

Creating decision-support methods for the planning of sustainable (urban) energy landscapes is a highly complex task. By defining five normative requirements, we draft a framework for such methodological developments. With the *Energy Zone Mapping tool* and the *ELAS-calculator*, we provide two examples how such strategic planning and assessment methods can be implemented by creating respective tools, and we provide evidence that these methods and tools make a difference in decision making. As both tools cover aspects of energy saving and energy sources, they make decision processes comprehensible. Both tools and their underlying methods promote the development of sustainable (urban) energy landscapes.

Many more of these decision-support methods and tools are necessary to cover all aspects of integrated planning of sustainable energy landscapes. Though many aspects of the tools are tailormade to the Austrian situation,* the underlying methods as well as the normative framework for strategic planning and assessment methods are generic. We hope that our chapter(s) inspire readers to use the methods, to adapt them to their own context, and to develop further strategic planning and assessment methods and share them with the international scientific community as well as with practitioners and include them in university education.

Acknowledgments

The research project "PlanVision—Visions for Energy Optimised Spatial Planning" was funded by the Austrian Climate and Energy Fund and carried out within the program "NEUE ENERGIEN 2020" (grant number 818916). We thank our fellow researchers Michael Weiss, Hermine Mitter, Georg Neugebauer, and Gerlind Weber from the Institute of Spatial Planning and Rural Development, Department of Spatial, Landscape and Infrastructure Sciences, University of Natural Resources and Life Sciences Vienna; Michael Narodoslawsky, Karl-Heinz Kettl, Michael Eder, and Nora Sandor from the Institute of Process and Particle Engineering of the Graz University of Technology; Horst Steinmüller, Barbara Pflüglmayer, Beatrice Markl, Andrea Kollmann, Christina Friedl, Johannes Lindorfer and Martin Luger from the Energieinstitut an der Johannes Kepler Universität Linz GmbH; as well as Karl Steininger and Veronika Kulmer from the Wegener Center für Klima und globalen Wandel of the Karl-Franzens-Universität Graz.

---

* For example, the subsidy criteria for biomass district heating.

The research project "ELAS—Energetic Long-term Analysis of Settlement Structures" was funded by the Austrian Climate and Energy Fund and carried out within the program "NEUE ENERGIEN 2020" (grant number 818915). Further cofinancing was provided by the Province of Upper Austria, Province of Lower Austria, and the Town of Freistadt. We thank our fellow researchers Michael Weiss, Hermine Mitter, and Georg Neugebauer from the Institute of Spatial Planning and Rural Development, Department of Spatial, Landscape and Infrastructure Sciences, University of Natural Resources and Life Sciences Vienna; Michael Narodoslawsky, Karl-Heinz Kettl, Nora Niemetz, and Nora Sandor from the Institute of Process and Particle Engineering of the Graz University of Technology; and Wolfgang Baaske and Bettina Lancaster from Studia Austria Schlierbach.

References

Argyris, C. 1993. *Knowledge for Action: A Guide to Overcoming Barriers to Institutional Change*. San Francisco, CA: Jossey Bass.

Birkmann, J. 2000. Nachhaltige Raumentwicklung im dreidimensionalen Nebel. *UVP-Report* 3/2000, 164–167.

Busch Lüty, C., Dürr, H.P., and Langer, H. (Eds.) 1990. Die Zukunft der Ökonomie: Nachhaltige Wirtschaft. München, Germany: Politische Ökologie. Sonderheft 1.

Greenwood, D. and Levin, M. 1998. *Introduction to Action Research—Social Research for Social Change*. London, U.K.: Sage Publications.

Hacking, T. and Guthrie, P. 2008. A framework for clarifying the meaning of triple bottom-line, integrated and sustainability assessment. *Environmental Impact Assessment Review*, 28, 73–89.

Innes, J. and Booher, D.E. 2000. *Collaborative Dialogue as a Policy Making Strategy*. Berkeley, CA: Institute of Urban and Rural Development, University of California. Working Paper Series, http://repositories.cdlib.org/iurd/wps/WP-2000–2005 (last accessed March 23, 2009).

Jha-Thakur, U., Gazzola, P., Peel, D., Fischer, T.B., and Kidd, S. 2009. Effectiveness of strategic environmental assessment—The significance of learning. *Impact Assessment and Project Appraisal*, 27(2), 133–144.

Krotscheck, C. and Narodoslawsky, M. 1996. The sustainable process index. A new dimension in ecological evaluation. *Ecological Engineering*, 6, 241–258.

Lawrence, D.P. 2000. Planning theories and environmental impact assessment. *Environmental Impact Assessment Review*, 20, 607–625.

Mayer, A.L. 2008. Strengths and weaknesses of common sustainability indices for multidimensional systems. *Environment International*, 34, 277–291.

Meadows, D.H., Meadows, D.L., and Randers, J. 1992. *Die neuen Grenzen des Wachstums. Die Lage der Menschheit: Bedrohung und Zukunftschancen*. Stuttgart, Germany: Deutsche Verlags Anstalt.

Narodoslawsky, M. and Stoeglehner, G. 2010. Planning for local and regional energy strategies with the ecological footprint. *Journal of Environmental Policy and Planning*, 12(4), 363–379.

Noble, B.F. 2000. Strategic environmental assessment: What is it? and what makes it strategic? *Journal of Environmental Assessment Policy and Management*, 2(2), 203–224.

Stoeglehner, G. 2003. Ecological footprint—A tool for assessing sustainable energy supplies. *Journal of Cleaner Production*, 11, 267–277.

Stoeglehner, G. 2010a. SUP und Strategie—Eine Reflexion im Lichte strategischer Umweltprobleme. *UVP-Report*, 23(5), 262–266.

Stoeglehner, G. 2010b. Enhancing SEA effectiveness: Lessons learnt from Austrian experiences in spatial planning. *Impact Assessment and Project Appraisal*, 28(3), 217–231.

Stoeglehner, G., Brown, A.L., and Kornov, L. 2009. SEA and planning: 'Ownership' of SEA by the planners is the key to its effectiveness. *Impact Assessment and Project Appraisal*, 27(2), 111–120.

Stoeglehner, G. and Narodoslawsky, M. 2008. Implementing ecological footprinting in decision-making processes. *Land Use Policy*, 25, 421–431.

Stoeglehner, G. and Narodoslawsky, M. 2009. How sustainable are biofuels? Answers and further questions arising from an ecological footprint perspective. *Bioresource Technology*, 100, 3825–3830.

Stoeglehner, G., Narodoslawsky, M., Baaske, W. et al. 2011a. ELAS—Energetische Langzeitanalyse von Siedlungsstrukturen. Projektendbericht. Gefördert aus Mitteln des Klima- und Energiefonds. Wien, www.elas-calculator.eu (last access: May 21, 2012).

Stoeglehner, G., Narodoslawsky, M., Steinmüller, H. et al. 2011b. PlanVision—Visionen für eine energieoptimierte Raumplanung. Projektbericht. Gefördert aus Mitteln des Klima- und Energiefonds. Wien, https://www.boku.ac.at/fileadmin//H85/H855/materialien/planvision/EndberichtPlanVision.pdf (last access: May 21, 2012).

Stoeglehner, G. and Neugebauer, G.C. in press. Integrating sustainability assessment into planning: Benefits and challenges. In: A. Bond, A. Morrison-Saunders, and R. Howitt (eds.) *Sustainability Assessment. Pluralism, Practice and Progress*. Routledge, London, U.K.

Stremke, S., Dobbelsteen, A., and Koh, J. 2011. Exergy landscapes: Exploration of second-law thinking towards sustainable landscape design. *International Journal of Exergy*, 8(2), 148–174.

Wallington, T., Bina, O., and Thiessen, W. 2007. Theorising strategic environmental assessment: Fresh perspectives and future challenges. *Environmental Impact Assessment Review*, 27, 569–584.

18

Assessment of Sustainability for the Danish Island of Samsø by Application of a Work Energy (Exergy) Balance: A Preliminary Assessment

Sven E. Jørgensen and Søren N. Nielsen

CONTENTS

18.1 Introduction

The chapter presents the primary result of a sustainability analysis that is performed for the Danish island of Samsø by the application of work energy capacity, also denoted as exergy or sometimes ecoexergy, to emphasize when we calculate exergy for natural systems. The theoretical background knowledge and the definitions that are necessary to understand the analysis and to interpret the analytical results were given in Chapter 11.

Samsø is an island situated in the Kattegat between the Danish parts of Zealand and Jutland. The area of the island is $114\,km^2$, and it is $26\,km$ long and $7\,km$ wide at the maximum width. The island has about 4100 inhabitants with about 7% working in the agricultural sector. The second most important sector is tourism.

In 1997, the Ministry of Energy announced a competition: which local area or island could present the most realistic and realizable plan for a 100% transition to self-sufficiency with renewable energy? Samsø won the competition in October 1997. Samsø has, therefore, during the last 15 years, introduced the application of alternative energy and, at the same time, made a significant effort to reduce the energy (or rather exergy) consumption. The energy

FIGURE 18.1
Solar panels used for heating. The district heating station is shown in the background.

consumption on Samsø is about 130 MJ/capita and about 50 MJ/ha. The average for Denmark is 160 MJ/capita and 200 MJ/ha, and due to an early and relatively massive introduction of green taxes on energy, Denmark has a low energy consumption compared with the GNP/capita of similar countries (see Jørgensen 2006). The alternative energy sources applied on Samsø today are mainly solar panels for heating, wind energy for the production of electricity, and straw for district heating. Figures 18.1 through 18.3 illustrate these three applications of alternative energy. Meanwhile, the Samsø Energy Academy was erected with the objective to arrange exhibitions, workshops, and courses about renewable energy and disseminate as widely as possible the gained knowledge about the use of alternative energy and energy saving.

The Samsø Energy Academy is situated at Strandengen, Ballen, where an administration building (see Figure 18.4) was constructed in accordance with ecological principles. Rainwater is used for flushing of toilets, and the building has a low energy consumption as it is very well insulated. Solar panels have been integrated on the roof and demonstrate this alternative energy technology. This partially covers the demand for electricity. The heating is supplied by the district heating station that uses straw (see Figure 18.3). The electricity consumption is low due to the selection of low-energy electrical devices and fittings. The electricity produced by the solar cells is supplemented by wind energy.

FIGURE 18.2
Windmills offshore from Samsø.

FIGURE 18.3
Storage of straw applied for district heating.

FIGURE 18.4
Samsø Energy Academy Building, Strandengen 1, Ballen, Samsø.

18.2 Energy Balance and the Principles for an Exergy Balance for Samsø

An overview of the use and production of energy today can approximately be presented as shown in Table 18.1. The numbers are from 2005 (Samsø Energy Academy 2007) and have only been improved since.

Table 18.1 shows that almost 90% of the energy consumption is covered by alternative energy. A few tractors use sunflower oil, but it is still a general problem to replace the use of fossil fuel by vehicles.

TABLE 18.1

Samsø: Energy Use and Production (TJ/Year)

Type of Energy	Consumption	Production
Electricity	286	386
Heating	140	66
Gasoline, including ferries	86	0
Total	512	452

As indicated in Chapter 11, it is necessary to supplement Table 18.1 with a work capacity balance (exergy or ecoexergy balance) to assess the exergy efficiency that is equal to the ratio of work performed in energy units with the total use of energy. The idea is that if it would be possible to get more work done with the same amount of energy, it would be possible to reduce energy consumption. The exergy (work capacity) balance should of course include the agricultural production. It is based on solar energy that in this context is considered free. The principles applied to set up the exergy balance are shown in Figure 18.5. Based upon these principles (for further explanation, see Jørgensen [2006]), the following quantifications can be developed:

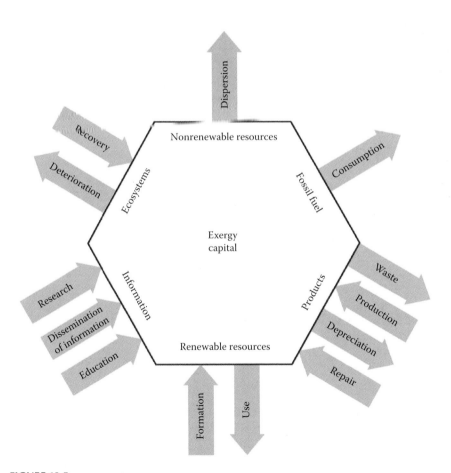

FIGURE 18.5
Principles applied for the development of an exergy (or work capacity) balance.

1. The *loss of exergy (work capacity) due to the consumption of fossil fuel* is found by the addition of chemical-free energy (the work capacity) of the fossil fuel and the loss of exergy due to the dispersion of the gases resulting from the chemical processes.

    The exergy loss due to dispersion of the components of fossil fuel is found by the following calculations: if we consider 1 g of coal that contains 1% of sulfur and 99% of carbon (coal contains also ash, but let us not consider it in our calculations), the exergy loss due to the dispersion can be determined by the following calculations:

$$0.01\left(\frac{8.314*300}{32}\right)\ln\left(\frac{0.01}{50\times10^{-9}}\right)+0.99\left(\frac{8.314*300}{12}\right)$$

$$\ln\left(\frac{0.99}{4\times10^{-4}}\right)=1617\,J\approx1.6\,kJ$$

where
    50×10^{-9} and 4×10^{-4} represent concentrations (expressed as ratios, i.e., no units) of sulfur dioxide-S and carbon dioxide-C in a typical town atmosphere
    8.314 is the gas constant R
    300 is the absolute temperature

The chemical exergy content of 1 g coal is about 32 kJ. The loss of exergy by dispersion is therefore only 5% of the loss directly of chemical exergy by burning coal. As all our calculations will have a greater uncertainty than 5% and the quality of coal may vary more than 5%, it seems acceptable not to include the dispersion exergy loss by the use of fossil fuel or, as an alternative, to multiply all exergy losses due to our consumption of fossil fuel by a factor of 1.05 to, by approximation, compensate for the exergy loss due to the dispersion of formed gases in the atmosphere. It means that the exergy lost by the use of fossil fuel becomes

    Loss of exergy by fossil fuel
       = Consumption in kg * F * 1.05 MJ   (18.1)

where F is the chemical energy content of the fossil fuel. F is 32 MJ/kg for coal and 42 MJ/kg for oil or natural gas.

2. *The deterioration of ecosystems.* Exergy can be used as an indicator for ecosystem health and can be used to express ecosystem services (Jørgensen 2010). Table 18.2 illustrates this approach, the calculations of ecosystem services based on exergy for various ecosystems. The annual net production contributes positively to the exergy balance,

TABLE 18.2

Work Capacity Used to Express the Ecosystem Services for Various Types
of Ecosystems[a]

Ecosystem	Biomass (MJ/m² Year)	Information Factor (β-Value)	Work Capacity (GJ/ha/Year)
Desert	0.9	230	2,070
Open sea	3.5	68	2,380
Coastal zones	7.0	69	4,830
Coral reefs, estuaries	80	120	960,000
Lakes, rivers	11	85	93,500
Coniferous forests	15.4	350	539,000
Deciduous forests	26.4	380	1,000,000
Temperate rainforests	39.6	380	1,500,000
Tropical rainforests	80	370	3,000,000
Tundra	2.6	280	7,280
Croplands	20.0	210	420,000
Grassland	7.2	250	18,000
Wetlands	18	250	45,000

[a] Calculated as biomass × the information factor.

while the deterioration of ecosystems contributes negatively to the
exergy balance and can be calculated from the following equation:

$$\text{Exergy lost by deterioration of ecosystems}$$
$$= \text{ha lost} * (\text{kg biomass/ha}) * \beta \text{ MJ} \qquad (18.2)$$

3. *The use of renewable resources* (the formation of renewable resources)
   is found separately by multiplication of the annual consumption
   of the various resources multiplied by the exergy content of each
   renewable resource. When, for instance, the annual loss of fishery
   in the Kattegat is 1000 ton, then the exergy is annually reduced by
   $10^9 \times 499 \times 18.7 \text{kJ} = 9.3 \times 10^{15}$ J; 499 is the β-value for fish (see also
   Chapter 11). The sustainability requires that the growth of fish bio-
   mass compensates for this loss of exergy. For a couple of decades in
   many marine ecosystems, including the North Sea and Kattegat, it
   has, unfortunately, not been the case due to overfishery.

4. *Loss of exergy due to depreciation of our products*: this concerns cars,
   houses, refrigerators, TVs, computers, etc. The exergy content of
   these products is approximately declining by 10% per year, which
   implies that the exergy content times 0.1 is lost per year. The number of these products per capita is known for most industrialized
   countries. The exergy requirement for these products is to a high
   extent a question of providing the materials used for the production.
   Table 18.3 gives the exergy requirement according to Hayes (1997)

TABLE 18.3

Exergy Requirements (MJ/kg Product) in Producing
Various Materials from Ore and Raw Source Materials

Product	From Ore	From Scrap
Glass	25	25
Steel	50	26
Plastics	162	112
Aluminum	250	8
Titanium	400	Not yet applied
Copper	60	7
Paper	24	15

Source: Hayes, E.T. *Implication of Material Processing,* EPA,
          Athens, GA, 1997.

for producing various materials from ore or scrap. An approximate
calculation for Samsø is presented later.

5. *Loss of exergy due to waste,* which is highly dependent on the waste-
   handling method. When landfills are used, the loss of exergy corresponds to the loss of land, and it can be calculated by the use
   of Equation 18.2. When incineration is applied, the corresponding saving of heating can be added as a positive contribution to
   the exergy balance. The loss of exergy due to our use of paper
   must, however, be added, as the exergy gained by incineration of
   paper is minor. It would correspond to about 20 MJ/kg of paper (it
   is approximately the average of the exergy requirements for paper
   listed in Table 18.3). The loss of exergy due to our application of
   paper is therefore

$$\text{kg paper/capita} * \text{The number of inhabitants} * 20 \text{ MJ} \qquad (18.3)$$

A comparison of the exergy requirements by the use of ore and scrap shows
clearly that reuse and recycling give double benefits: the use of our limited
resources is reduced and exergy is saved. Notice that particularly for aluminum the difference in energy requirement by use of scrap instead of ore is
considerable.

The gain of exergy can be calculated by adding the following contributions:

1. *Exergy of products.* When the products are used, they are depreciated
   (Section 18.2). When the product is produced, exergy is gained as
   production in kilogram times the exergy requirement per kilogram
   (see Table 18.3). It is of course presumed that the various energy
   forms are coved by point 1 shown earlier.

2. *The formation of renewable resources.* It is calculated in the same way as the losses, that is, the annual growth is multiplied with the exergy of the considered renewable resource.

3. *Repair of products* that give them higher exergy values. This is in most industrialized societies minor, as repair of products is applied less and less due to the high costs of labor.

4. *Increase of our knowledge*/our information capital.

5. *Recovery of ecosystems.*

A continuous steady annual net loss of exergy or work capacity is of course from a long-term viewpoint completely unacceptable because it would mean that we are reducing the sum of human and natural capital year after year, which will imply that life conditions for mankind are steadily deteriorating. If we use exergy as sustainability index = (gain of exergy/loss of exergy), it is of course unacceptable that the sustainability index for a longer period of time is considerably less than 1.00.

Our perception of the evolution is that it has increased the order and exergy on earth; however, the core question is whether mankind by its massive change of the environment, including the change of nature to man-made systems (towns, factories, roads, and so on), has started a decline of order and exergy, which sooner or later will result in a major crisis for life on earth. The content of exergy in a natural system is usually many times the content of a technological system (see Table 18.2). It implies that what is named progress for mankind may be a step backward for the exergy content of the earth. The decline of exergy can continue some decades or maybe even some centuries, but it cannot continue on a long-term basis because the basic for human activities will thereby be significantly reduced.

18.3 Exergy Balance for Samsø

The exergy balance for the island of Samsø has been calculated from the principles listed earlier and illustrated in Figure 18.5. At this stage, only a first rough balance has been carried out. A more detailed balance will be developed during the coming year. The first draft balance is, however, very illustrative for the possibilities to ensure a sustainable development on the island. It is not expected that the more detailed balance will change the general conclusions, but it will, of course, change the numerical analysis slightly. Therefore, the first rough balance will be presented and discussed in the following.

The application of fossil fuel corresponds to heating minus the heating produced + the gasoline used (see Table 18.1). These quantities are multiplied by 1.05 to account for the dispersion of gases as a consequence of the use of fossil fuel. Electricity can be converted directly to exergy. Samsø has a management of nature that with good approximation yields a status quo situation. The fishery is minor, and, therefore, there is no contribution to loss or gain of exergy due to renewable resources.

Samsø has about 4100 inhabitants, and if the number of tourists and the number of summer houses are considered, it would with a good approximation be reasonable to count with an average of 6000 inhabitants that have invested in durable consumer goods. When we apply a good Danish average number, an average family (2.25 persons) has 1800 kg of durable consumer goods. By application of exergy requirements in Table 18.3 and presuming that durable consumer goods are mainly made of steel, the depreciation (10% per year) will correspond to (6000 * 1800 * 50 * 0.1/2.25) MJ = about 24 TJ.

The waste treatment is still under discussion, but the general application of about 0.2 ha/year agricultural land for landfills for 6000 inhabitants will probably be replaced by incineration in the near future. According to Table 18.2, the application of landfills will correspond to 84 TJ/year.

Exergy of the agricultural production is based upon 18.7 kJ/g, which is a good average of the exergy content of agricultural products. It is also the value that we use for the exergy content of detritus.

It is furthermore presumed that the consumption of plastic is 2 kg/year/capita and that of paper 250 kg/year/capita. These will give (see earlier text and Table 18.3) the following exergy consumption:

$$\text{Exergy consumption (plastic and paper)}$$
$$= ((2 * 162) + (250 * 20)) * 6000 \text{ MJ/year} = 33 \text{ TJ/year}$$

The exergy lost by the consumption of glass can be calculated similarly:

$$\text{Exergy consumption (glass)} = 12 * 25 * 6000$$
$$= 1.8 \text{ TJ/year, corresponding to an annual use of glass on 12 kg/capita.}$$

The exergy or work capacity balance is shown in Table 18.4. The agricultural production contributes as shown significantly to the positive side of the balance, which of course is rooted in the input of the "free" solar energy. Without the positive exergy contribution from agriculture, the balance results would be negative.

There are several possibilities to improve the balance. First of all, the 84 TJ lost for the treatment of waste by landfills could be eliminated, which has

TABLE 18.4

Exergy (Work Capacity) Balance for Samsø (TJ)

Item	Consumption	Production
Electricity	286	386
Heating	145	66
Gasoline + diesel	91	
Food	27	
Milk		2
Potatoes		150
Vegetables		112
Grain		271
Pigs		106
Pesticides	1	
Waste	84	
Durable consumer goods	24	
Plastic and paper	33	
Glass	5	
Total	696	1093

been planned to be realized during probably 2012. Reuse or recycling of plastic and paper would also reduce the exergy lost. It is hardly possible to reduce the loss of exergy due to durable consumer goods, but it is of course possible to cover the heating 100% or close to 100% by the use of alternative energy. It would reduce the exergy consumed by about 80 TJ.

18.4 Discussion and Conclusions

The exergy balance shows that the development on Samsø is sustainable due to the use of alternative energy and due to the agricultural production. It is, however, possible to improve the balance as discussed earlier by a better waste treatment, total use of alternative energy for heating, and wider use of recycling for paper and plastic. By these three initiatives, it would be possible to reduce the exergy consumption to about 520 TJ. It may also on a long-term basis be possible to replace the use of gasoline and diesel with the use of electricity for vehicles and ferries.

It is presumed that the natural ecosystems are not changed (reduced or enlarged), which is in accordance with information from the municipality. But as shown in Table 18.2, ecosystems have a very high ecoexergy. The global loss of sustainability is mainly rooted in the deforestation of the rainforest and the use of fossil fuel (see Jørgensen, 2006). It is therefore very important for the future exergy balance of Samsø that the ecosystems are maintained or if the areas of some ecosystems are reduced that other ecosystems are enlarged, for instance, by increasing the area of forests.

The experience of the use of work capacity balance (ecoexergy balance) on the Danish island Samsø has shown that

1. It is relatively easy to set up the balance
2. Ecoexergy is a very useful sustainability indicator
3. The balance can be used as a management tool by giving quantitative information on how the balance and the general energy efficiency can be improved

References

Hayes, E.T. 1997. *Implication of Material Processing*. EPA, Athens, GA.
Jørgensen, S.E. 2006. *Eco-Exergy as Sustainability*. WIT Press, Southampton, U.K., 220pp.
Jørgensen, S.E. 2010. Ecosystem services, sustainability and therrmodynamic indicators. Ecological Complexity, 7, 311–313.
Samsø Energy Academy. 2007. Samsø, a Renewable Energy Island, 10 years of Development and Evaluation, 60pp.

19

Carbon Emission Intensity and Areal Empower Density: Combining Two Systemic Indicators to Inform the Design and Planning of Sustainable Energy Landscapes

Riccardo M. Pulselli, Pietro Romano,
Michela Marchi, and Simone Bastianoni

CONTENTS

19.1 Introduction

In the last decade, methods of environmental accounting have been performed, combined with geographic information systems (GIS), in order to develop a spatial analysis of man-made and natural processes and their impacts within a region. Among others, the *greenhouse gas inventory* (hereafter GHG for greenhouse gases) accounts for emissions and removals of GHG, in terms of equivalent CO_2 (t CO_{2eq}), and the emergy evaluation measures the intensity of environmental resource use by a population settled in an area, in terms of equivalent solar energy (solar emergy Joules—seJ).

In this chapter, the carbon emission (measured in t CO_{2eq} year^{-1} km^{-2}) and the areal empower density (seJ year^{-1} km^{-2}) are presented and discussed as indicators of environmental performance. Measuring and visualizing the concentration of CO_2 outflows (net carbon emission) and the intensity of energy and material inflows (areal empower density) that feed structures and processes are operations that intend to contribute for a deeper under-standing of the behavioral organization of a given region. Both the indicators provide complementary information on the effects of anthropogenic and natural functions on the environment at a local and even wider (landscape) scale, until the global.

In a broader sense, indicators such as the carbon emission and the empower density—as well as the exergy destruction, the entropy production, and a few others—concur to reveal a kind of "thermodynamic profile" of the region depending on land uses, human settlements, processes, technologies, and landscape structures. The core research topic is to investigate how to increase (spatial) accuracy and reliability of the present assessment methods and indicators and how to use and interpret their results, made spatially explicit through the use of GIS, and thus improve our capacity to inform energy con-scious planning and design in the context of sustainable development.

19.2 Thermodynamics-Based Framework in Urban Studies

Abel Wolman introduced the concept of urban metabolism in 1965 (Wolman, 1965). He highlighted how some operations in urban systems closely resem-ble a set of related metabolic processes. A key contribution in this sense was provided by Ilya Prigogine, the father of evolutionary thermodynamics, and winner of the Nobel Prize in 1977. Prigogine and Stengers (1979) formulated the concept of dissipative structures as thermodynamic systems, open to energy and matter, that self-organize toward higher complexity and orga-nization. Prigogine stated (1997): "The simplest example of dissipative struc-tures that we can evoke by analogy is the city." In recent years, a number of researchers (Pulselli and Tiezzi, 2009; Stremke and Koh, 2010; Stremke et al., 2011, among others) in the field of ecology, urban studies, and landscape architecture have been treating cities, with their surrounding environment, as if they were organisms and have analyzed the processes and mechanisms that form their dissipative structures. The urban metabolic system has been interpreted as a mechanism for processing low-entropy resources (exergy input) and producing high-entropy wastes (exergy destruction). The sec-ond law of thermodynamics states that during any process, exergy (order, work capacity) is destroyed and entropy (disorder) is produced. Identifying constraints imposed by the second law of thermodynamics to energy (and entropy) flows from sources to sinks and deriving ecological concepts and strategies have the capacity to expand the frame of reference in planning of the human physical environment and provide insights into how to adapt human behavior to renewable energy sources (Stremke, 2010).

In this thermodynamic framework, the investigation of the general functioning of a regional system requires a systemic approach capable of detecting the properties of the whole as generated by the combined actions of a number of agents operating in the system, such as cities, industries, farms, and ecosystems. All these systems basically maintain their organization by processing environmental resources and generating wastes. Environmental accounting methods and systemic indicators, combined with GIS, can be developed as tools for understanding and representing the behavioral organization of a regional system based on the energy and material inflows and outflows that feed natural and man-made processes.

19.3 Approach and Methods

Among a number of possible methods, the *GHG inventory* (Section 19.3.1), based on the IPCC standard procedure (2006), and the *Energy evaluation* (Section 19.3.2), based on the H.T. Odum energy systems theory (1971, 1996), were selected and processed with reference to a case study, the Region Abruzzo (Section 19.4). Indicators, namely, *carbon emission* (in t CO_{2eq} year^{-1} km^{-2}) (Section 19.4.1) and *areal empower density* (in seJ year^{-1} km^{-2}) (Section 19.4.2) were assessed. The information obtained through flow assessment and indicators was made visible through the use of GIS as a tangible physical phenomenon at the landscape scale that would be otherwise invisible (Section 19.5).

The elaboration of these indicators combined with procedures in the field of geographic information sciences allowed for measuring the intensity of environmental resource use (energy and material inflows) and GHG emission (material outflow), relative to specific land uses. In a wider sense, with the energy and material inflow assessment (expressed through the areal empower density), we intended to provide a measure of the "low-entropy" inputs, while, with GHG emissions (carbon emission), we intended a quantitative interpretation of entropic dissipation ("high entropy" outputs) or a portion of it. Moreover, the intensity of both inflows and outflows and their variations in space, once visualized on maps, let us achieve both a qualitative and quantitative understanding of systems' dissipative behavior within the region.

19.3.1 GHG Inventory

The GHG inventory is an instrument for monitoring variations in emissions and removals of natural and man-made GHG. The inventory estimates direct GHG emissions, namely, emissions of carbon dioxide (CO_2), methane (CH_4), nitrous oxide (N_2O), and others (hydrofluorocarbons, perfluorocarbons, and sulfur hexafluoride—SF_6), the global warming potentials (GWPs) of which are available in the IPCC Fourth Assessment Report (2007). In physical chemistry, the GWP measures the long-term contribution of a gas to global warming and is weighted according to the atmospheric lifetime of the gas and its capacity to absorb infrared radiation emitted by the Earth. The GHG emissions and the absorption by forests (the capacity of CO_2 uptake depends on specific climatic and geographical conditions) are provided in tons of carbon dioxide equivalent (t CO_{2eq}).

The IPCC inventory is basically conceived as a standard GHG emission assessment method of nations. Nevertheless, in recent years, IPCC inventories were implemented to be performed at the local level in order to investigate the effects of local activities with regard to global climate change (Wilbanks and Kates, 1999). Local and regional policy makers in many nations are increasingly asking what they can and should be doing to respond to the challenge of GHG emissions reduction (Carney and Shackley, 2009). Byrne et al. (2007) provided an overall review of a number of policies, strategies, and frameworks that have emerged at national, regional, and local levels to guide climate protection, trying to identify the benefit linked to each emission reduction program. Rabe (2007) stated that climate change is a challenge of multilevel governance, creating opportunity for local policy experimentation.

In this framework, district-level GHG emission assessments aim to improve spatial accuracy and develop potential sound mitigation strategies at manageable smaller scales (Garg et al., 2001; Dalvi et al., 2006). A site-specific information on GHG emission would allow for meeting national and international reduced carbon emission targets (i.e., Kyoto Protocol Actions; European Commission Directive) and provide matter for a burden sharing policy at the subnational, regional scale (D.L. 03/03/11, no. 28 in Italy).

Methods have been provided since the late 1990s (i.e., USEPA, 1998) that combine environmental indicators, such as GHG emission and air quality, with GIS in order to make results spatially explicit and thus to inform regional planning and management practices. Diem and Comrie (2001) investigated the potentiality of GIS for enhancing an existing spatially aggregated anthropogenic emissions inventory in Tucson (Southern Arizona) and discussed the ozone-specific management implications through a spatially disaggregated emissions inventory. Dennis et al. (2002) presented a spatial analysis of acreage burned, fuel loading, and emission factors in Texas. Despite fire emissions being a relatively small fraction of the total emission, GIS was used for investigating the spatial and temporal distribution of atmospheric particulate matter (PM) from wildfires, rangeland fires, agricultural operations, and slash burns. Gaffney (2002) investigated the opportunity of gradually developing a GIS-based emission inventory in California, following an incremental approach that was conceived to be progressively upgraded. Global position systems (GPS) were investigated to be used for locating sources of emissions at the very local level (Alexis, 2002). Sozanska et al. (2002) show a spatial inventory of N_2O emissions from agricultural soils in Great Britain using a simple regression model depending on variation in soil management and climate within a GIS framework. Koerner and Klopatek (2002) presented a GIS-based analysis of soil carbon dioxide emissions and other natural and anthropogenic CO_2 emissions in the metropolitan region of Phoenix (Arizona), focusing on the effect of urbanization on the whole regional pattern. Smiatek (2005) developed GIS relational database management system (RDBMS) functions to process land-use data in south Germany. In Jones et al. (2007), outcomes form an energy and environmental impacts prediction model of South Wales (United Kingdom) were elaborated through GIS at the urban scale (Cardiff, United Kingdom). Zhang et al. (2008) calculated urban-scale emission of atmospheric pollutants from both fossil fuel consumption and industrial production processes through air quality monitoring stations in Hangzhou (Zhejiang Province, South China). An anthropogenic emission inventory of gaseous pollutants was presented by Markakis et al. (2010) integrating GIS with structural query language (SQL) program to provide sets of disaggregated emissions. Zhao et al. (2011) mapped total emissions in Huabei region (a part of eastern China) with reference to cells in a spatial grid. Spatial analysis of GHG emissions was performed through GIS based on land-use changes in Italy by Castaldi et al. (2005) and Pennati et al. (2009).

Wobber and Fleming (2002) provided a guideline of how an inventory of GHG emissions could be used to estimate the potential impacts on climate change at the regional level. GHG emission calculators have recently emerged in order to be applied as regional management tools. Models were developed for evaluating GHG emissions of urban and regional systems as well as specific processes (Titheridge et al., 1996; Jones et al., 2000; Fleming et al., 2001; Barrett et al., 2005; Baggott et al., 2006). Among others, the Greenhouse Gas Regional Inventory Protocol (GRIP) software package (Carney and Shackley, 2009) proved to be a valuable tool for eliciting detailed stakeholder data, idea, and opinions and exploring the challenges of meeting a given CO_2 reduction target. The GRIP methodology, developed at the University of Manchester to be applied on regional systems, allows for processing a GHG inventory that basically follows the IPCC guidelines. Emphasis is given to the quality of data that is managed and represented (through different colors) with respect to its spatial accuracy.

Not far from the GRIP approach, the IPCC methodology performed in this chapter, based on the "2006 IPCC Guidelines for National Greenhouse Gas Inventories," was developed at the regional scale. Quality of data is the critical point as it is in most of the existing environmental accounting methods applied to regional systems. In general, a desirable condition is that data are collected at the local level and then aggregated through a bottom-up procedure. Unfortunately, due to the limitation of the current monitoring systems, data are often collected at a wider scale (e.g., that of a national administration) and need to be scaled down from an upper to a lower level (top-down).

Data availability, including site-specific emission and absorption factors, affects spatial accuracy and reliability of outcomes from a GHG inventory.

From a theoretical viewpoint, GHG emissions determine an exergy loss. On the other hand, carbon uptake by trees determines an exergy gain, due to the increase of biomass and information contained in the forest areas. Net GHG emissions thus correspond to an exergy destruction taken as the difference between an exergy input that increases the system's organization (CO_2 absorption) and the exergy of the waste to which the input exergy is dissipated (Rosen and Scott, 2003).

19.3.2 Emergy Evaluation

Emergy (spelled with an "m") is defined as the quantity of solar energy used, directly or indirectly, to obtain a product by a process or to renew a resource that has been consumed (Odum, 1971, 1996). The units of emergy are solar emergy Joules or solar emjoules (seJ).

The emergy evaluation assigns an "environmental value" to resources in proportion to the "dimension" of the transformation processes producing man-made and natural systems. This value represents the energy used throughout a series of processes to obtain a system product. Its assessment includes energy and material transformations that have previously occurred, moving back through a chain of processes from a final grade product to its primary sources (solar energy).

Similarly to any other cost measure, the emergy used up is no longer available to drive further transformations: it is embodied in the products, generally in the form of upgraded quality and hierarchical role (Sciubba and Ulgiati, 2005). As a consequence, emergy can be related to exergy (and entropy), but differs significantly. Exergy is defined as the maximum amount of work that can be produced by a flow of matter or energy as it comes to equilibrium with a reference environment. It is interesting in this context that exergy measures the energy needed to break down to thermodynamic equilibrium (Jørgensen et al., 1995; Wall and Gong, 2001). Exergy accounts for the resources that are still available to potentially nurture a system process, while emergy accounts for the whole of energy and resource flows that have been used to supply a system process. An increase in resource use (emergy) is therefore supposed to determine a decrease of available energy (exergy).

In absolute terms, emergy and exergy give only a partial indication of the environmental cost associated with the organization of a system process. An emergy/exergy ratio can be interpreted as an indicator of a (eco)system efficiency. It corresponds to the amount of environmental resources (given in solar energy equivalent) per unit exergy, the latter being a measure of organization (the distance from the thermodynamic equilibrium). If the ratio has an increasing trend (ignoring oscillations due to normal biological cycles), it means that natural selection is taking the system along a path that will lead it to a lower level of organization (Bastianoni and Marchettini, 1997; Tiezzi, 2003).

On the other hand, the exergy/empower ratio represents the work capacity of an emergy flow. This indicates system efficiency in transforming direct and indirect solar energy inputs into organization, reflecting the state (i.e., internal complexity) of a system per unit input (Bastianoni et al., 2006). The application of emergy/exergy ratio can provide a more effective and comprehensive measure of the environmental impact of a system process than emergy and exergy considered alone.

Emergy evaluations of territorial systems have been performed for the assessment of environmental resource use in nations (Campbell, 1998; Campbell et al., 2004, 2005), regions (Higgins, 2003; Pulselli et al., 2007, 2008), and urban systems (Ascione et al., 2009; Su et al., 2009; Zhang et al., 2009a–c).

Emergy flows per unit time, namely, empower, are given in seJ year^{-1} and represent the intensity of environmental resource use by a process or system. In regional systems, the areal empower density (unit: seJ year^{-1} km^{-2}) shows the intensity of emergy flows into a unit area. This value depends on the type of land use, processes, technologies, natural, or man-made systems settled within an area.

19.4 Carbon Emission and Areal Empower Density: The Case of the Abruzzo Region

The Abruzzo region is in the central part of the Italian peninsula, covering an area of 10,789 km^2, with a population of 1,305,500 and a population density of 121 km^{-2}, lower than the Italian average of 195 km^{-2}. The regional system has an easily readable structure, being organized in a highly urbanized and productive area along the seacoast, with its core in the metropolitan area of Pescara-Chieti, and around the inner towns of L'Aquila, Teramo, and Avezzano. The highest peaks of the Apennines, Gran Sasso, and la Maiella are only 60 km from the Adriatic coast, and parks and protected areas cover 2970 km^2, namely, Parco Nazionale d'Abruzzo, La Maiella, Gran Sasso, and Sirente-Velino (Figure 19.1).

19.4.1 Carbon Emission: The Assessment of CO$_2$ Net Emissions

Here, we show the GHG inventory of the Abruzzo region (Italy) for the year 2006, compiled according to the 2006 IPCC Guideline for National Greenhouse Gas Inventories. Table 19.1 shows direct GHG emissions generated in the Abruzzo region in 2006.

Among the main GHG reported in the table, the preponderance of CO$_2$ emissions (around 76%) was largely due to direct combustion of fossil fuels (energy sector). Methane (21%) is the main GHG in the wastes sector, produced by anaerobic processes of organic degradation in solid urban wastes

(a)

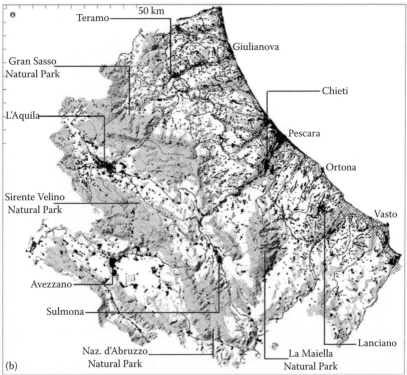

FIGURE 19.1
(a) Map of Italy and (b) map of the Abruzzo region (scale approximately 1:1.500.000).

TABLE 19.1

Total Emissions (Tons) of Greenhouse Gases
in the Abruzzo Region (2006)

Direct Greenhouse Gases	t Gas	t CO_{2eq}	%
CO_2	8,513,746	8,513,746	77
CH_4	90,422	2,260,558	20
N_2O	1,142	340,186	3
Other gases		7,350	0.07
CO_{2eq}		11,121,840	100

consigned to landfills. A minor percentage of this gas was also emitted by cattle breeding where it was largely generated by enteric fermentation in rumi-
nants. N_2O (3%) was the main GHG produced in the agricultural sector as
manure, organic and synthetic fertilizer, and residual matter from cultivation.

The energy sector is generally the major part of a GHG inventory in industrialized countries. In the Abruzzo region, it contributes to 70% of total GHG
emissions. This includes the combustion of fossil fuels (transport, heating,
industry, agriculture) and the production and use of electricity (i.e., emissions from the local thermoelectric production and due to the use of electricity imported from the national grid) (Table 19.2).

TABLE 19.2

Greenhouse Gas Emissions by Sector
in the Abruzzo Region in 2006

Sector	t CO_{2eq}	%
Energy (fuels and electricity)	**7,777,483**	**70**
Transport	3,230,930	29
Residential	2,207,491	20
Industrial	2,099,825	19
Agriculture	239,237	2
Industry (main production)	**621,501**	**6**
Wastes	**2,060,515**	**19**
Landfill	1,938,975	17
Wastewater	82,642	0.7
Compost	38,898	0.3
Agriculture	**662,342**	**6**
Carbon loss (fires and crops)	197,491	2
Cattle breeding (stable)	256,025	2
Cattle breeding (pasture)	28,625	0.3
Organic fertilizers	97,185	1
Synthetic fertilizers	83,016	0.7
Total Emissions	**11,121,840**	**100**

Industrial production in which materials are transformed physically or chemically (energy production excluded) releases GHG into the atmosphere. This is based on the amount of primary materials, carbonates, and fluorinated gases used by the main industries operating in Abruzzo that are cement, glass, ceramic, paper, and electronics factories. This is around 6% of the total emission.

We estimated emissions of CO_2, CH_4, and N_2O from wastes treated by the following processes: disposal in landfills, biological treatment (digestion and composting), and wastewater treatment (waste incineration with cogeneration of electricity was not detected within the region). This is around 18% of the total emission.

Emissions from the sector of agriculture, forests, and other land uses can be divided into the following categories: (1) changes in land use, (2) changes in the stock of carbon in biomass (i.e., hectares of forest lost in forest fires and forestry uses, removals due to tree crops in orchards and vineyards), (3) cattle breeding and manure management (manure and other organic fertilizers), and (4) direct and indirect emissions from nitrogen fertilizers. This is around 6% of the total emission.

Absorption of CO_2 by forests and perennial crops was accounted as variation in the stock of carbon in biomass. We used the gain–loss approach proposed by the IPCC (2006) based on annual increases in biomass and growing stock by forest type.

As a result, the carbon emission of the Abruzzo region in 2006 was 11,121,840 t CO_{2eq}. This was mainly due to domestic activities (20%), industrial productions (25%), wastes (19%), transportation (29%), and agriculture (6%). The carbon emission per person corresponds to 8.52 t CO_{2eq} inhab^{-1}. Considering the absorption of 2,125,344 t CO_{2eq} by forests (19%), the net emission is around 8,800,000 t CO_{2eq}, that is 6.89 t CO_{2eq} inhab^{-1} (Figure 19.2).

Table 19.3 shows values from the assessment of GHG emissions in the Abruzzo region (2006) compared to the Tuscan region (2007), to the Italian national inventory (2006), and to the European inventory (2006).

19.4.2 Areal Empower Density: The Assessment of Environmental Resource Use

Results from an emergy evaluation applied to the Abruzzo region were published and widely discussed in Pulselli (2010). The empower was detected equal to 8.71E + 22 seJ year^{-1}. The emergy evaluation highlighted that 73% of the resources used derives from external sources (F), including fuels (15%) and goods and products, especially food, imported from the market (58%). The classification in categories of resources has also highlighted the measure to which the system draws from renewable R (2%), such as direct solar irradiation, rain precipitation, and geothermal heat, and nonrenewable N (25%) local resources, such as extracted materials and water. The areal empower density of the Abruzzo region was 8.07E + 18 seJ year^{-1} km^{-2}. Nevertheless,

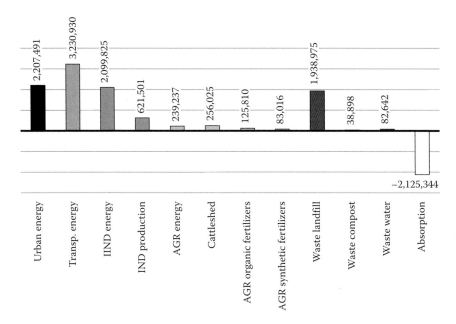

FIGURE 19.2

GHG emissions and absorption (t CO_{2eq}) for the Abruzzo region in 2006.

TABLE 19.3

Greenhouse Gas Emissions Assessment of Abruzzo Region (2006), Tuscan Region (2007), Italy (2007), and EU (2006)

	Abruzzo Year 2006	Tuscany[a] Year 2007	Italy[b] Year 2006	Europe[c] Year 2006
Inhabitants	1,305,500	3,638,211	58,751,711	731,000,000
Area (m²)	10,789,000	22,994,000	301,340,000	10,180,000,000
Gross emissions (t CO_{2eq})	11,121,840	35,472,387	562,982,420	5,117,000,000
t CO_{2eq}/inhabitant	8.52	9.75	9.58	7.00
t CO_{2eq}/m²	1.03	1.54	1.87	0.50
Removal by trees (t CO_{2eq})	−2,125,344	−9,232,978	−89,804,030	−401,000,000
% Abatement of gross emissions	−19	−26	−16	−8

[a] IRSE—Tuscan Region, 2007. Regional GHG Inventory 2007. Direzione Generale delle Politiche Territoriali e Ambientali.

[b] SINAnet, 2006. Italian Greenhouse Gas Inventory 1990–2009. National Inventory Report. Available at: http://www.sinanet.isprambiente.it/it/sinanet/serie_storiche_emissioni/CRF%201990–2009/view

[c] EEA, 2010. Annual European Union greenhouse gas inventory 1990–2008 and inventory report 2010. Submission to the UNFCCC Secretariat, Copenhagen.

this is an average value that was supposed to be not homogeneous but variable in space according to specific land uses.

Considering that emergy evaluation applied to territorial systems is still performed within certain margins of customization—that is both a limitation and an advantage—it is not useful here to compare results from other regions that would require a discussion on diversities in the methodological approaches followed and the baseline. Nevertheless, emergy has the characteristics to allow for a deeper understanding of the mutual relationship among different parts of a regional system, such as provinces and municipalities, different sectors of activity (residential, industrial, and agricultural), and different classes of resources (renewable–nonrenewable; local–imported). In this case, emergy values were scaled down from the regional to the municipal level (315 municipalities within the region) and then elaborated through GIS in order to provide a geographical representation of emergy intensities and visualize their spatial configuration within the region.

19.5 Integrated Spatial Analysis Based on Indicators through GIS

Both the indicators, the carbon emission and the areal empower density, provide values on the general environmental performance of the whole regional system and, based on their criteria of assessment, make possible to analyze the contribution of different sectors of activity and processes. These can be implemented through a spatial analysis in order to provide a more accurate description of the territory and find out the direct relation between indicators and landscape structures. A procedure of data geoprocessing through GIS was therefore elaborated in order to achieve a qualitative interpretation of the indicators and their variation in space as well as a quantitative estimation of their values relative to land uses.

The operation of geoprocessing is based on the localization of activities and land uses. This allowed for detecting the varying intensity that energy and material inflows (in terms of empower) and material outflows (in terms of CO_{2eq}) develop in the region. Essentially, this geographic representation has revealed an unusually shaped landscape that is a spatially explicit representation of the behavioral organization of the population settled in the region with its own social, economic, and cultural system.

In Figure 19.3, maps of GHG emissions are shown. Patterns represent at a glance the distribution of GHG emissions related to land uses, such as urban (Figure 19.3a), industrial (Figure 19.3b), and agricultural (Figure 19.3c). In particular, GHG emissions associated with the urban sector (includes energy use for the transport sector that was plotted spatially at the location

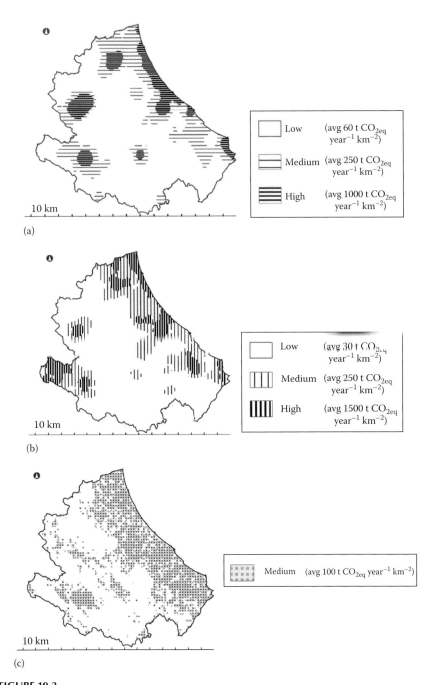

FIGURE 19.3
Patterns of GHG emissions associated with the (a) urban, (b) industrial, and (c) agricultural sectors in the Abruzzo region (2006) and

(continued)

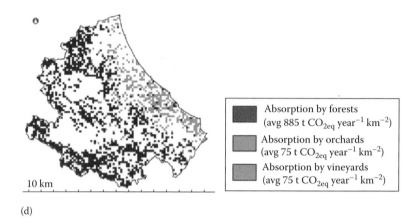

(d)

FIGURE 19.3 (continued)
Patterns of GHG emissions associated with the (d) pattern of CO_2 absorption by forests, orchards, and vineyards.

of dwellings) are mostly localized in a defined high-intensive area of about 1087 km^2 (10% of the region) with an average intensity of 1000 t CO_{2eq} km^{-2}. The medium intensive area covers 2727 km^2 (25% of the region) with an average of 250 t CO_{2eq} km^{-2}. GHG emissions due to the industrial sector are mostly concentrated in a few bounded areas, with industrial districts, of 1250 km^2 (12%) with an average intensity of almost 1500 t CO_{2eq} km^{-2}. The medium intensive area covers 2800 km^2 (26% of the region) with an average of 250 t CO_{2eq} km^{-2}. GHG emissions associated to the agricultural sector are localized in a wider area of over 5000 km^2 (43%) with an average low intensity of almost 100 t CO_{2eq} km^{-2}. Figure 19.3d shows the location of CO_2 absorption through forests (2559 km^2, equal to 27.3% of the region) with an average absorption of 885 t CO_{2eq} km^{-2} and agricultural fields such as orchards and vineyards (773 km^2, equal to 7.2%) with an average 75 t CO_{2eq} km^{-2} absorption.

In Figure 19.4a, GHG emissions from the different sectors of activity described earlier were overlaid to each other in order to give a disaggregated description of gross emissions in a unique representation. The map in Figure 19.4b combines GHG outflows and carbon inflows and thus shows a synthetic representation of CO_{2eq} net emissions throughout the region as a balance between effective GHG emissions and the CO_2 absorption by woods, forests, orchards, and vineyards. This identifies different roles, with respect to environmental impact and climate change, associated with different land uses. The areas with a high intensity of CO_{2eq} outflows (red pixels), which mainly include urban centers and industrial districts, identify a circumscribable zone within the coastal plain and a few areas around the inland towns (high-intensive net emission: 3500 t CO_{2eq} km^{-2}; low-intensive net emission: 500 t CO_{2eq} km^{-2}). Areas with CO_2 inflows (high net absorption: 250 t CO_{2eq} km^{-2}; low net absorption: 100 t CO_{2eq} km^{-2}), which means the

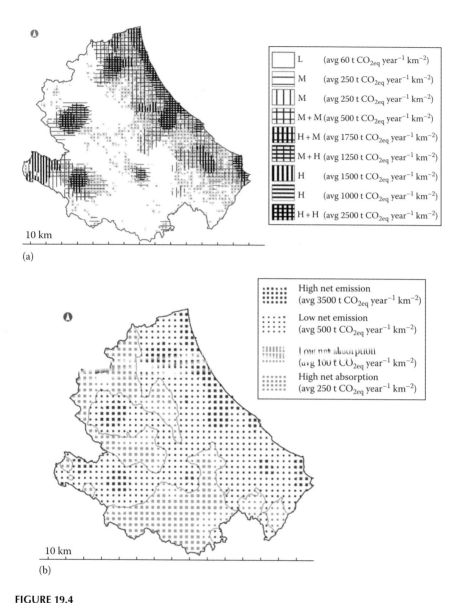

FIGURE 19.4
(a) Overlap of GHG emissions associated to the urban, industrial, and agricultural sectors and (b) map of total GHG net emission and absorption intensity in the Abruzzo region (2006).

removal of CO_2, correspond to the natural parks and protected areas and give a clear information about the local consistence of natural systems and the importance of ecosystems' services.

Figure 19.5 shows a configuration emerged from the assessment of the flows of energy and matter converging on the territory and measured in

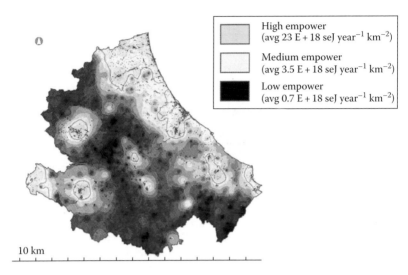

High empower
(avg 23 E + 18 seJ year^{-1} km^{-2})

Medium empower
(avg 3.5 E + 18 seJ year^{-1} km^{-2})

Low empower
(avg 0.7 E + 18 seJ year^{-1} km^{-2})

10 km

FIGURE 19.5
Areal empower density in the Abruzzo region (2006). Areas classified from high intensive (yellow) to low intensive (blue).

emergy terms (data elaborated from Pulselli [2010]). It reflects through a graduated colored scale (from yellow to blue), the intensity of use, and exploitation of environmental resources. Also, this emergy-based geography is nonhomogeneous and is different in the various local areas depending on the intensity and type of resource utilized. Considering three main intervals of empower—from a higher range, to medium, to lower—some areas can be circumscribed within defined boundaries in order to highlight congruencies in urban development. The map shows distinct areas that were drawn based on the spatial distribution of empower values. Low-intensive areas correspond to an average areal empower density around 0.7E + 18 seJ year^{-1} km^{-2}, medium areas to 3.5E + 18 seJ year^{-1} km^{-2}, and high-intensive areas have an average areal empower density of 23E + 18 seJ year^{-1} km^{-2}, covering less than 15% of the region. The first and the latter corresponds to protected areas (natural parks, mountain regions, and forests) and highly urbanized areas (urban and industrial centers), respectively. Medium values generally correspond to cultivated areas or those in which agriculture and urban sprawl coexist.

Figure 19.6 is a combined representation of carbon emission and areal empower density aimed at providing a qualitative interpretation of outcomes. In the map, the empower is in the color scale from yellow (high density) to blue (low) and the carbon emission in pixels from red (net emission) to green (net absorption). Although the two indicators refer to different units and intensities, the overlaid patterns show a basic coherence. The whole configuration revealed landscape structures—such as cities, industries, and farms—as elements that absorb and

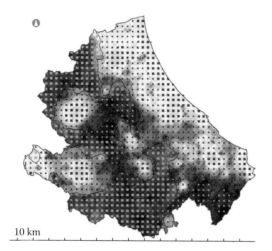

10 km

FIGURE 19.6
Overlaid patterns of GHG net emission–absorption (red-green pixels, respectively) and areal empower density (yellow-blue from high to low intensive).

process environmental resources (energy and material inflows) and proportionally release carbon dioxide and other GHG in the atmosphere (material outflow). The latter can be interpreted, in a wider sense, as a partial representation of the entropic output depending on the behavioral organization of the region.

19.6 Conclusion

The combined use of environmental indicators, carbon emission, and areal empower density has expressed the difference among the various land uses with reference to the net GHG emission and the environmental resource use as indices of environmental performance. These were assessed in terms of tons of equivalent carbon dioxide (CO_{2eq} year^{-1} km^{-2}) and solar emergy Joules (seJ year^{-1} km^{-2}), respectively.

The assessment of indicators and their elaboration through GIS allowed for a deeper analysis, visualized in the form of spatial patterns with varying intensity, for a quantitative and qualitative understanding of the less evident but essential aspects of the life of the regional system. With reference to the case study, outcomes contributed to reveal a kind of "thermodynamic profile" of the Abruzzo region. A whole configuration emerged as the representation of a general behavioral organization that depends on land uses, human settlements, processes, technologies, and landscape structures. This systemic approach helps to figure out the relations that the various territorial spheres establish with the sources of resources they feed through (mainly through the empower density) and with external sinks (through the carbon emission).

The measure of both GHG emission and emergy flows is aimed at identifying and quantifying the impact of differently anthropized areas—such as urban and productive centers, farms, and natural systems—and to define their roles with respect to the entire region. In a thermodynamic framework, regional systems look to behave like dissipative structures. In particular, the high organization and dynamism of cities and industrial districts determine a high intensive use of environmental resources (low entropy input), measured through the empower, and a high entropy output. The carbon emission can be interpreted as a factor that provides an entropy increase as well as an exergy destruction. In general, high values of carbon emission correspond to areas with the highest empower values. From a sustainability viewpoint, these should be compensated by the presence of other areas, with different vocations, that provide renewable resources and absorb CO_2. These areas with a lower empower and a function of CO_2 absorption by natural systems (woods and forests) were clearly identified in the region. Their capacity, in terms of CO_2 removals, was estimated around 19% of the total GHG emissions within the region.

Traditionally, the aim of systemic indicators is to suggest the political and management solutions necessary for achieving a more sustainable way of life. Based on their systemic nature, both the indicators presented in this chapter were conceived to provide information about the state of a system, such as the Abruzzo region. Planners and designers of sustainable energy regions are expected to use the information provided by these, and a few other indicators, in order to foresee and monitor the effects of specific actions and policies on the whole regional sustainability. Based on a comprehensive balance, variations induced by specific phenomena, whether they were an emerging economic trend or a particular policy imposed by the administration, can be detected and measured as perturbations in a complex system at the landscape scale.

The areal empower density and the carbon emission, and their elaboration in the form of spatial patterns, show two complementary aspects of the region and inform on its life and sustainability. Based on this information, that is becoming more and more accurate, planning policies and strategies can be developed relative to different land uses and behaviors, and their effects can be easily measured and monitored in time. A progressive transition from an electricity production or a transport system that uses fossil fuels to renewable sources, a ruled program that limits the extraction of materials from quarries and mines in time, an efficient system of waste treatment and a social campaign aimed at promoting an ecological consciousness, a reduction in the use of synthetic fertilizers, and a more efficient use of biomass from agricultural and ecosystems, shorter process chains, restrictive constrains to urban sprawl in favor of the renewal of existing volumes, and passive techniques for energy saving in buildings are examples of diffuse actions aimed at decreasing the use of nonrenewable resources and GHG emissions. A "business-as-usual" scenario, based on current trends, and a "sustainable energy landscape" scenario, in which nonrenewable resources are progressively substituted by renewables and ecosystem functions—such as CO_2 removal by woods, forests, and urban green areas—are preserved and even strengthened, can be compared to each other. Sustainability indicators, elaborated through GIS, are promising tools that combine quantitative and qualitative information on regional systems' behavior and organization and allow for a more conscious planning and design of sustainable energy landscapes of the future.

Acknowledgments

This study was funded by the Parks, Territory, Environment, and Energy Management Sector of the Abruzzo region. The authors thank the Councilor Franco Caramanico and the Director Antonio Sorgi for kindly assisting us. This research project was conducted under the scientific supervision of Professor Enzo Tiezzi whose scientific heritage is today a fundamental key for interpreting and developing our interdisciplinary works.

References

Alexis, A. 2002. Use of GIS and GPS as a QA tool in emission inventory. Presented at the *11th USEPA-Annual Emissions Inventory Conference*, Atlanta, GA, 9pp.

Ascione, M., Campanella, L., Cherubini, F., and Ulgiati, S. 2009. Environmental driving forces of urban growth and development. An emergy-based assessment of the city of Rome, Italy. *Landscape and Urban Planning* 93: 238–249.

Baggott, S.L., Brown, L., and Cardenas, L. 2006. *UK Greenhouse Gas Inventory, 1990–2004*. AEA Consulting, Didcot, Oxfordshire, U.K.

Barrett, J., Birch, R., Cherrett, N., and Wiedmann, T., 2005. Exploring the application of the ecological footprint to sustainable consumption policy. *Journal of Environmental Policy and Planning* 7: 303–316.

Bastianoni, S. and Marchettini, N. 1997. Emergy/exergy ratio as a measure of the level of organization of systems. *Ecological Modelling* 99(1): 33–40.

Bastianoni, S., Pulselli, F.M., and Rustici, M. 2006. Exergy versus emergy flow in ecosystems: Is there an order in maximizations? *Ecological Indicators* 6: 58–62.

Byrne, J., Hughes, K., Rickerson, W., and Kurdgelashvili, L. 2007. American policy conflict in the greenhouse: Divergent trends in federal, regional, state and local green energy and climate change policy. *Energy Policy* 35: 4555–4573.

Campbell, D.E. 1998. Emergy analysis of human carrying capacity and regional sustainability: An example using the State of Maine. *Environmental Monitoring Assessment* 51: 531–569.

Campbell, D.E., Brandt-Williams, S.L., and Meisch, M. 2005. *Environmental Accounting Using Emergy: Evaluation of the State of West Virginia.* U.S. EPA, Narragansett, RI.

Campbell, D.E., Meisch, M., DeMoss, T., Pomponio, J., and Bradley, M.P. 2004. Keeping the books for environmental systems: An emergy analysis of West Virginia. *Environmental Monitoring Assessment* 94: 217–230.

Carney, S. and Shackley, S. 2009. The greenhouse gas regional inventory project (GRIP): Designing and employing a regional greenhouse gas measurement tool for stakeholder use. *Energy Policy* 37: 4293–4302.

Castaldi, S., Carfora, A., Vigliotti, M., and Valentini R. 2005. A GIS based application of the IPCC protocol for the estimation of N_2O emissions from agricultural ecosystems of the Italian region. *15th Meeting of the Italian Society of Ecology*, Torino, Italy, 7pp.

D.L. 03/03/11, No. 28. Decreto Legge in recepimento della direttiva 2009/28/CE relativa al burden sharing regionale in Italia, 12pp.

Dalvi, M., Beig, G., Patil, U., Kaginalkar, A., Sharma, C., and Mitra, A.P. 2006. A GIS based methodology for gridding a large-scale emission inventories: Application to carbon-monoxide emissions over Indian region. *Atmospheric Environment* 40: 2995–3007.

Dennis, A., Fraser, M., Anderson, S., and Allen, D. 2002. Air pollution emissions associated with forest, grassland, and agricultural burning in Texas. *Atmospheric Environment* 36: 3779–3792.

Diem, J.E. and Comrie, A.C. 2001. Allocating anthropogenic pollutant emissions over space: Application to ozone pollution management. *Journal of Environmental Management* 63: 425–447.

Fleming, P., Webber, P., Chadwick, H., and Devine-Wright, P. 2001. Greenhouse gas inventory. In: *Changing by Degrees: The Potential Impacts of Climate Change in the East Midlands.* Shackley, S., Kersey, J., Wilby, R., and Fleming, P. (Eds.). Ashgate, Aldershot, U.K., pp. 185–219.

Gaffney, P. 2002. Developing a statewide emission inventory using geographic information system (GIS). Presented at the *11th USEPA—Annual Emissions Inventory Conference*, Atlanta, GA, 12pp.

Garg, A., Bhattacharya, S., Shukla, P.R., and Dadhwal, V.K. 2001. Regional and sectoral assessment of greenhouse gas emissions in India. *Atmospheric Environment* 35: 2679–2695.

Higgins, J.B. 2003. Emergy analysis of the Oak openings region. *Ecological Engineering* 21: 75–109.

IPCC, 2006. IPCC guideline for national greenhouse gas inventories. In: *Methodology Report Prepared by the National Greenhouse Gas Inventories Programme.* Eggleston, H.S., Buendia, L., Miwa, K., Ngara, T., and Tanabe, K. (Eds.), 5 vols. IGES, Japan.

IPCC, 2007. Climate change: The scientific basis. In: *Report Prepared for Intergovernmental Panel on Climate Change by Working Group I.* Solomon, S., Qin, D., Manning, M., Chen, Z., Marquis, M., Averyt, K.B., Tignor, M., and Miller, H.L. (Eds.). Cambridge University Press, Cambridge, U.K., p. 996.

Jones, P., Patterson, J., and Lannon, S. 2007. Modelling the built environment at un urban scale—Energy and heath impacts in relation to housing. *Landscape and Urban Planning* 83: 39–49.

Jones, P., Williams, J. et al. 2000. Planning for a sustainable city: An energy and environment prediction model. *Journal of Environmental Planning and Management* 43(6): 855–872.

Jørgensen, S.E., Nielsen, S.N., and Mejer, H. 1995. Emergy, environ, exergy and ecological modelling. *Ecological Modelling* 77: 99–109.

Koerner, B. and Klopatek, J. 2002. Anthropogenic and natural CO_2 emission sources in an arid urban environment. *Environmental Pollution* 116: S45–S51.

Markakis, K., Poupkou, A., Melas, D., Tzoumaka, P., and Petrakakis, M. 2010. A computational approach based on GIS technology for the development of an anthropogenic emission inventory of gaseous pollutants in Greece. *Water Air Pollution* 207: 157–180.

Odum, H.T. 1996. *Environmental Accounting: Emergy and Environmental Decision Making*. Wiley, New York.

Odum, H.T. 1971. *Environment, Power and Society*. Wiley, New York.

Pennati, S., Castellani, V., and Sala, S. 2009. CO_2 budget estimation and mapping at a local scale. In: *Ecosystems and Sustainable Development VII*. Brebbia, C.A. and Tiezzi, E. (Eds.). WIT Press, Southampton, U.K., pp. 21–32.

Prigogine, I. 1997. *The End of Certainty: Time, Chaos and the New Laws of Nature*. Free Press, New York.

Prigogine, I. and Stengers, I. 1979. *La Nouvelle Alliance. Métamorphose de la Science*. Gallimard, Paris, France.

Pulselli, R.M. 2010. Integrating public policy and geographic information systems for monitoring resource use in the Abruzzo region (Italy). *Journal of Environmental Management* 91. 2349–2357.

Pulselli, R.M., Pulselli, F.M., and Rustici, M. 2008. Emergy accounting of the Province of Siena: Towards a thermodynamic geography for regional studies. *Journal of Environmental Management* 86: 342–353.

Pulselli, R.M., Rustici, M., and Marchettini, N. 2007. An integrated framework for regional studies: Emergy based spatial analysis of the Province of Cagliari. *Environmental Monitoring and Assessment* 133: 1–13.

Pulselli, R.M. and Tiezzi, E. 2009. *City Out of Chaos. Urban Self-organization and Sustainability*. WIT Press, Southampton, U.K.

Rabe, B. 2007. Beyond Kyoto: Climate change policy in multilevel governance systems. *Governance: An International Journal of Policy, Administration and Institutions* 20(3): 423–444.

Rosen, M.A. and Scott, D.S. 2003. Entropy production and exergy destruction: Part I—Hierarchy of Earth's major constituencies. *International Journal of Hydrogen Energy* 28: 1307–1313.

Sciubba, E. and Ulgiati, S. 2005. Emergy and exergy analyses: Complementary methods or irreversible ideological options? *Energy* 30: 1953–1988.

Smiatek, G. 2005. SOAP-based web services in GIS/RDBMS environment. *Environmental Modelling and Software* 20: 775–782.

Sozanska, M., Skiba, U., and Metcafe, S. 2002. Developing an inventory of N_2O emissions from British soils. *Atmospheric Environment* 36: 987–998.

Stremke, S. 2010. *Designing Sustainable Energy Landscapes: Concepts, Principles and Procedures*. PhD thesis, Wageningen University, Wageningen, the Netherlands.

Stremke, S., Dobbelsteen, A. van den, and Koh, J. 2011. Exergy landscapes: Exploration of second-law thinking towards sustainable landscape design. *The International Journal of Exergy* 8(2): 148–174.

Stremke, S. and Koh, J. 2010. Ecological concepts and strategies with relevance to energy conscious spatial planning and design. *Environment and Planning B: Planning and Design* 37: 518–532.

Su, M.R., Yang, Z.F., Chen, B., and Ulgiati, S. 2009. Urban ecosystem health assessment based on emergy and set pair analysis: A comparative study of typical Chinese cities. *Ecological Modelling* 220: 2341–2348.

Tiezzi, E. 2003. *The Essence of Time*. WIT Press Publications, Southampton, U.K., p. 123.

Titheridge, H., Boyle, G. et al. 1996. Development and validation of a computer model for assessing energy demand and supply patterns in the urban environment. *Energy and Environment* 7: 29–40.

United States Environmental Protection Agency (USEPA). 1998. A GIS-based modal model of automobile exhaust emissions. Report no. EPA-600/-98-097.

Wall, G. and Gong, M. 2001. On exergy and sustainable development—Part 1: Conditions and concepts. *Exergy, an International Journal* 1(3):128–145.

Webber, P. and Fleming, P. 2002. Emissions models and inventories for local greenhouse gas assessment: Experience in the East Midlands region of the UK. *Water, Air & Soil Pollution: Focus* 2(5–6): 115–126.

Wilbanks, T.J. and Kates, R.W. 1999. Global change in local places: How scale matters. *Climatic Change* 43: 601–628.

Wolman, A. 1965. The metabolism of the city. *Scientific American* 213: 179–190.

Zhang, L.X., Chen, B., Yang, Z.F., Chen, G.Q., Jiang, M.M., and Liu, G.Y. 2009a. Comparison of typical mega cities in China using emergy synthesis. *Communications in Nonlinear Science and Numerical Simulation* 14: 2827–2836.

Zhang, Q., Wei, Y., Tian, W., and Yang, K. 2008. GIS-based emission inventories of urban scale: A case study of Hangzhou, China. *Atmospheric Environment* 42: 5150–5165.

Zhang, Y., Yang, Z., and Yu, X. 2009b. Evaluation of urban metabolism based on emergy synthesis: A case study for Beijing (China). *Ecological Modelling* 220: 1690–1696.

Zhang, Y., Yang, Z., and Yu, X. 2009c. Ecological network and emergy analysis of urban metabolic systems: Model development, and a case study of four Chinese cities. *Ecological Modelling* 220: 1431–1442.

Zhao, B., Wang, P., Ma, J.Z., Zhu, S., Pozzer, A., and Li, W. 2011. A high-resolution emission inventory of primary pollutions for the Huabei region, China. *Atmospheric Chemistry and Physics Discussions* 11: 20331–20374.

Part IV

Education

20

Designing Sustainable Energy Islands: Applying the Five-Step Approach in a Graduate Student's Studio in the Netherlands

Renée de Waal, Sven Stremke, Rudi van Etteger, and Adri van den Brink

CONTENTS

20.1 Introduction

More and more educational programs for art, architecture, urbanism, and landscape architecture pay attention to planning and design of sustainable landscapes and cities and sustainable energy landscapes in particular. In this book, examples are given from the University of Minnesota and the Royal Institute of Art in Stockholm, who focus, respectively, on improving the sustainability of their own campus (Chapter 21) and improving the sustainability of cities worldwide (Chapter 22). Within the Netherlands, the master program of Landscape Architecture and Planning at Wageningen University devotes substantial efforts to the planning and design of sustainable energy landscapes. Earlier in this book, Stremke introduced the five-step approach to the planning and design of such landscapes (see Chapter 5). Here, we discuss how this approach was applied in the setting of the 2011 graduate students' studio at Wageningen University.

For several years, the transition to renewable energy has been a hot topic in society. The drivers behind renewable energy transition are the adverse effects of fossil fuels on the climate, resource depletion, and the fact that the world's main resources are concentrated in a small number of politically unstable countries. There is a broad consensus on a strong and bidirectional relationship between space and renewable energy. Various authors assert that both the landscape architecture and the planning discipline can play a significant role in the transition to renewable energy (see, e.g., Ghosn 2009, Schöbel 2009, van Hoorn et al. 2010, Noorman and de Roo 2011). In particular, it is believed that landscape design and planning can, and should, contribute significantly to the development of sustainable energy landscapes, namely landscapes that are well adapted to renewable energy sources without compromising other landscape services, landscape quality or biodiversity (Stremke 2010, p. 1).

The relationship between renewable energy and landscape is an important research focus of the Landscape Architecture Group at Wageningen University. Over the past years, this research has been connected with education through, for instance, MSc studios as well as BSc and MSc thesis projects. Each year, our group organizes the MSc studio* to train students in a current, relevant, and complex assignment in landscape architecture, planning, and research. The 2011 studio was dedicated to the planning and design of two sustainable energy islands in the Dutch delta region—Goeree-Overflakkee and Schouwen-Duiveland—with local, provincial, and national authorities acting as commissioner to the students.

Studios on energy landscapes were also given in 2007 and 2009, with different study areas and commissioners, as well as different assignments and educational approaches. By giving the 2011 studio a clear methodological structure (i.e., the five-step approach), we aimed to provide the students with an efficient and result-oriented way of learning while, at the same time, provide the commissioners with interesting and useful results. This presented a change compared to the previous studios, in which the students developed their own approaches. The question that is discussed in this chapter therefore is: What difference did the application of the five-step approach in the studio make to the results, and how did the students, commissioners, and teachers evaluate its application?

---

* Also referred to as "Atelier," see, e.g., Stremke et al. (2011).

In Section 20.2, we present how we (the authors and teachers of the studios) incorporated the five-step approach into the overall structure of the studio. In Section 20.3, we explore the similarities and differences in the outcomes for the two islands by providing examples of the students' work. In Section 20.4, we compare the structure and outcomes of the 2011 studio with those of the 2007 and 2009 studios. We conclude this chapter by reflecting on the application of the five-step approach in the educational setting of a graduate studio on sustainable energy landscapes.

20.2 The 2011 "Sustainable Energy Islands" Studio

20.2.1 About the Studio

The 3 month studio is a compulsory course in the Landscape Architecture and Planning MSc program. It is worth 18 ECTS credits.* About 60% of the participants are students of landscape architecture, about 30% study spatial planning, and about 10% study socio-spatial analysis. About half of the student body comes from outside the Netherlands, that is, from Asia, Africa, North and South America, and many European countries. In the studio, students form multidisciplinary teams to exchange knowledge and learn how to work together in a result-oriented way. To make the assignment realistic, students work on a project commissioned by a body outside the university.

Each studio is tutored by teachers from the Landscape Architecture, the Land-Use Planning, and the Cultural Geography chair group, as well as external experts, and a team of coaches who support the intercultural and interdisciplinary teamwork. Because the majority of the students in the 2011 studio had little knowledge of renewable energy prior to the course, we invited experts, researchers, and landscape architects to give a series of lectures on energy landscapes. Two books† were suggested to the students to help them with, for instance, energy calculations.

Each studio is structured in three phases, regardless of the topic. Studios commence with an analysis phase that is followed by a phase of scenario making and/or envisioning on the regional scale. During these two phases, students work in interdisciplinary teams. In order to also train students in individual competences relevant to their specialization (landscape architecture, planning, or research), the third phase consists of an individual project. The students have to formulate their own assignments, based on the results

---

* ECTS stands for European Credit Transfer System. One credit equals a study load of 25–30 h.
† Namely, *Sustainable Energy—Without the Hot Air* by MacKay (2009) and *Renewable Energy Resources* by Twidell and Weir (2006).

of the previous two phases. Every year, we take the students on field trips and on a 1 week international excursion.

In the 2011 studio, we visited the study area twice and stayed for a total of 5 days. During the field trips, the students experienced the power of natural forces in the landscape—how wave power erodes the coast, how the wind shapes dunes, and how powerfully the sun shines—which further encouraged them to think about the relationship between landscape and renewable energies. For additional inspiration, we took the students to the island of Samsø (Denmark) and the Western Harbor area in Malmö (Sweden). The inhabitants of Samsø have managed to become energy self-sufficient by making use of locally available renewable energy sources. Although the Danish island has a far lower population density than the two Dutch islands,* this example of successful energy transition in a similar rural landscape inspired both the students and the commissioners. The Western Harbor area in Malmö provided an energy-conscious reference for the more urban parts of the two Dutch islands.

20.2.2 Specific Assignment

During the 2007 and 2009 studios on renewable energy, we explored how two Dutch regions—Zuid Limburg and Arnhem–Nijmegen—could be transformed such that they would become energy self-sufficient on the basis of local renewable sources. Based on these studios, other commissioned projects (e.g., Broersma et al. 2011), and reference studies abroad (Waal and Stremke 2011), we concluded that a full transition to renewable energy is technically feasible in the Netherlands, especially in rural areas with relatively low population density.

However, the Netherlands is still far from reaching the targets for renewable energy provision. According to EU targets, the Netherlands should aim for a 14% share of renewable energy in 2020 (van Hoorn et al. 2010). The share of renewables in 2010, however, was no higher than 3.8% (CBS 2011). Despite the great potentials for renewable energy in the Netherlands (see, e.g., WWF 2011) and technological innovation, difficulties occur in implementation. There are also other problems, primarily of economic, political, and social nature (see e.g., Pasqualetti 2011).

Given these preconditions, we asked the students to rethink the sustainability of renewable energy in the physical environment and, subsequently, to design *sustainable* energy landscapes rather than *renewable* energy landscapes. The assignment of the 2011 studio postulated that renewable energy provision should not compromise landscape quality, biodiversity, food production, and other life-supporting ecosystem services, people's physical safety, and social cohesion (Stremke et al. 2011).

---

* Samsø has a population density of 36 inhabitants/km^2 compared to 184/km^2 on Goeree-Overflakkee and 148/km^2 on Schouwen-Duiveland.

20.2.3 Context and Study Areas

The 2011 studio was commissioned by national, regional, and local governments. Between 2008 and 2011, the Dutch Ministry of Infrastructure and Environment funded the *Mooi Nederland* ("Beautiful Netherlands") innovation program, which aimed to enhance the spatial quality of, for instance, renewable energy landscapes (Ministerie van I&M 2011). Via this program, we got in contact with representatives from the island of Goeree-Overflakkee (see Figure 20.1). Representatives of the ISGO*—a collaboration between the

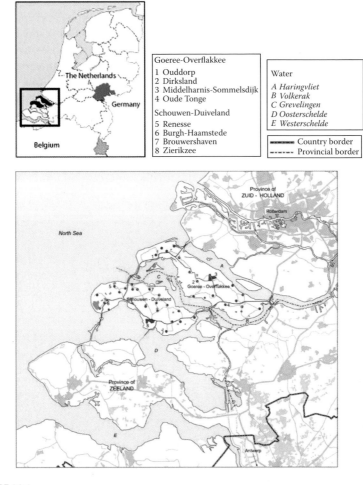

FIGURE 20.1
The islands Goeree-Overflakkee and Schouwen-Duiveland and their location within the Netherlands.

* ISGO stands for Intergemeentelijk Samenwerkingsverband (Intermunicipal collaboration) Goeree-Overflakkee.

island's four municipalities—want to increase renewable energy provision to counteract the economic recession and population decline and to mitigate the adverse effects of climate change in the long run. The island offered the students a case to study the perception of renewable energy in these local communities and to develop scenarios, visions, policy recommendations, and designs for energy transition in a real setting. Because Goeree-Overflakkee would have been a small study area for the 60 students that had registered for the course, we contacted the municipality of the neighboring island Schouwen-Duiveland (see Figure 20.1). The municipality was willing to facilitate the studies of about 30 students working on the same assignment that had been put forward by the representatives of Goeree-Overflakkee.

The island of Goeree-Overflakkee is part of the province of Zuid Holland. The island of Schouwen-Duiveland lies just to the south of Goeree-Overflakkee but constitutes a part of the province of Zeeland. The four municipalities on Goeree-Overflakkee are in the process of being merged into one municipality. The island of Schouwen-Duiveland already consists of one municipality, as a result of recent reforms. The population density of Goeree-Overflakkee is 184 inhabitants/km^2 and of Schouwen-Duiveland 148 inhabitants/km^2. Compared to the average population density of the Netherlands (almost 400 inhabitants/km^2), both islands are quite rural. Both islands consist of four major landscape types, namely, agricultural, recreational, urban, and coastal landscapes (see Figures 20.2 through 20.5).

FIGURE 20.2
Large-scale agricultural landscape on Goeree-Overflakkee. (Photo by author, 2011.)

FIGURE 20.3
Lake Grevelingen, with the recreational landscape of the Port Zélande holiday village in the distance. (Photo by author, 2011.)

FIGURE 20.4
Street in the village of Dirksland on Goeree-Overflakkee. (Photo by Loes van Schie, 2011.)

FIGURE 20.5
Coastal dune landscape near the village of Renesse on Schouwen-Duiveland. (Photo by author, 2011.)

20.2.4 Structure of the Studio around the Five-Step Approach

To enable the students to compose innovative and yet comparable proposals for the development of the two islands, we offered them the five-step approach as it was put forward by Stremke (see also Chapter 5). Briefly, the five steps comprise the following:

 I. Analysis of the current landscape (including historical developments) and energy systems

 II. Inventory of near-future developments that have to be taken into account

 III. Exploration of possible far-futures in the form of scenarios

 IV. Envisioning and illustration of a set of desired futures

 V. Identification and elaboration of concrete energy-conscious interventions

In order to subdivide the 2011 studio into the three standard phases, we combined step I with step II in phase 1 and step III with step IV in phase 2. Step V was dealt with in phase 3 (see Figure 20.6). Before the start of the studio, we divided the 60 students equally over the two islands and formed (per island) four subgroups of seven to eight students.

Besides offering a methodological framework, we also partly determined the substance of the first two phases. During phase 1, each of the subgroups

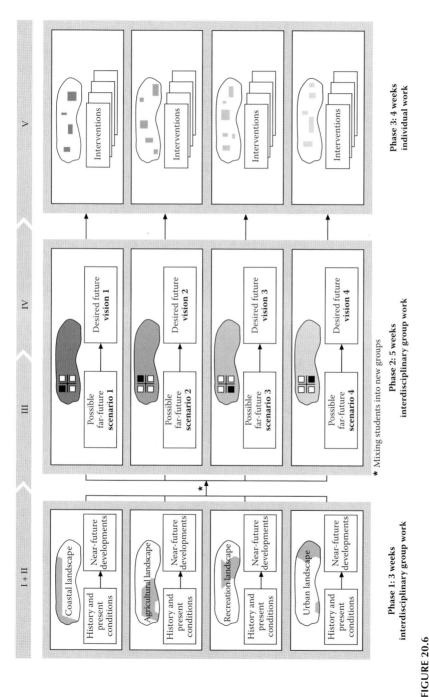

FIGURE 20.6

Structure of the 2011 studio around the five-step approach.

had to study one of the four landscape types that are found on the islands. For each of these landscape types, students had to study both historical and current conditions, the potentials and limitations for renewable energy, as well as near-future developments.

At the outset of phase 2, students were divided into four new groups, allowing them to share the knowledge they had acquired of one of the landscape types. Students were asked to study and concretize existing socioeconomic scenario studies in order to illustrate possible futures for their island. The scenario study contrasts a civil society with a central government and an open global economy with a closed regional economy resulting in four scenarios, namely Strong Europe, Global Economy, Regional Government, and Regional Civil Society.* After a broad exploration of one possible future, each subgroup had to envision a desired future, taking into account the conditions described in their respective scenario.

In phase 3, the students had to focus on either a particular topic or a specific area. During this phase, more than 50 proposals for sustainable energy landscapes were put forward, each one containing several physical interventions and/or policy recommendations.

Although the studio was structured as described above, the students worked iteratively throughout steps III, IV, and V. During the tutoring, the students consulted the teachers and each other, continually revising and improving their proposals. Some of the students used the field trips to test their ideas on site and to discuss them with local actors.

20.3 Results of the 2011 Studio

A total of 60 students worked on this project full time for a period of 12 weeks, thus investing about 28,800 h of work. The resulting body of work is way too large to be discussed within the constraints of this chapter. Nevertheless, surprising similarities and differences were found when we compared the results from both islands. Here, we give a brief account of these similarities and differences by highlighting some of the students' results. All of the results discussed in this section can be found in the original student reports, and most of it in the 2011 studio book entitled *Beyond Fossils. Envisioning desired futures for two sustainable energy islands in the Dutch delta region* (Stremke et al. 2011).

20.3.1 Results from Steps I and II: Analysis of Current Conditions and Near-Future Developments

Based on their analysis, the students concluded that the two islands are rather similar. Both islands are the result of several islands growing together due

---

* Largely based on http://www.welvaartenleefomgeving.nl/scenario.html

to land reclamation. On both islands, one finds large-scale agricultural land on the reclaimed areas (Figure 20.2) and more small-scale agriculture on the older parts. On both islands, the coastal areas have always played an important role in the defense against the sea. In recent centuries, the two islands have become increasingly important tourist destinations (Figure 20.3). Most tourist activities are concentrated in the coastal landscape, mainly along the North Sea side, although, over the last years, the rural areas received more and more tourists. Finally, nature areas are present on both islands in the form of salt marshes and dune landscapes (Figure 20.5).

As for renewable energy, the potential for solar, wind, tidal energy, and biomass is equally good on both islands. The potential of the latter three is excellent, compared with the rest of the Netherlands.* Extracting heat from geothermal sources and heat/cold storage are two of the many possibilities for the built environment (Agentschap NL 2011). Further, blue energy† has a potential on some locations on Goeree-Overflakkee (Post 2009).

20.3.2 Results from Steps III, IV, and V: Scenarios, Envisioning, and Concrete Interventions

Despite the earlier similarities, the students identified unique characteristics of both the physical and the institutional landscapes on the two islands. These dissimilarities are one possible explanation for the different proposals that were put forward later in the studio. Although the islands are products of similar geological formation processes, Schouwen-Duiveland is older and therefore smaller scaled than its neighbor. The towns on the island exude a stronger historical atmosphere, compared to those on Goeree-Overflakkee. This is why larger (and probably more visible) energy-conscious interventions were proposed for Goeree-Overflakkee than for Schouwen-Duiveland.

The difference between the islands' water systems appeared to be another factor that led to differences between the proposals. The estuaries around Schouwen-Duiveland hold salt or brackish water, whereas the waters north and east of Goeree-Overflakkee contain freshwater. The Haringvliet and the Volkerak, respectively, are closed off from the North Sea by the Haringvliet dam (see Figure 20.1). Due to increasing problems with the water quality, there are plans to reopen the Haringvliet dam. Farmers on Goeree-Overflakkee fear that if the dam is opened, saltwater seepage will negatively affect current agriculture practice. This led some groups to envision cultivation of salt-resistant energy crops on Goeree-Overflakkee. On Schouwen-Duiveland, however, saltwater has already penetrated the groundwater system; a fact that caused even more students to address this challenge. If, on the contrary, the Haringvliet

---

* See, for example, Wolters-Noordhoff Atlasproducties (2008).
† Also referred to as osmotic power—energy that can be derived from the difference in salt concentration between saltwater and freshwater.

FIGURE 20.7
Bird's eye view of the blue energy plant (white arrow) and the surrounding landscape to the
east of Goree-Overflakee. Project by Koen Verhoeven and Karlijn Looman.

remains closed off from the North Sea, the region constitutes one of the best
potential sites for blue energy generation in the Netherlands. Two of the students took that as a starting point to design a blue energy plant. In the proposal,
they makes use of the existing Volkerak sluices, which presents a man-made
barrier between freshwater (in the Haringvliet) and saltwater (in the Volkerak)
(see Figure 20.7).

Concerning wind energy, the students designed a range of solutions for
the two islands. Three of the four Schouwen-Duiveland groups envisioned
offshore wind energy, whereas only one of the four Goeree-Overflakkee
groups proposed this for their island. However, the wind park proposed
for Goeree-Overflakkee was by far the biggest of the four proposals. This
difference between the proposals cannot be explained by the wind energy
potential, since it is equally good. Rather, it is possible that the Goeree-
Overflakkee students knew of the existing ideas for offshore wind energy
and that the proposed spot is the only place where an offshore wind park
would be permitted (according to current legislation). When it comes to
onshore wind energy, the opposite occurred and again this is not a consequence of differences in wind speeds. Both the number and the capacity of
turbines are much larger for Goeree-Overflakkee than their counterparts for
Schouwen-Duiveland. This might be because Goeree-Overflakkee is part
of Zuid Holland, a quite urbanized and densely populated province. The
rural island of Goeree-Overflakkee needs to accommodate a large share of

the province's capacity for wind energy generation. Schouwen-Duiveland has a different position in this respect, because it is part of the province of Zeeland. This province needs to accommodate a large share of wind energy too, but in Zeeland, the population is more evenly distributed in space and so is the pressure to install wind turbines.

This difference can also be recaptured when comparing the Global Economy vision* for both islands. For Goeree-Overflakkee, the ambition for renewable energy generation is much higher than for Schouwen-Duiveland (300% vs. 200% of the current consumption). This is because the Goeree-Overflakkee students felt more need to serve the needs of the nearby and densely populated Randstad.† This resulted for Goeree-Overflakkee in fewer but major energy interventions such as a large-offshore wind park, about 70 onshore wind turbines, a blue energy plant, and a biofuel production plant (see Figure 20.8). The vision for Schouwen-Duiveland, in contrast, proposes much smaller and more varied energy interventions, such as smaller wind turbines, solar energy, road heat, biomass, and heat/cold storage (see Figure 20.9). Another difference is that the Global Economy vision for Goeree-Overflakkee aims for good connections with the Randstad area, for instance, by proposing a new train connection powered by wind energy. The Global Economy vision for Schouwen-Duiveland, in contrast, proposes some new greenhouse clusters, but is not concerned with improving the island's connections with the rest of the Netherlands.

The previous examples concern proposals for renewable energy systems. The students, however, were required to focus not only on technical and spatial aspects but also address social, cultural, and aesthetic aspects. How to combine renewable energy with the existing leisure and tourism activities was one of the key research questions defined by the commissioners. For both islands, students dealt with this challenge in quite similar ways. The great majority of proposals reveal a strong belief that sharing information and providing education via revelatory design will increase the acceptance of and the sense of urgency about renewable energy. In some cases, this was underpinned by referring to literature from environmental aesthetics and environmental psychology. A number of students are convinced that energy-conscious interventions will attract people to the islands if the projects are innovative and unique. In this way, an increase in renewable energy provision can coincide with the increased attractiveness of the two islands for day-trippers and tourists. During the individual phase, students designed, for example, bike paths connecting prominent renewable energy facilities and, an "energy experience" route in the village of Middelharnis (Figure 20.10).

---

* The "Global Economy" vision combines a global and open economy with a civil society; see also Section 20.2.4.
† The "Randstad" is the conurbation in the west of the Netherlands including the major cities of Amsterdam, Rotterdam, The Hague, and Utrecht.

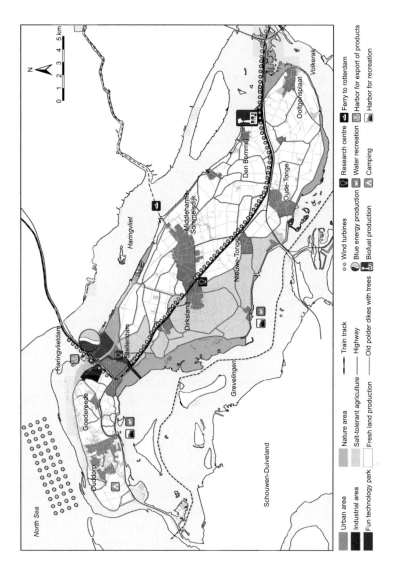

FIGURE 20.8
Global Economy vision for Goeree-Overflakkee.* (Redrawn on basis of plan in Stremke, S. et al. (Eds.), *Beyond Fossils. Envisioning Desired Futures for Two Sustainable Energy Islands in the Dutch Delta Region*, Wageningen University, Wageningen, the Netherlands, 2011.)

---

* Only the interventions proposed by the students are explained in the legend; other items on the map are existing.

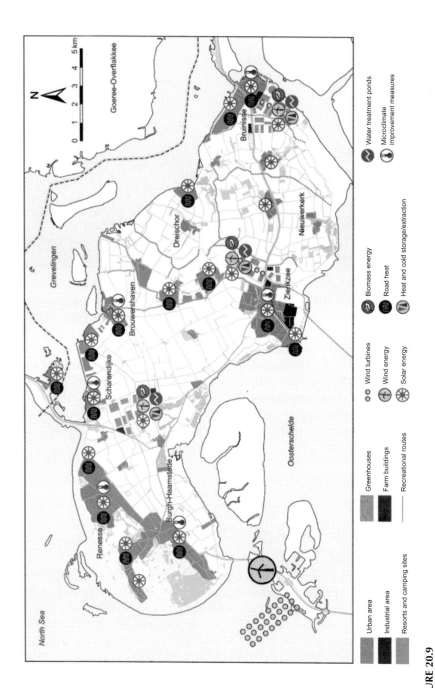

FIGURE 20.9

Global Economy vision for Schouwen-Duiveland. (Redrawn on basis of plan in Stremke, S. et al. (Eds.), *Beyond Fossils. Envisioning Desired Futures for Two Sustainable Energy Islands in the Dutch Delta Region*, Wageningen University, Wageningen, the Netherlands, 2011.)

FIGURE 20.10
Renewable energy provision and ecological water treatment can be experienced from recreational infrastructure on the island of Schouwen-Duiveland. Project by Zuzana Jancovicova.

Besides stimulating awareness and acceptance through energy-revelatory design, all groups concluded that the involvement of stakeholders throughout the transition process is of paramount importance. Among the various strategies, cooperatives were suggested to take the lead in new wind energy projects. Such cooperatives could help to spread more equally both the drawbacks of energy transition (e.g., nuisance) and the benefits of renewables.

20.4 Using the Five-Step Approach Compared to Two Earlier Studios

The 2007 and 2009 studios not only focused on different study areas but, more importantly, also employed frameworks other than the five-step approach (see Table 20.1). Since all three studios dealt with energy transition, it still seemed worthwhile to compare and discuss our experiences with the different approaches. For each of the three studios, a book has been put together to illustrate the structure of the respective studio and present the best students' projects. These books (Etteger and Stremke 2007, Helm et al. 2010, Stremke et al. 2011) and the original student reports provided the basis for the below comparison of the following three studios.

TABLE 20.1

General Information, Structure, and Methodological Approach of the 2007, 2009, and 2011 Studios

General Information	2007 Studio	2009 Studio	2011 Studio
Study area	Zuid Limburg	Arnhem–Nijmegen	Goeree-Overflakkee and Schouwen-Duiveland
ECTS	18 (12 weeks)	18 (12 weeks)	18 (12 weeks)
Number of students	9	22	60
International students	Yes	Yes	Yes
Disciplines	Landscape architecture	Landscape architecture	Landscape architecture
		Land-use planning	Land-use planning
		Socio-spatial analysis	Socio-spatial analysis
Commissioner	No, but the studio was related to the SREX[a] research project	Yes; Arnhem–Nijmegen City Region	Yes; Ministry of Infrastructure and Environment, Province of Zeeland, ISGO and Municipality of Schouwen-Duiveland
Teachers [b]	1, 2	1, 3	1, 2, 3, 4, 5, 6
Title of studio book	*ReEnergize South Limburg: Designing Sustainable Energy Landscapes*	*Sustainable Energy Landscapes: Biomass and other opportunities for an energy neutral City Region Arnhem–Nijmegen*	*Beyond Fossils: Envisioning desired futures for two sustainable energy islands in the Dutch delta region*
Structure of the studio			
Phases	Phase 1 group work; 3 weeks	Phase 1 group work; 3 weeks	Phase 1 group work; 3 weeks
	Phase 2 group work; 5 weeks	Phase 2 group work; 5 weeks	Phase 2 group work; 5 weeks
	Phase 3 individual; 4 weeks	Phase 3 individual; 4 weeks	Phase 3 individual; 4 weeks
Five-step approach applied as described in this chapter?	No	No	Yes
Comparable steps taken?			
Step I (analysis of current conditions)	Yes (approach determined by teachers)	Yes (approach determined by teachers)	Yes (approach determined by teachers)

(continued)

TABLE 20.1 (continued)

General Information, Structure, and Methodological Approach of the 2007, 2009, and 2011 Studios

General Information	2007 Studio	2009 Studio	2011 Studio
Step II (inventory of near-future developments)	No	No	Yes
Step III (socioeconomic scenarios)	No	Yes (defined by students)	Yes (preselected by teachers)
Step IV (composition of energy visions)	Yes	Yes	Yes
Step V (energy-conscious interventions)	Yes	Yes	Yes

ᵃ www.exergieplanning.nl
ᵇ Each number refers to one staff member.

20.4.1 The 2007 Studio "ReEnergize Zuid Limburg"

The 2007 studio focused on the region of *Zuid Limburg* bordering Belgium and Germany in the south of the Netherlands (see Figure 20.11). It is the only region in the Netherlands where black coal used to be mined; the last mine was closed in the 1970s. Being an energy region is therefore familiar to the inhabitants, though nowadays this no longer has any substance. In 2007, the region provided only 2% of its own energy. The goal of the 2007 studio was to find out whether the region could become self-sufficient based on locally available renewable energy sources. Issues such as economic short-fall, soil erosion, increased need for water storage, and the rebuilding of a local identity were also addressed. This required, similar to the 2011 studio, an integration of energy-conscious interventions with other issues.

The students worked in the same three phases as in 2011 (see Table 20.1), but the methodological approach was less predetermined. Phase 1 comprised an analysis of the entire region and the energy assignment. The students did this by unraveling reality in a *matterscape, powerscape,* and *mindscape.* Respectively, these scapes refer to the material and political landscape and to the perceived landscape as matter of imagination and appreciation (Jacobs 2006). For phase 2, the region was divided into three subregions: *Parkstad* ("park city"), *Maasvallei* ("Meuse valley"), and *Heuvelland* ("hill area"). Each group created a proposal for one of the aforementioned energy landscapes. In phase 3, every student elaborated a detailed design based on the group proposal. There were no scenarios included in the 2007 studio.

The results of this studio show great consistency throughout the phases. The visual presentation of the analysis, the proposed energy landscapes, and the detailed designs are both convincing and attractive. The focus lies on interventions in the matterscape, realizing energy-conscious interventions, increasing water storage, and improving landscape connectedness (both physically and visibly). Landscape perception is mainly addressed from an architectural perspective. Although students interviewed a few inhabitants, the proposals do not seem to involve inhabitants in the planning and design processes. Looking back on this studio, we hypothesize that it was the "bypassing" of the scenario step, among other reasons, that prevented more involvement of stakeholders in the overall planning and design process.

20.4.2 The 2009 Studio "Energy Neutral City Region Arnhem–Nijmegen"

The study area of the 2009 studio was the City Region Arnhem–Nijmegen (see Figure 20.11). Arnhem and Nijmegen are two municipalities in the east of the Netherlands including relatively large cities of the same name. Similar to ISGO in 2011, the City Region acted as commissioner to the 2009 studio. The assignment was to envision how the area could

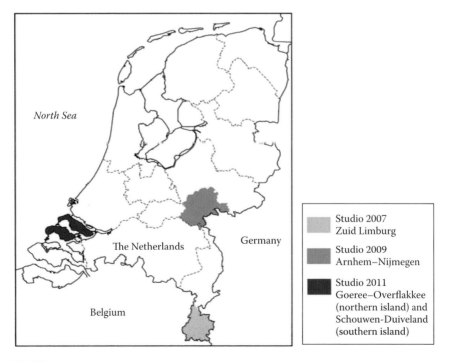

FIGURE 20.11
Study areas of the 2007, the 2009, and the 2011 studio.

become energy self-sufficient, with a focus on biomass as the main renewable energy source.

The studio was structured in the same way as in 2007 and 2011, but the students were divided into four groups (compared to three groups in 2007 and eight groups in 2011). Not only were matterscape, powerscape, and mindscape analyzed, but a fourth group studied latest energy technologies and mapped renewable energy potentials in the study region.

At the outset of the second phase, students had to develop their own methodological approach—a strategy that differed from 2007 to 2011. After studying and discussing possible approaches for a couple of days, the students proposed to work with scenarios and to create visions. The scenarios were inspired by (but not consisted with) the same existing socioeconomic scenario study that was utilized in 2011. However, the students created a coherent set of storylines for each scenario, taking exogenous and endogenous developments into account. Based on these storylines, the students developed four visions for a sustainable energy region Arnhem–Nijmegen: *Autarkic Rural, Autarkic Urban, Global Rural,* and *Global Urban.*

It turned out that asking the students to develop their own methodological approach was a rather tough task. It weakened their self-confidence and eventually led to a complaint being lodged with the tutors. Based on the published results, the strategy chosen for the 2009 studio proved to be less inspirational, compared with the five-step approach employed in 2011. Looking at the results of phase 2, it appeared that the 2009 students were able to create a sensible method in the form of well-organized and mutually synchronized scenarios. However, because they fail to explicitly use existing scenario studies, the results were less credible to outsiders compared to the 2011 scenarios, which built upon a well-known national scenario study. Although (or because) those national socioeconomic scenarios are perhaps not the most recent scenario study, they are (by now) well known to many policy makers as well as planning and design professionals. A drawback that, in terms of communication, often appeared to be a great advantage. Students combined their four scenarios and visions in a single report but did not assess the robustness of their proposed interventions. The individual plans, designs, and studies of 2009 are comparable both in quality and extensiveness with those in the other two studios.

20.5 Discussion and Conclusion

By integrating the five-step approach into the 2011 studio *Sustainable Energy Islands*, we aimed to provide a clear methodological structure that would allow students to work both efficiently and in a result-oriented manner, while providing room for architectural imagination. Another important objective was to generate results that were both interesting and useful for the commissioners of the studio. The main question raised in this chapter is: What difference did the application of the five-step approach in the studio make to the results, and how did the students, commissioners, and teachers evaluate its application?

To evaluate the application of the five-step approach in the studio, we asked the students to participate in a quick assessment at the end of the course. They voted on 11 statements—prerequisites for a meaningful approach to long-term planning and design at the regional scale—which are described in depth by Stremke et al. (2012). Overall, students were quite positive about the approach (see Figure 20.12).

Looking at the results in more detail, one can notice that the students, in spite of their overall appreciation of the five-step approach, consider statement 3—enabling active stakeholder participation—as a weak point. Within the scope of this educational project, they indeed went out on the street to make enquiries and arranged interviews with various actors and experts on the islands. However, they felt that they could only ask citizens to inform them rather than take an active role in the planning and design process—to "take a pencil and sketch a plan." This is probably due to the difficulties maintaining intense, two-way communication with stakeholders (beyond the group of commissioners) throughout the course of the studio.

The students were also critical about statement 9—the possibilities to assess the robustness of energy-conscious interventions. Without drawing any further conclusions, we should like to indicate that this might have been due to time constraints at the end of the studio. Many (but not all) students simply did not manage to perform a robustness analysis in the allotted time. Assessing the robustness of individual energy-conscious interventions is a task of great complexity that would have required intense collaboration between students who had just spent 4 weeks focusing on their individual projects. If an in-depth robustness analysis of the different energy-conscious interventions and policy recommendations is to be carried out, one must teach them how to do so, facilitate the collaboration between students during the final phase of the studio, and allot sufficient time for this task.

According to the teachers,* the students worked efficiently and in a result-oriented manner during the 2011 studio, especially when compared with the 2009 studio where students spent a couple of days on discussing possible methodological frameworks. The aforementioned difficulties of the students to realize stakeholder participation and to identify most robust interventions are acceptable from a teaching point of view as these do not present learning goals of the studio. Having said this, we like to stress that the lack of stakeholder participation should, however, be addressed in future studios, regardless of the chosen methodological framework.

---

* Including all teachers, also the ones who do not coauthor this chapter.

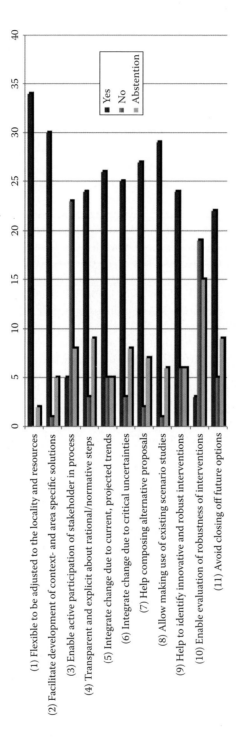

FIGURE 20.12
Student evaluation of the five-step approach as methodological framework in the 2011 studio. The length of the three different bars indicates the number of student votes.*

* Only two-thirds of the 60 students of the 2011 studio participated in this assessment.

All commissioners of the 2011 studio appreciated the student proposals. The representatives of the ministry noted "The research, scenarios, energy visions, and designs created in the atelier show that renewable energy production can add to the identity and spatial quality of the place. Even in rural areas with strong identities—like Goeree-Overflakkee and Schouwen-Duiveland—it is possible to increase the renewable energy production in a way that suits the islands and their inhabitants" (Bisschop and van Zelm van Eldik 2011 in Stremke et al. (2011); p. 16). The commissioner from Goeree-Overflakkee appreciated in particular the out-of-the-box thinking and controversial visions. He pointed out that "The scientific way of working with scenarios differs from our governmental way of working, and resulted in wild and ambitious plans and visions from which it is hard to distil feasible, desired developments. But then again, this way of working opened doors for the students that we, public servants, cannot open because of political or policy restrictions. It also provided us with plenty of possible futures within the scenarios; especially appealing was the way the students worked with various possibilities for energy production" (Seriese 2011 in Stremke et al. 2011; pp. 151–152).

On Goeree-Overflakkee, the studio generated a lot of positive attention for sustainable developments that resulted in the wish to consider the realistic potential of the plans and designs in more detail. A journalist from a regional newspaper interviewed teachers and students before she visited some sites on Schouwen-Duiveland together with the students (Pankow 2011). In addition, and quite surprisingly to us, the commissioners from both islands indicated that the studio might lead to more cooperation between Goeree-Overflakkee and Schouwen-Duiveland in the future. Based on our evaluations, we can ascertain that the goal of the 2011 studio—to provide the commissioners with both interesting and useful results—has been met.

When comparing the results of the 2011 studio between the two islands, it appeared that applying the five-step approach resulted in both similarities and differences. Applying the five-step approach led not only to solid and comparable results, but also allowed the identification of particularities. This is exactly what we as teachers are hoping for when we ask students, on the one hand, to base their proposals on a grounded understanding of relevant theories and the site while, on the other hand, to employ architectural imagination and human ingenuity.

The comparison of the 2011 studio with the studios in 2007 and 2009 showed that working in interdisciplinary teams led to richer proposals for sustainable energy landscapes, compared to working with students of landscape architecture alone. Relatively more attention was devoted to stakeholders in the 2009 and 2011 studios; this can be regarded as typical contributions from spatial-planning students. Working with concepts such as landscape experience and identity and researching the assumptions made for planning and design are valuable contributions from the students of socio-spatial analysis.*

---

* The interdisciplinary character of the studio is a consequence of a change in the organization and not a consequence of applying the five-step approach.

Although publications on the five-step approach (Stremke 2010, Stremke et al. 2012) indicate that interdisciplinary team work is beneficial to long-term regional planning and design, it should be emphasized even more in the description of the five-step approach and probably become a prerequisite.

When comparing the 2011 studio to that of 2009, it appeared that a structuring methodological framework enhances the efficiency of students' work and increases the depth of their results. Working with scenarios helped to stretch imagination and to explore innovative energy-conscious interventions that go well beyond studies that are limited to the analysis of today's physical reality. Working with existing socioeconomic scenarios, moreover, helps to preserve resources and to prevent students from losing the bond with the real world.

To conclude, we recommend applying the five-step approach not only in research (as discussed by Stremke in Chapter 5) but also in educational settings such as graduate studios on the planning and design of sustainable energy landscapes. The five-step approach has an important structuring value when working with a large group of students on the regional scale, when the topic of the studio is new to the students, and/or when a commissioner is awaiting inspiring yet useful results.

Acknowledgments

We thank the participants of the 2007, 2009, and 2011 studios, and especially those students whose illustrations are presented in this chapter. Moreover, we thank all commissioners for their great enthusiasm and their contributions to the studios. We are grateful to our colleague Adrie van't Veer for helping us with the figures. Last but not least, we thank the referee of this chapter for his useful comments.

References

Agentschap, N.L. 2011. *WarmteAtlas Nederland*. (Online) Available at: http://agentschapnl.kaartenbalie.nl/gisviewer/viewer.do?code=311bc9828c8015a87e9d3dd8fd179ed8 (accessed on May 30, 2012).

Bisschop, N. and Zelm van Eldik, D. van. 2011. The identity of energy landscapes. In: Stremke, S., Etteger, R. van, Waal, R. de, Haan, H. de, Basta, C., and Andela, M. (Eds.). *Beyond Fossils. Envisioning Desired Futures for Two Sustainable Energy Islands in the Dutch Delta Region*. Wageningen, the Netherlands: Wageningen University, pp. 15–16.

Broersma, S., Fremouw, M. A., Dobbelsteen, A. A. J. F. van den, Stremke, S., Waal, R. de, and Klap, K. 2011. *Duurzame energiebeelden voor de Veenkoloniën op basis van energie-potentiekartering en netwerkanalyse*. Research report. Delft, the Netherlands: TU Delft.

CBS. 2011. *Hernieuwbare Energie in Nederland 2010* (Renewable energy in the Netherlands in 2010). Den Haag, the Netherlands: Centraal Bureau voor de Statistiek.

Etteger, R. van and Stremke, S. (Eds.) 2007. *ReEnergize Zuid Limburg. Designing Sustainable Energy Landscapes.* Wageningen, the Netherlands: Wageningen University.

Ghosn, R. (Ed.) 2009. *Landscapes of Energy.* Cambridge, MA: Harvard University Press.

Helm, S. van den, Knaap, W. van der, and Telgen, S. van (Eds.) 2010. *Sustainable Energy Landscapes. Biomass and Other Opportunities for an Energy Neutral City Region Arnhem–Nijmegen.* Wageningen, the Netherlands: Wageningen University.

Hoorn, A. van, Tennekes, J., and Wijngaart R. van der. 2010. *Quickscan energie en ruimte* (Quickscan Energy and Space). Den Haag, the Netherlands: Planbureau voor de Leefomgeving.

Jacobs, M. 2006. *The Production of Mindscapes: A Comprehensive Theory of Landscape Experience.* PhD thesis. Wageningen, the Netherlands: Wageningen University.

MacKay, D. J. C. 2009. *Sustainable Energy—Without the Hot Air.* Cambridge, U.K.: UIT.

Ministerie van I&M. 2011. *Mooi Nederland. 2,5 Jaar Innovatie en Waardecreatie.* Den Haag, the Netherlands: Ministerie van Infrastructuur en Milieu (Ministry of Infrastructure and the Environment).

Noorman, K. J. and Roo, G. de (Eds.) 2011. *Energielandschappen—de 3de generatie; over regionale kansen op het raakvlak van energie en ruimte* (Energy Landscapes—The 3rd Generation; About Regional Chances at the Interface of Energy and Space). Assen, the Netherlands: Provincie Drenthe en Rijksuniversiteit Groningen.

Pankow, A. 2011. Schouwen-Duiveland: Zelfvoorzienend Energielandschap. *Provinciale Zeeuwse Courant*, May 25, 2011, p. 19.

Pasqualetti, M. J. 2011. Social barriers to renewable energy landscapes. *The Geographical Review*, 101(2): 201–223.

Post, J. W. 2009. Blue energy: *Electricity Production from Salinity Gradients By Reverse Electrodialysis.* PhD thesis. Wageningen, the Netherlands: Wageningen University.

Schöbel, S. 2009. Windenergie, Gesellschaft und Landschaft (Wind energy, society and landscape). *Garten + Landschaft*, 3: 15–19.

Seriese, L. 2011. Reflection on the atelier from the perspective of ISGO. In: Stremke, S., Etteger, R. van, Waal, R. de, Haan, H. de, Basta, C., and Andela, M. (Eds.) 2011. *Beyond Fossils. Envisioning Desired Futures for Two Sustainable Energy Islands in the Dutch Delta Region.* Wageningen, the Netherlands: Wageningen University, pp. 151–152.

Stremke, S. 2010. *Designing Sustainable Energy Landscapes. Concepts, Principles and Procedures.* PhD thesis. Wageningen, the Netherlands: Wageningen University.

Stremke, S., Etteger, R. van, Waal, R. de, Haan, H. de, Basta, C., and Andela, M. (Eds.) 2011. *Beyond Fossils. Envisioning Desired Futures for Two Sustainable Energy Islands in the Dutch Delta Region.* Wageningen, the Netherlands: Wageningen University.

Stremke, S., Kann, F. van, and Koh, J. 2012. Integrated visions (Part I): Methodological framework for long-term regional design. *European Planning Studies*, 20(2): 305–319.

Twidell, J. and Weir, A. D. 2006. *Renewable Energy Resources.* London, U.K.: Taylor and Francis.

Waal, R. de and Stremke, S. 2011. Güssing, Jühnde and Samsø: Three European energy landscapes. In: *International Federation of Landscape Architects (IFLA), Scales of Nature. 48th IFLA World Congress*, Zürich, Switzerland, June 27–29, 2011. Switzerland: Stadt Zürich.

Wolters-Noordhoff Atlasproducties. 2008. *De Bosatlas van Nederland.* Groningen, the Netherlands: Wolters-Noordhoff Atlasproducties.

WWF. 2011. *The Energy Report. 100% Renewable Energy by 2050.* Gland, Switzerland: World Wide Fund for Nature (formerly World Wildlife Fund).

21

Toward the Zero+ Campus: Multidisciplinary Design Pedagogy in the United States

Barry Lehrman, Loren Abraham, Mary Guzowski,
Lance Neckar, Derek Schilling, and Elizabeth Turner

CONTENTS

Each building is a unique ecosystem within the larger ecosystems of landscape… Ecologically designed buildings and institutions afford a chance to make such relationships explicit, thereby becoming part of the educational process and research agenda organized around the study of local resource flows, energy use, and environmental opportunities (Orr 2006, p. 181).

21.1 Introduction

The *Zero+ Campus Design Project* (2010–2012) in the College of Design at the University of Minnesota engaged students, staff, and administrators in research and teaching focused on increasing campus beauty and sustainability through the integration of building and landscape performance-based design.

Founded in 1851, the University of Minnesota is older than the State of Minnesota (established 1858) and so has a unique charter among the land-grant public education institutions in the United States. The Twin Cities campus is located adjacent to downtown Minneapolis on the bluffs above the Upper Mississippi River just below St. Anthony Falls (Figure 21.1); an ecologically sensitive site on a major flyway and crossed by other ecological corridors. Inspired by this setting, the University has a long tradition of using environmental issues in the classroom and progressive policies to guide campus operations to mitigate impacts.*

The 392 acre (159 ha) Twin Cities campus is a powerful teaching analog for larger urbanized settings; providing design students with a site having a single owner, a clear regulatory structure, and access to performance and other longitudinal data about the site, structures, and users. The campus serves over 52,000 students with 22,500 faculty and staff who are engaged in a comprehensive range of research activities. The 161 buildings[†] consume about 8.8 trillion BTUs (9.3 billion MJ) of energy annually, including about 380 million KWh (1.4 billion MJ) of electricity.

The University of Minnesota Board of Regents implemented an updated sustainability policy in 2004, stating that the University "is committed to incorporating sustainability into its teaching, research, and outreach and the operations that support them... The University shall strive to be a model in the application of sustainability principles to guide campus operations" (University of Minnesota Board of Regents 2004). Since adapting this policy, the University has indeed become a leader in implementing programs that significantly have enhanced the sustainability of campus operations. In 2011, the University of Minnesota ranked the 4th best college in the United States and the #1 large public university by the Sustainable Endowment Institute (2011). The University of Minnesota continues to raise the sustainability bar and to grapple with the real and urgent challenges of integrating ever more ambitious sustainability goals and performance metrics into campus planning.

---

* In the 1970s, for example, the University pioneered green roofs and earth-sheltered buildings, along with other energy efficiency measures for cold climates.
† 17.19 million ft²/1.6 million m² gross.

FIGURE 21.1
Location of the University of Minnesota and the Church Street Corridor.

However, there are no published efforts to date about the University of Minnesota, or elsewhere, that look into improving the holistic performance of campuses through exploring the linkages between systems, such as the water–energy nexus and the interface between buildings and landscapes. Implementing sustainable projects at the University has been limited by accounting practices that separate upfront capital costs from the operating costs of projects, along with disciplinary and administrative silos that are common institutional obstacles at other campuses, municipalities, and large organizations.

American university and college campuses traditionally are dominated by their iconic landscape—the English inspired academic quad (Harvard Yard) and/or the New World educational parkland that surrounds the buildings (e.g., Jefferson's University of Virginia) that represent the cultural center of the community (Gumprecht 2007). As demand for higher education increases and scientific advances foster new building typologies, campus landscapes are often compromised or undermined as density increases.

Of the 47 courses listed on the website of the Association for the Advancement of Sustainability in Higher Education (AASHE) that are currently offered on the topic of campus sustainability at American/Canadian universities, 24 are categorized as environmental studies (AASHE n.d.). The *Zero+ Campus Project* appears to be unique in North America in that architecture and landscape architecture faculty are jointly developing and leading our courses and that we are looking at the campus from both a design and operations perspective.

21.1.1 Zero+ Campus Design Project

To assist the University in achieving ever-higher levels of sustainable performance in campus operations, faculty and graduate students in the School of Architecture and Department of Landscape Architecture have developed a four-pronged approach (Figure 21.2) to design education for the integration of energy and hydrological performance across scales that are known as the *Zero+ Campus Design Project* (Lehrman et al. 2011).

Working in the context of sustainability-driven decision-making processes at the institutional level, the *Zero+ Campus Project* holistically combines (1) curriculum development with (2) research/tool development, (3) a stakeholder

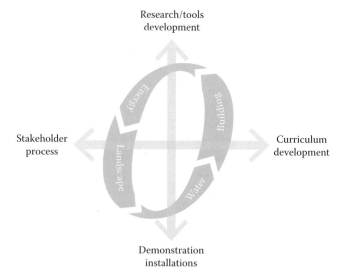

FIGURE 21.2
Zero+ Campus Project diagram.

process, and (4) demonstration installations. The research focuses on critical problem identification, strategic design, and planning processes and developing integrated modeling and tool capacity. The curriculum provides students with complex design challenges framed through current University projects that address well-defined project types and issues that are both site- and system-specific with broader applicability to the University campus and metropolitan settings.

For the courses and demonstration installations, the authors worked with the staff of the University Capital Planning and Project Management (CPPM) to identify project sites with specific regard to the energy and water issues. In the site-selection process for our design studios, a range of project typologies were initially identified based on upcoming capital projects. The projects selected were a research laboratory, student housing, and classroom buildings. The institution's designers, planners, and facility managers were engaged in a series of workshops at the launch of the project that identified obstacles and opportunities, along with short- and long-term goals. The stakeholders were also invited to participate in the courses taught and proved to be essential to provide a comprehensive professional practice setting.

University of Minnesota administration provided the project team with a wealth of performance data for each of the buildings and campus landscape areas that the students were investigating. Data at both the campus level and for specific buildings included water (potable water, stormwater, and sanitary sewage), solid-waste volumes/weights, transit and bicycle ridership, and energy (electricity, steam, chilled water, and natural gas) data (quantities and utility fees) for the previous 5–10 years. Spatial data provided included AutoCAD files of the buildings and landscapes, geographic information system (GIS) files of infrastructural systems, soils/geologic formations, and surface topography.

Institutional challenges to implement best practices on the University campus were identified through a series of stakeholder meetings with University staff (2010–2011). These meetings revealed significant institutional obstacles to implement the highest level of sustainable performance. Not surprisingly, short-term economic constraints of project and maintenance budgets undermine the ability to pursue long-term benefits and savings. This and other constraints at the University of Minnesota seem fundamentally grounded in the dominant cultural values that privilege short-term economic performance at the expense of other issues of sustainability (even lifecycle costs). Compounding this societal emphasis on upfront costs are the externalities of the economic and political trend in the United States to reduce the public support of higher education.*

---

* For example, with Minnesota public projects, development of landscapes and the spaces between buildings only receive funding when tied to a capital project like a building, though there is widespread agreement that better outdoor spaces enhance both the core mission and branding in the context of well-designed sylvan American college campuses (UMN 2009; SRF Consulting Group et al. 2001; Gumprecht 2007).

The *Zero+* concept stems from *net-zero* energy design (NSTC 2008; Hernandez and Kenny 2010), where a building or site annually generates enough resources (i.e., energy, water, and clean air) to operate without external resources or approach a net life-cycle impact of zero.* Net-zero positive performance advances the broader concept of *regenerative design* (Lyle 1996) and the *Living Building Challenge* (ILBI 2010) to enable projects to produce more resources than they consume, leading to net annual gains.

In *Zero+ projects*, building and landscape design integrate energy and hydrological performance in mutually dependent systems across the scales of the built environment.

Zero+ design exceeds the target of carbon/energy neutrality, which exclusively focuses on limiting greenhouse gas (GHG) emissions through renewable energy provision and high efficiency, while not mitigating other environmental impacts of the built environment. These positive environmental improvements can be quantified as generating surplus *ecosystem services* (Millennium Ecosystem Assessment 2005; Gomez-Baggethun et al. 2010), through regulating and provisioning the environment.

The project presented in this chapter focuses on the broader context of the built environment (including landscapes) as both a generator and consumer of energy and other resources. Generating energy sustainably via renewable resources is a critical task for the survival of our society, but equal among other mandates for reducing anthropogenic environmental impacts and preventing the transgression of additional planetary boundaries (Rockstrom et al. 2009). *The Zero+ Campus Design Project* (in the following referred to as *Zero+ Project*) looks at opportunities for users and stakeholders in managing and maintaining the built environment to find efficiencies across ecotechnical systems, not just maximizing the benefits of specific resources.

Teaching sustainable design requires more than providing methods for students to "do less bad" but instead mandates teaching them how to "do more good" within a rapidly changing understanding of earth sciences, anthropogenic impacts, human behavior, and advancing technology. The methodology from the *Zero+ project* is intended to be applicable in professional practice settings on a wide range of buildings and landscapes typologies.

21.2 Coursework and Pedagogy

The *Zero+ Project's* courses have aimed, in this spirit, to go beyond the teaching of fundamentals of energy and environmental performance metrics. Three of the four-pronged project objectives challenge students to succeed

---

* Including energy embedded in the extraction, fabrication, transportation of building materials, and then in the demolition and disposal of the building.

in designing buildings and related landscapes that have net-positive energy and water use with strategies that integrate water and energy across scales, all in the institutional context of promoting an understanding of sustainability decision-making process.

On the research side, the project team is developing more *designer-friendly* tools (Attia et al. 2009) to enable the modeling of the water–energy nexus and is creating a holistic methodology for modeling these performance objectives in the larger context of ecosystem services in the urban campus environment. These tools are directly tied to providing students the means to achieve the *Zero+* goals.

The faculty and graduate research assistants on the project team have, so far, offered four courses as part of the *Zero+ Campus Project* (Figure 21.3). Each course investigated different aspects of the integration of landscape and architectural design strategies to reduce campus emissions, energy, water, and waste:

1. *Zero+* Campus Design Studio (7 weeks, four credits, Architecture) (Figure 21.4)
2. Video as an Ecological Design Tool Catalyst (1 week, one credit, Architecture)
3. Optimizing the Building-Landscape Interface Multidisciplinary Studio (3 weeks May-Term, three credits, Architecture and Landscape Architecture)
4. Whole Building Analysis Seminar (15 weeks, three credits, Architecture)*

Each of the *Zero+ Project's* courses has sought to

1. Develop integrated building/landscape design proposals that improve ecological design performance and beautify the campus
2. Identify and apply advanced performance metrics and integrated modeling practices
3. Develop integrated concepts and strategies for the reduction of
   a. Energy, water, and resource consumption and waste generation
   b. Stormwater runoff and the associated mitigation of pollution, flooding, and energy consumption
   c. Urban heat island effect and related threats to human and ecological health

---

* A fifth integrative Architecture and Landscape Architecture seminar was underway as this chapter was being written that is testing the technically advanced tools through the comparative evaluation of complex campus design project proposals.

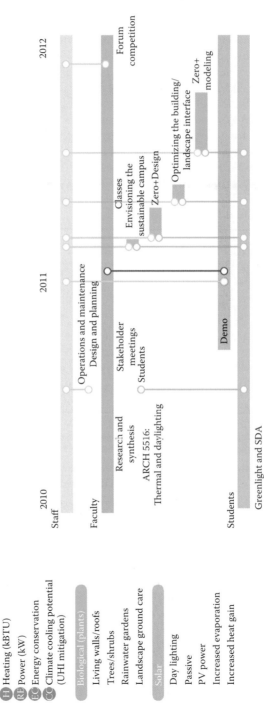

FIGURE 21.3
Zero+ curricular strategy and timeline.

FIGURE 21.4
Zero+ laboratory project. (Rendering by Chris Brenny and Pablo Villamil.)

21.3 Optimizing the Building/Landscape Interface Multidisciplinary Design Studio

The May-Term studio *Optimizing the Building/Landscape Interface*, which presents the focus of this section, enrolled four architecture and five landscape architecture students.* The students were organized into three cross-disciplinary teams that each took a unique approach to the project based on their personal interests. Prior to the course, Abraham and Guzowski developed a robust pedagogy for teaching the design of net-zero/carbon neutral architecture (Guzowski and Abraham 2009) that formed the departure point for developing the course (see Figure 21.3).

The May-Term course provided iterative opportunities to apply theoretical knowledge to the design process, as is fairly typical in most architecture and landscape architecture studios. What was unique in this studio was the interaction between modeling different systems at multiple scales plus the multidisciplinary mix of students and instructors. On another quadrant of collaboration, Neckar had taught several American Institute of Architects award-winning research collaborative urban design studio courses with Architecture faculty over his 20-year teaching career. In a still more unusual pedagogical and disciplinary crossover that enriched the course, Landscape Architecture Lecturer Vincent de Britto enrolled as a "student" in the course, providing leadership and expertise to one of the integrative teams.

---

* Two of the landscape architecture students were also architects by training.

FIGURE 21.5
Sketch of Church Street. (Sketch by Emily Lowery and Steven Grootaert.)

The May-Term studio had an ambitious scope, strategically focused on the Church Street Corridor (Figure 21.5) through the center of the campus. The eight buildings adjacent to Church Street and their associated landscapes span over 85 years of campus development and so provide a representative range of energy performance scenarios and opportunities for renovation and new design interventions. The landscape along Church Street is a seemingly random mixture of asphalt pavement with concrete curbs, intermittent concrete sidewalks, a pedestrian mall of concrete pavers with no curbs, turf grass, and hardwood tree studded pastoral parklands, foundation plantings of mixed shrubs/perennials/annuals. Church Street is bisected by "Scholars Walk" commemorating the University's 150th anniversary. There is, however, no integration between this ceremonial landscape and the utilitarian character of Church Street. Notably, there are also a few large trees shading the street, several of which predate the university.

Tate Physics Laboratory, an early-mid twentieth century building that is scheduled for renovation, was studied by most teams as the catalyst for reshaping the Church Street landscape and related infrastructure on the campus. All projects successfully modeled institutional change within the current pragmatic context of capital building project-based programming and budgets.

The May-Term met a total of 12 times for 3 h each session over the 3 weeks. Each week had a specific focus (also see Figure 21.6):

1. WEEK 1: Investigating the baseline provided the introduction to the *Zero+* methodology and site through a series of lectures (case studies, performance metrics, baseline information, and science of the water/energy nexus), a 2 h walking tour of Church Street, a panel discussion, and several work sessions. A panel discussion by Facilities Management staff that shared energy, water, and waste practices and obstacles to higher performance from their perspective. The

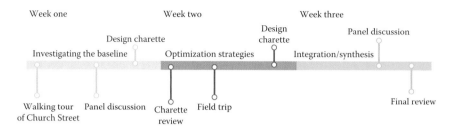

FIGURE 21.6
May-term studio steps and activities.

week concluded with a design charette to catalyze the formulation of *Zero+* strategies for the specific sites.

2. WEEK 2: Optimization strategies started with an informal review of the concepts generated in the first charette and lectures on integration linkages and methods. The class took a field trip to the green roof of the St. Paul Fire Department Headquarters (designed by Abraham) and to the Minnesota Science Museum's Zero Energy House. At the end of the week, the students participated in a second design charette.

3. WEEK 3: Integration/Synthesis started with a design review and a lecture on "Exemplary Campus Projects and Processes." A panel discussion organized by Elizabeth Turner brought together the campus sustainability administrators from the University of Minnesota Morris, Carleton College, St. Olaf College, Macalester College, Augsburg College, and Gustavus Adolphus College—for the first time according to the participants—to provide a regional perspective across a range of campus scales. The course wrapped up with a formal presentation by each team to a jury comprised of the *Zero+ Project* team and staff from CPPM.

21.3.1 Exercise 1: Establishing the Performance Baseline

During the first week of the studio, students were asked to assess the current performance of the built campus environment in the historic context as the focus of the inventory and analysis process. Defining the baseline conditions challenged the students to define the intended sustainability goals beyond the broad goal of *Zero+* and to balance competing demands for space and resources onsite to achieve these goals. All participants had prior coursework where they learned the core issues of sustainable design and developed their own definitions of what a sustainable landscape or building needed to be. The site inventory also included a study to define bioclimatic forces and ecotechnical systems that influence the site; investigated the poetic/pragmatic design opportunities for deploying new ecotechnical systems; considered the architectural and environmental implications of site, climate, and place; and investigated processes and tools for bioclimatic analysis and design for a *Zero+* campus. The performance baselines created by the students were not a snapshot of one moment, but instead looked across long-term trends with multiple data points. The learning objective for this exercise was to enable students to identify essential systems of linkages, feedbacks, benefits, and impacts to feedback loops across scales and time.

In the process of defining the baseline, students found that improving the performance of one specific criterion often had a negative impact on other equally important metrics (Figures 21.7 and 21.8). For example, increasing the amount of glazing in a building can not only improve the daylighting of the interior, but also increase the heat loss in winter that needs to be offset, for example, by larger infrastructural changes, such as the construction of geothermal systems or the introduction of solar arrays or green roofs. Identifying such reciprocal benefits and impacts across scales requires iterative parametric studies that can build both intuitive and a more a holistic understanding of the built environment.

The first week's charrette asked students to evaluate the ecotechnical system(s) embedded around the Church Street Corridor at multiple scales, to define the design parameters that impact the baseline performance of these systems, to create a kit of parts (Figures 21.9 and 21.10), and to create a narrative about these systems. After the intensive 3 h charrette sequence of diagramming, writing, and drawing in plan and section, the students prepared refined versions of the work. Students were asked to select two of the many ideas for further development in the following weeks.

21.3.2 Exercise 2: Optimization

The second week focused on exploring the interconnections between systems, site, and time (Figure 21.11); defining performance metrics that value poetics and qualitative benefits*; exploring deployment and implementation strategies; and defining strategies to achieve *Zero+* performance through an iterative optimization process.

The second week's charrette explored the implications and potential interconnections between systems to provide resilience through redundancy and synergies, balanced with creating high performance and efficient systems (Figure 21.12). Beyond the goals of achieving *Zero+* energy and water performance, was the implicit charge to design a resilient system that could maintain high performance even in the worst-case scenarios of climate change and, as Heinberg of the Post Carbon Institute says, "peak everything" (Heinberg 2007).

21.3.3 Exercise 3: Integration and Synthesis

Synthesis, like most design processes, is an iterative process. To wrap up the studio, the assignment was to integrate and synthesize performance metrics

---

* For example, of association, recreation, or interaction.

Energy analysis-base case

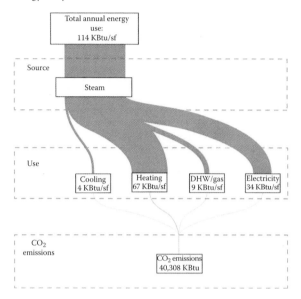

Design case

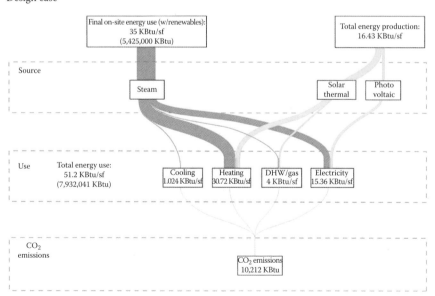

FIGURE 21.7
Energy optimization. (Drawing by Dawn Keeler.)

Water balance analysis-base case

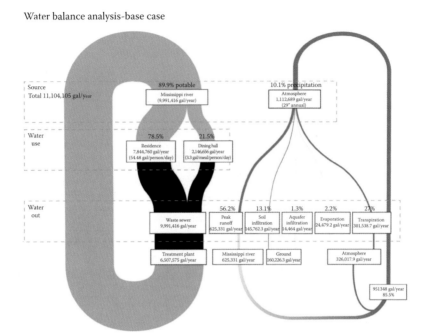

Design case

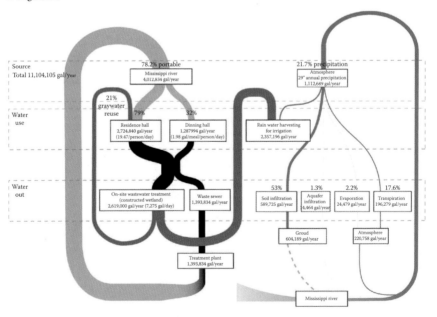

FIGURE 21.8
Water optimization. (Drawing by Dawn Keeler.)

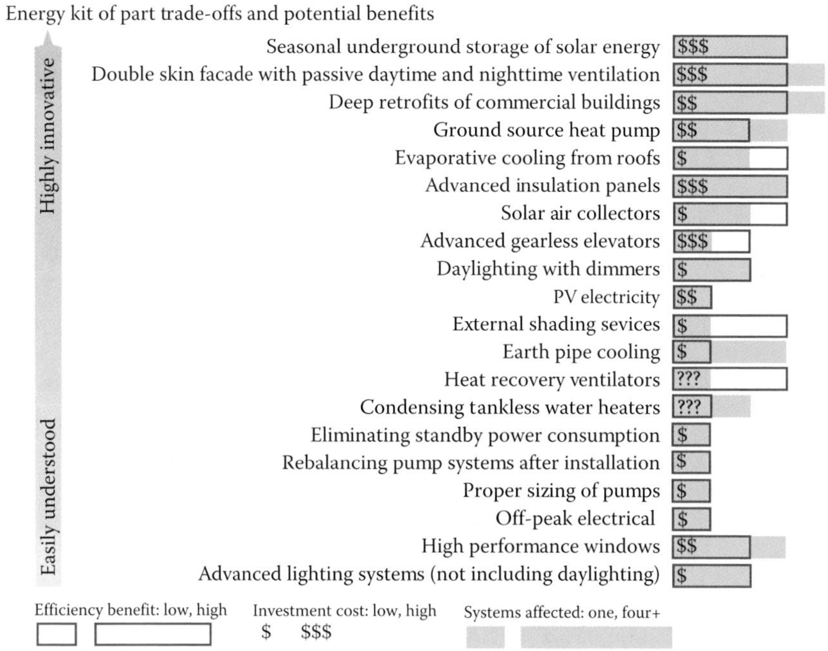

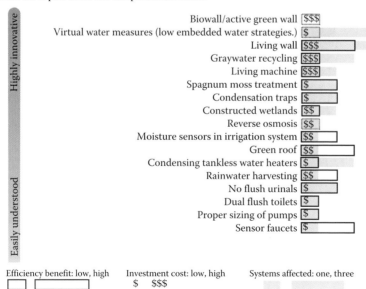

FIGURE 21.9
Energy kit of parts. (Drawing by Barett Steenrod.)

FIGURE 21.10
Water kit of parts. (Drawing by Barett Steenrod.)

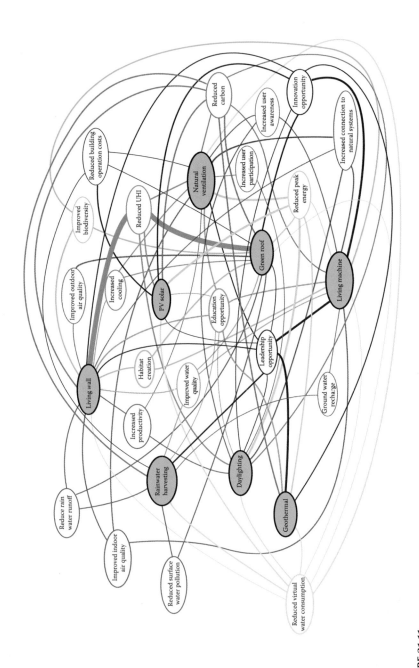

FIGURE 21.11
System interrelationships. (Drawing by Thea Holmberg-Johnson.)

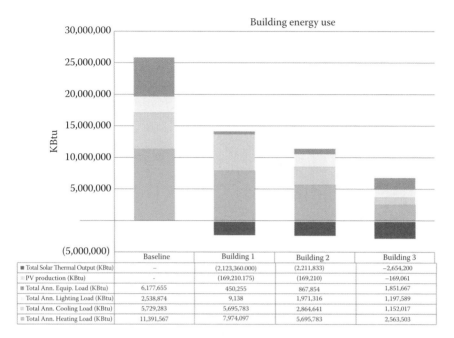

	Baseline	Building 1	Building 2	Building 3
■ Total Solar Thermal Output (KBtu)	–	(2,123,360.000)	(2,211,833)	−2,654,200
▨ PV production (KBtu)		(169,210.175)	(169,210)	−169,061
■ Total Ann. Equip. Load (KBtu)	6,177,655	450,255	867,854	1,851,667
Total Ann. Lighting Load (KBtu)	2,538,874	9,138	1,971,316	1,197,589
▨ Total Ann. Cooling Load (KBtu)	5,729,283	5,695,783	2,864,641	1,152,017
■ Total Ann. Heating Load (KBtu)	11,391,567	7,974,097	5,695,783	2,563,503

FIGURE 21.12
Building performance optimization (output from Zero+ calculator).

into a multifunctional design and system. For optimizing buildings and landscapes to achieve *Zero+* performance, integration requires more than just fine-tuning the interface and physical connections between systems.* Most integration approaches are iterative and require several cycles to achieve the desired level of performance. Through integration, there is the potential to improve the performance of the entire system and create new capacity that exceeds the individual abilities of the subsystems.

The student-integration processes operated differently in each group. Some groups paid more attention to the urban campus scale and some to the building and the immediate landscape. At the campus scale, one emerging interest manifested itself in integrating diverse issues and scales from bicycle and transit modes and utilities into the social fabric of campus life. Other efforts generated both graphic representations and explored performance metrics or at least referred to them in their strategic approaches (Figure 21.13).

One team focused on several blocks of Church Street to develop the "Bike Pasture" (Figure 21.14), one of the ongoing demonstration projects of the *Zero+ Project*. The Bike Pasture was initially proposed in a student-led design workshop before the May-Term as a way of reimagined bike parking being purposefully integrated into landscape planning—a place to pasture our

---

* Design professionals and engineers typically define the process of system integration as just attempting to make subsystems work harmoniously.

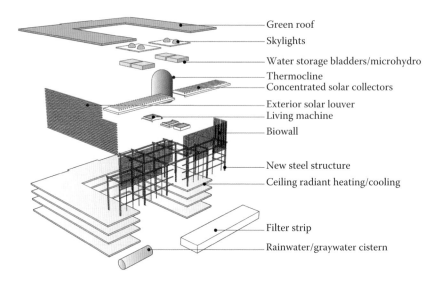

Green roof
Skylights
Water storage bladders/microhydro
Thermocline
Concentrated solar collectors
Exterior solar louver
Living machine
Biowall
New steel structure
Ceiling radiant heating/cooling
Filter strip
Rainwater/graywater cistern

FIGURE 21.13
Tate Laboratory components axon. (Drawing by Emily Stover.)

FIGURE 21.14
Bike Pasture rendering. (Rendering by Emily Lowery.)

modern-day metal horses and refresh ourselves = (a human energy take on *renewable energy landscapes*). The Bike Pasture aims to be beautiful with or without bikes, provide an integrated social space, and increase infiltration through permeable paving and plantings. The outcomes of the studio helped solidify the design parameters of the demonstration project and were used to secure funding from the President's office to begin construction in 2012. The "real-life" nature of this demonstration project provides the opportunity for students to research sustainable materials and strategies while balancing environmental and social benefits with a real budget. This project will provide landscape architecture students a chance to interact with design strategies while giving the University community a place to enjoy.

21.4 Zero+ Performance Modeling and Optimization Tools

On the research and teaching quadrants (see Figure 21.2), the May-Term experience sharpened the need to have better integrated and more *designer-friendly* tools to provide measurable simulation/optimization capacities for energy and water use.

21.4.1 Modeling Landscape Performance

At the start of the *Zero+ Project*, we conducted an inclusive review of the energy and other modeling tools and assessed their viability for addressing landscape performance (see Figure 21.15). In response to this,

Tool name	Features/interface key: s = Supported features, p = Partially supported, blank = Feature not supported/unknown

Tool name	Norm. text output	Norm. text input	Predictive	Text/spreadsheet based	3D modeling-based	GIS-based	2D CAD/BIM-based	Scripting language	Location/climate	Land use/building type	Demographics	Building footprint/area	Transportation parameters	Materials/assemblies	Impervious/pervious surfaces	Vegetation types/land cover	Potable H20/sewage volume	Runoff volume/hydrograph	Evapotranspiration	Water quality/pollution load	Habitat/vegetation	Energy use	GHG emissions/air pollution	Embodied energy/LCA	Demographics trends	Land value/Economics
Athena Impact Estimator for Buildings	s		p	s				C++	s	s				s				p			s		s	s	s	
Community viz	s		p	p	s	s	s	C++	s	s	s	s					p					p	p		s	s
Energyplus/ECOTECT analysis	p	p	s		s		s	Fortran 2003	s	s		s		s			s					s	s			
Energyplus/openstudio	p	p	s		s			C	s	s		s		s			s					s	s			
Facility Energy Decision System (FEDS)	s			s					s	s		s					p					s	s	s		s
Holistic city	s								s	s		s	s		s		s					s	s			
IES VE-Toolkit,VE-PRO	p		s	s	s		s		s	s		s		s			s					s	s			
Index and Cool Spots			p		s				s	s		s										s	s			
i-Tree suite (eco, streets, and hydro)	s	s	s					Visual basic	s							s		s	s	s		s	s			
i-Tree Green Roof Tool (GBO, Mini-model)	s	s	s					Visual basic				s			s	s	s									
UPlan	s					s		Python	s	s						s			s	s	s	s	p		s	
UrbanSIM	s				s		s		s	s	s	s													s	s
HydroCAD	s	s	s					Excel	s					s			s		s	s	s					
Andrew Edwins Water Balance Model	s	s	s	s					s					s		s	s	s	s	s						

Note: Data from the website of the tool developer or as noted from Condon, 2009 or the building energy software tools directory (BESTD). http://apps1.eere.energy.gov/building/tools_directory/subjects.cfm (accessed november 2010.)

FIGURE 21.15
Suitability of available tools for modeling landscape performance.

Abraham and Schilling created a plugin for use with *Google Sketchup*™ that works in concert with the *IES Virtual Environments*™ for the purpose of quickly generating *shoebox* models that approximate the building size and type under consideration for design optimization and other analyses. Functionality for this tool emulated methods and concepts introduced in *Energy 10* developed in the 1990s by NREL (Balcomb and Hayter 2001) but improved on them significantly by utilizing the 3D modeling of *Google Sketchup* as the design interface and engaging the superior analysis capabilities of *IES Virtual Environments*. Abraham also developed a *Patchwork Calculator* tool to tie together the results from various analyses and allow for the aggregation and integration of simulation results from a number of other assessment tools into the final set of integrated site water balance, utility cost, life cycle cost, carbon footprint, and other key performance metric calculations needed by students, designers, and decision-makers in the early design phases.

Optimization and further performance simulations can be created using a combination of programs (Crawley et al. 2008), including *Google Sketchup*, the proprietary *Zero+ Shoebox* Plugin (created by Abraham and Schilling), the *IES VE Suite* including its *Sketchup plugin*, and the *Zero+ Patchwork Calculator* (Figure 21.16).* Performance optimization criteria considered issues such as envelope-insulation values, window-to-floor area ratio, daylighting design

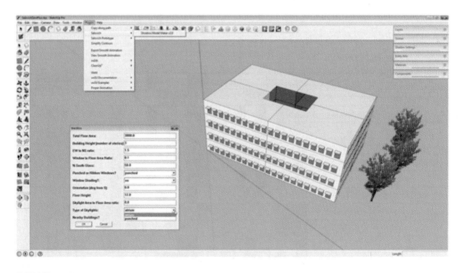

FIGURE 21.16
Screenshot of the Shoebox Plugin for Google Sketchup™.

---

* The programs were used to inform the design for the proposed renovation and addition to the Tate Physics Laboratory and Church Street Corridor.

parameters, benefits of green roof, and integration of various renewable energy strategies using a single-variable approach.*

From the myriad of available choices, students in the May-Term evaluated three design proposals including net-zero energy and zero-runoff scenarios as compared to the baseline building. Their performance evaluations focused on the critical metrics identified in the earlier stakeholder workshops and were the focus of their design process.

21.4.2 Calculating Water–Energy Nexus

Quantifying the water–energy nexus is a multistep process that starts with estimating building water use and site water balance, that is, rainfall, evapotranspiration, infiltration, and runoff (Figure 21.17). From this data, combined with potential cistern capacities, potential rainwater-harvesting yields can be calculated.† These calculations can be made using standard weather data (such as TMY3 files), surface area as well as orientation and then applied toward the building loads with a qualified coefficient of performance value. Additionally, there are external inputs such as the energy needed to treat and deliver potable water to the site and energy to treat the wastewater effluent from the site and the waste heat energy expended in these processes. We had data on the quantity of high-pressure steam delivered to the buildings along Church Street (including the amount of fuel used) for heating and cooling from the centralized steam plant, the water usage, and general data about the energy consumed by the local water and sewer utilities that serve the campus.

To calculate the change in water use based on the student's designs, we used the *Urban Water Model version 2 Tool*—a spreadsheet calculator that was developed by graduate student Andrew Edwins in 2009 and modified by Schilling and Abraham for the course. This tool was adapted to take outputs from and provide inputs back to the *Zero+ Patchwork Tool*, which is then able to calculate water-use reduction in the building as well as the associated source energy and carbon emissions associated with the water usage and wastewater discharged. Another feature of the tool allows the user to provide wastewater treatment onsite as an alternative to discharging the wastewater. In this case, the energy and associated impacts on GHG emissions are transferred to the site and building. Evapotranspiration and shading effects of landscape elements were also calculated using stand-alone tools such as *i-Tree* (created by the US Forest Service).

---

* Future versions of this tool may incorporate the capability to perform optimizations using a *multivariate approach* (Thomasson 2007) combined with *fuzzy logic* to limit the considerable number of simulations and therefore processor time that would otherwise be required (Ellis et al. 2006).

† In cases where living envelope systems are contemplated, this harvested rainwater can be evapotranspiration converting the water into energy, that is, cooling potential.

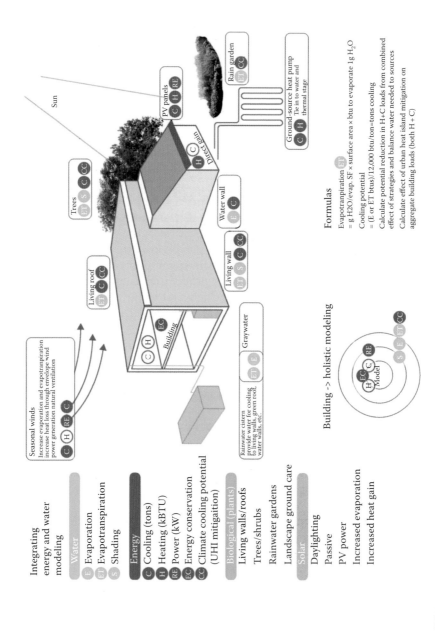

FIGURE 21.17
Modeling the water–energy nexus.

21.5 Lessons Learned

Developing and teaching the May-Term studio within the context of a research project has both deepened and broadened the research inquiry. The lack of teaching tools in both architecture and landscape architecture and the inquiry of students point toward a range of further research questions, the resolution of which promise to more fully realize sustainable energy landscapes, campuses, and *Zero+* places.

21.5.1 Design Education Lessons

Integrated multidisciplinary education is core to educating design professionals who can create *Zero+* buildings, landscape, campuses, and cities. Over the 3 weeks of the May-Term, and in other architecture courses, the instructors have observed growth in the students' ability to integrate complex system across multiple scales, to identify and apply performance metrics specific to the site, and to represent graphically (and beautifully) the design, systems, and performance metrics. Much of the growth came from collaboration between the students across disciplines, including the following lessons:

1. Aesthetic and pragmatic integration: There is insufficient language to adequately communicate sustainability and ecological processes within the usual design context. In design schools, there is a gap between selecting performance goals and the mandate to achieve those goals in the context of aesthetics and poetics.

2. Challenge of collaboration across disciplines: Cross-disciplinary collaboration in the context of disparate theory and training bases in architecture and landscape architecture adds complexity to the integration challenge. Further, the *disciplinary* focus of professional accreditation standards for architecture and landscape architecture in the United States practically restricts the ease of interdisciplinary collaboration in studios.* Interdisciplinary courses are, however, increasingly important in professional practice and should be supported—even mandated—by all professional design-accreditation organization.†

3. Calibrate learning curves across disciplines: The professions of architecture and landscape architecture are at different points

---

* Few design schools include engineering and ecology or hydrology, and stringent disciplinary requirements tend to limit electives outside the core topics and to discourage integrative course offerings.

† Please also see Chapter 20, where De Waal et al. discuss their experiences with another cross-disciplinary studio at Wageningen University in the Netherlands.

on the learning curve when it comes to modeling sustainability performance. Architecture has several decades of experience in teaching and using simulation tools for high-performance architectural design with tools providing a range of capacities including daylighting and illumination, energy and water use, acoustics, and fire spread and blast damage. Landscape architecture has limited experience with performance simulation beyond hydrologic modeling. Therefore, landscape architecture, which operates explicitly at all scales and increasingly even inside buildings, needs to incorporate performance-based design and planning tools (i.e., *SITES*, *CityGreen*, or *Seattle Green Factor*) with architecture-focused tools (i.e., *EcoTect* or *IES VE*) to develop the disciplinary capacity to teach and apply these tools. Greater integration of tools across disciplines within design education is sorely needed.

4. Systems thinking and ecological literacy: Teaching system thinking and ecological literacy is improving, but still needs to be nurtured in all fields. The boundaries between quantification of performance and qualitative design processes also need to be dissolved. In practice, design firms often delegate the performance modeling to specialists, which limits the holistic integration of sustainable features into projects. In academic settings, performance modeling is often taught as an elective and not fully integrated into the design studio. Integrating beauty with quantitative performance is crucial for widespread deployment and acceptance of high-performance buildings and sites.

5. Aesthetics of sustainable landscapes: Native planting designs face a reputation for "weedy" gardens full of plants that do not fulfill society's desire for neatness (such as manicured turf and topiary shrubs). Photovoltaic panels and wind turbines are perceived as being technological and thus in direct opposition to ideals of natural beauty when they occupy the landscape outside of industrial settings. These negative perceptions can be overcome only by providing designers with the tools to integrate performance and design excellence on all projects. A new landscape design language that combines native flora in more manicured appearance settings with renewable energy technologies can also increase public acceptance, such as Walter Hood's *Solar Strand* project in Buffalo, New York.

6. Fundamental challenges are present in designing a new curriculum aimed at integration of landscape and architectural design objectives and targeting tool development and use for *Zero+* emissions, energy, water, and waste. First, the definition of new metrics is needed for measuring degrees of improvement to meet the *Zero+* objective. Second, further research is needed to adapt and develop tools with

which to simulate, compare, and optimize various integrated design interventions across scales and issues. Once these initial challenges are overcome, then these protocols need to be tested in a multidisciplinary setting and with stakeholders, before being deployed in the wider professional practice.

21.5.2 Decision-Making Tools Lessons

The decision-making process needs to be changed to enable the implementation of *Zero+* design features.

1. Up-front costs are the primary institutional decision driver, while operating costs and utility bills are rarely included in capital spending decisions. Because capital costs and operating costs are accounted for through a separate funding processes (bonds versus income), neither return on investment calculations nor payback times are considered by the university when looking at sustainability options for capital projects.

2. By providing cost-benefit analysis of different systems and strategies in the predesign phase, there may be more incentives for the University administration to select higher-performance building and landscape strategies in the future. As demonstrated by the students in the May-Term studio and other courses, there is a wealth of performance indicators that can be collected and analyzed about up-front and life-cycle costs (see, e.g., Loftness et al. 2005).

3. Another significant dichotomy is that the *desire* for additional space that is not coupled to actual *need* for new space—bigger is not always better. Better analysis of existing operations and forecasting for future space needs can help to evaluate the trade-offs of improved quality versus increased quantity of space and buildings.

4. In interviews and workshops, the team quickly learned that the key stakeholder is in need for clear evidence of the long-term financial viability of sustainable strategies and indirectly, the benefits that support the educational and sustainability values of the institution. In some cases, such economic metrics and tools exist, but in other cases, there is a definite gap (Keysar and Pearce 2007), which provides an opportunity to develop new financial sustainability analysis tools, perhaps integrating both life-cycle analysis and ecosystem services.

21.5.3 Need for New Tools

As we look into the near future, we can see an opportunity to expand the concept of the *Zero+ Patchwork Tools* to create an interface-driven tool providing links between tools with more extensive capabilities, layers of information, and external knowledge bases.

1. There is great potential for a holistic urban modeling tool to facilitate the integration of GIS with Building Information Models (BIM) and discrete energy and hydrologic analysis applications.
2. At the site-specific scale, BIM-based tools are integrating a powerful range of energy and water-use modeling.
3. At the urban scale, GIS is already providing hydrologic and energy modeling at the resolution of individual parcels, but not within the parcels. The design driver behind this type of tool would be the same as the tools we introduced earlier, namely to meet the simulation/optimization assessment and performance evaluation needs and abilities of stakeholders and students. For now, this need remains an aspiration for future funding and development.

21.6 Conclusion

Optimizing the performance of building and landscape interface represents a paradigm shift that has the potential to significantly contribute to climate change mitigation and adaptation efforts through systems integration and by establishing closed-loop urban systems at optimal scales. A more effective integration of performance across water, energy, and ecological systems is a long-term process that will require a high level of cross-disciplinary collaboration and new performance metrics and tools. Some large interdisciplinary firms (such as Arup) have already piloted sophisticated proprietary integrated models that can provide insight into the next generation of tools and evaluative modeling tools (Kirkpatrick 2007).

Tempering the promise of innovations of such performance modeling programs like *IES VE* and GIS plug-ins, this modeling requires a computational leap to calculate data-intensive algorithms compared to the current practice of building and landscape modeling and thus represent the need for investing in both software and hardware. Modeling integrated landscape performance of campuses, neighborhoods, and even entire cities represents a new range of services that promise to grow the practice of landscape architecture.

The future of sustainable landscape architecture is not designing single-function works of infrastructure, such as renewable energy "brightfields" or biomass plantations, but in generating multifunctional places that integrate a range of ecosystem services, energy, and water systems, and other sustainability features into the inhabited built environment. If humanity is to fully realize the potential of ecocities and living buildings to avert catastrophic climate and ecological changes, then a new generation of design practitioner needs to be educated to collaborate in new ways and to apply the performance-based tools yet to be developed.

Midstream in the *Zero+ Campus Design Project*, the challenges referenced above suggest that there are significant obstacles to success on all four quadrants of these efforts (Figure 21.2). However, the experience to date also suggests that broadened discourse and integrated design teaching and research can yield breakthroughs on all four of the project's well-defined trajectories. In summary, the *Zero+ project* models a process that promises the advancement of design education and practice toward the creation of new, beautiful, and sustainable design solutions for buildings and landscapes. And not a moment too soon.

Acknowledgments

This chapter could not have been completed without the dedicated assistance of our graduate research assistants: Derek Schilling, Elizabeth Turner, Dawn Keeler, and Julian Lemon. Orlyn Miller, Director and Monique MacKenzie, Capital Planner of CPPM were members of the teaching/research team and our primary "client." Jay Denny, Kathryn Bene, and Dana Donnatucci of Facilities Management and other University staff shared their wisdom and provided megabytes of data to the *Zero+ Project*. The students of the May-Term studio and other *Zero+* courses shared their design strategies that illustrate this chapter. The Salovich *Zero+ Project* was funded by the Ann Salovich bequest, administered by the University Provost, Thomas Sullivan. We also thank Jon Foley (Director, Institute on the Environment, University of Minnesota), and Thomas Fisher (Dean, College of Design) for their support and advocacy on the project's behalf.

References

AASHE, Courses on Campus Sustainability | Association for the Advancement of Sustainability in Higher Education (AASHE). *AASHE.org*. Available at: http://www.aashe.org/resources/courses-campus-sustainability (accessed on February 3, 2012).

Attia, S. et al. 2009. "Architect Friendly": A comparison of ten different building performance simulation tools. In *Building Simulation*. Eleventh International IBPSA Conference. Glasgow, Scotland, U.K.

Balcomb, J.D. and Hayter, S.J. 2001. Hourly Simulation of Grid-Connected PV Systems Using Realistic Building Loads. In American Solar Energy Society (ASES) National Solar Conferences Forum. Washington, DC: National Renewable Energy Laboratory.

Condon, P.M., Duncan C., and Nicole M. 2009. *Urban Planning Tools for Climate Change Mitigation.* Policy Focus Report/Code PF021. Cambridge, MA: Lincoln Institute of Land Policy.

Crawley, D.B. et al. 2008. Contrasting the capabilities of building energy performance simulation programs. *Building and Environment*, 43(4), 661–673.

Ellis, P.G. et al. 2006. Automated multivariate optimization tool for energy analysis. In *SimBuild*. IBPSA SimBuild. Cambridge, MA.

Gomez-Baggethun, E. et al. 2010. The history of ecosystem services in economic theory and practice: From early notions to markets and payment schemes. *Ecological Economics*, 69(6), 1209–1218.

Gumprecht, B. 2007. The campus as a public space in the American college town. *Journal of Historical Geography*, 33, 72–103.

Guzowski, M. and Abraham, L. 2009. Integrated Luminous and thermal design: A cold climate approach to zero-energy carbon-neutral design education. In *PLEA 2009. 26th Conference on Passive and Low Energy Architecture*. Quebec City, CA.

Heinberg, R. 2007. *Peak Everything: Waking Up to the Century of Declines* FEP Torn., New Society Publishers, Gabriola Island, British Columbia, Canada.

Hernandez, P. and Kenny, P. 2010. From net energy to zero energy buildings: Defining life cycle zero energy buildings (LC-ZEB). *Energy and Buildings*, 42(6), 815–821.

ILBI, 2010. Living Building Challenge 2.0: A visionary Path to a Restorative Future. Available at: https://ilbi.org/lbc (accessed on December 1, 2011).

Keysar, E. and Pearce, A.R. 2007. Decision support tools for green building: Facilitating selection among new adopters on public sector projects. *Journal of Green Building*, 2 #3(Summer) 1–19.

Kirkpatrick, N. 2007. Dongtan eco city and integrated resource management modeling. In Nightingale Associates Sustainability Conference. Warwickshire, Coventry, U.K.: Nightingale Associates. Available at: http://vimeo.com/15770219 (accessed on November 29, 2011).

Lehrman, B., Turner, E., and Schilling, D. 2011. Zero+ Campus Project.

Loftness, V. et al. 2005. Building Investment Decision Support (BIDSTM): Cost-Benefit Tool to Promote High Performance Components, Flexible Infrastructures and Systems Integration for Sustainable Commercial Buildings and Productive Organizations. Available at: http://www.aia.org/aiaucmp/groups/ek_public/documents/pdf/aiap080050.pdf (accessed on November 28, 2011).

Lyle, J.T. 1996. *Regenerative Design for Sustainable Development*, John Wiley and Sons, New York.

Millennium Ecosystem Assessment, 2005. *Ecosystems and Human Well Being: Synthesis*, Washington, DC.: Island Press.

NSTC, 2008. *Federal Research and Development Agenda for Net-Zero Energy, High Performance Green Building Report*, Washington, DC.: Office of Science and Technology Policy. Available at: http://www.bfrl.nist.gov/buildingtechnology/documents/FederalRDAgendaforNetZeroEnergyHighPerformanceGreenBuildings.pdf

Orr, D.W. 2006. *Design on the Edge: The Making of a High-Performance Building*, The MIT Press, Cambridge, MA.

Rockstrom, J. et al. 2009. A safe operating space for humanity. *Nature*, 461(7263), 472–475.

SRF Consulting Group et al. 2001. University of Minnesota Exterior Design Standards. Available at: http://www.cppm.umn.edu/assets/pdf/exterior_design_standards. pdf (accessed on December 1, 2011).

Sustainable Endowment Institute, 2011. Report Card 2011—The College Sustainability Report Card. Available at: http://www.greenreportcard.org/report-card-2011/ schools/search/235 (accessed on January 26, 2012).

Thomasson, F. 2007. Envelope Optimization and Control Using Fuzzy Logic. *Practical Process Control by Control Guru*. Available at: http://www.controlguru. com/2007/060607.html (accessed on February 17, 2012).

UMN, 2009. University of Minnesota Twin Cities Campus: Discover Connect Sustain. Available at: http://www.cppm.umn.edu/assets/pdf/2009_BOR_ mp.pdf (accessed on December 1, 2011).

University of Minnesota Board of Regents, B. of R., 2004. Sustainability and Energy Efficiency Policy. Available at: http://www1.umn.edu/regents/policies/ administrative/Sustain_Energy_Efficiency.pdf (accessed on January 26, 2012).

"Resources": An Educational Approach to Address Future Urban Uncertainties

Henrietta Palmer and Michael Dudley

CONTENTS

22.1 Introduction

The uncertainties that epitomize our time are to a great degree based on the
fact that we are rapidly leaving the fossil-based era. This is occurring without a roadmap to guide us, as the ideas of modernism once did. The city as
an oil-based phenomenon can be read as a particularly insecure minefield,
yet implored to provide a place for an increasing human population. How
can we navigate in this archipelago of unexpected events and new situations,
while simultaneously securing the fuel necessary for the journey and a seat
for all that are on the same trip?

22.1.1 Urbanity as Uncertainty

The media critic Marshall McLuhan proposed that observing the present
provides clues to seeing our future. This seems particularly relevant for
studies of cities. The city as a place of social interaction and negotiation has
always been the place where humans have tried to deal with the unexpected
and unpredictable. In the legible traces left after our built environments, we
can discern stories that explain both individual and societal responses to
specific political contexts. Existing environments already provide the seeds
to future solutions. To find these, germinate and position them in a coherent
discourse concerning our urban predicament is one of the explicit goals of the
postmasters program "Resources" at the Royal Institute of Art in Stockholm.

Another one of our fundamental tasks is to pose the question: for *whom*
are we building the future city? The present global dichotomy of urban
growth, which consists on the one hand of the *gated community* and on
the other of the *slum*, lulls one privileged part of the population in comfort and security, while the rest are left all the more vulnerable. How can
urban solutions, be they for energy, food production, or protection from the
effects of climate change, act in concert while providing a more just city?
What possibilities do we entreaty to the concept of urbanity? What kind of
city dare we dream of?

22.1.2 "Resources"

Resources began in 2005 as a one year, postmasters research program on
the theme of future cities in a time of shifting conditions. The program is
one of three run at *Mejan Arc*, which is a forum for advanced urban studies at the Royal Institute of Art in Stockholm. Each year, *Resources* selects
a theme connected to a specific city or geography and posits an overarch-
ing question. Within this context, we collect a multidisciplinary group of
participants that provide us with a broad knowledge base and enable us
to develop solutions beyond the familiar borders of our professions. Since
the challenges our urban future faces are nested and complex, it logically

follows that this complexity has to be approached in an interdisciplinary way. A mixed group has the strength of bringing diverse knowledge and professional experience together.

Through the years, our resolve to look beyond the western world has grown all the more strong. This is partly due not only to the fact that urban growth is primarily occurring outside of the western hemisphere, but also because cities not yet homogenized by western modernism provide us with a rich depository for a postfossil society. The research in *Resources* is therefore based on the conviction that finding a path forward requires looking beyond a western conception of urban life, one which is deeply rooted in oil dependency. Consequently, the depletion of fossil resources necessitates a fundamental reconsideration of urbanity and urban form.

Key concepts for *Resources* include the terms *ecology* and *resilience*. The former we use in its widest sense, meaning that it includes economic, natural, and man-made components of a particular environment. The latter we use in order to discuss the city's ability to cope with change without compromising its resources. We believe that only through studying varying urban landscapes can we discover and generate new solutions that can contribute to a global discourse on the future of cities.

22.2 Cities and Energy: Studios 2007–2009

22.2.1 Shanghai: Beyond Oil (2007)

During the years 2007–2009, we ran a series called *Cities and Energy*, in which we investigated the three globally dominant economies: China, United States, and India. One city from each country—Shanghai, Los Angeles, and Pune—formed the basis for a case study. Each year built upon the previous, leading to more complex and fundamental questions. We took as our point of departure the cardinal question—how will the city be *beyond oil*?

In many regards, this particular take on the urban environment was liberating. *Beyond oil* became, in opposition to the more elusive term sustainability, a physically manifest limitation. Without oil, no airplanes or other oil-powered transport, no artificial fertilizers, and no plastics of the type we are so familiar with—all of these having direct consequences for our lifestyles. The setting for *Resources.07* was Shanghai, a year before the World Expo was to be held, and the city declared that it would be *the last fossil city and the first renewable one.*

Interest in China accelerated in the Swedish press, and China's urban growth became a hot topic during the year. Particularly apparent for the student group was, despite a clear understanding of issues concerning energy and the environment, that in China, there was an obvious resistance

FIGURE 22.1
The inner city landscape of Shanghai is rapidly transforming from a traditional two-story Lilong structure to a 30-story city of gated compounds. (Photo: Henrietta Palmer.)

to continue with what we considered to be sustainable solutions because many of them were too closely associated with their recent communist past. Here, sustainability should arise in a modern and high-tech guise, while traditional and low-tech solutions were politically stigmatized. The year in Shanghai was our first international study, and the solutions generated were primarily based on synergies and localization of systematic solutions. Still, the viability of consumption as a primary societal driving force in a post-oil situation became unavoidable. How could a city as consumption-driven as Shanghai dare to posit itself as leaving the fossil era (Figure 22.1)?

22.2.2 Los Angeles: Beyond Desire (2008)

The following year, in *Resources.08*, we focused on energy and consumption. If our oil dependency is driven by consumption, and our consumption is driven by our desires—for welfare, status, or power—then perhaps is *desire* worth examining? In our search for a relevant case study, the United States in general and Los Angeles in particular were chosen.

Literally springing from oilfields, Los Angeles is replete with the traces of capitalism. Nowhere is the monumentalization of desire more palpable or "motopia" so emblematically embodied. Los Angeles is the place for extreme lifestyles and secured utopias. However, *The City of Style* is also a place where alternative cultures and progressive grassroots movements seem to originate

FIGURE 22.2
The L.A. River presents us with a piece of complex infrastructure, that effectively divides Downtown L.A from East L.A. while still connecting the grand mountain range to the north with the beaches of the Pacific to the south. (Photo: Måns Tham.)

and flourish, fulfilling other kinds of dreams. Ultimately, the invention of alternative patterns of consumption must be based on other aspirations and longings. Our research converged on the Los Angeles River, along the banks of which the city's first citizens settled. Today, the river is a concrete ditch, a storm sewer, a relic of modernist planning, and a veritable wall of infrastructure dividing the poor areas of East L.A. from the tentatively awakening Downtown (Figure 22.2).

22.2.3 Pune: Beyond Development (2009)

The modern notion of development defines growth as an increase in GDP, that is, increased consumption. Solving the *Gordian Knot* of continuous growth while reducing greenhouse gas emissions is on the agenda of every politician. Despite this, no one has yet succeeded to strike the decisive blow. Countries that manage to combine low emissions with relatively high economic growth, such as Sweden, have done so by outsourcing production of goods to low-wage countries and by dependent on carbon-emitting energy production in neighboring countries such as Russia and Germany. To define development in terms of growth is obviously problematic. The term development must be augmented and based on other criteria than merely growth, without loosing the essential idea of providing improved living conditions for all the more. An analysis of development theory became therefore central to our research in India.

In *Resources.09*, we chose to work with the city of Pune, a medium-sized Indian city southeast of Mumbai with a population of approximately four million and experiencing expansive growth. Just as in most of the cities in India, Pune lacks comprehensive infrastructure systems for energy, water supply, sewage, and transport. At the same time, private car ownership is increasing, as are shopping malls and segregated, so-called ecocommunities built with green, high-tech solutions. The dilemma of poverty inherent to current developing urban environments in India stands in stark contrast to a nation in a dynamic state of economic growth—a split that is painstakingly apparent in Pune with its growing numbers of IT-centers and large percentage of slum areas (Figure 22.3).

From a planning perspective, Pune challenges western precepts. The impact of British colonialism is not particularly present in the planning of

FIGURE 22.3
This middle-class family lives in a run-down housing complex in inner-city Pune. This common situation demonstrates the complexities of the Indian society in which material wealth is not necessarily measured in the same way as in western society. It also shows how different the expectations of the growing middle class are, compared to their western counterparts, when it comes to demands on public infrastructural services. (Photo: Michael Dudley.)

the city, and the city has grown according to its own logic, which demands another understanding—another viewpoint. Compelling ourselves to develop other ways of *seeing* is a conscious strategy within *Resources*. A genuine discussion of urbanity is predicated on the creation of new images and alternative maps. The Indian city forces us all to question assumptions concerning urbanity, infrastructure, and lifestyles. As difficult as it is to understand the complexity confronting India's urban poor, it is equally difficult to relate to a member of India's growing middle class—racing toward its future in a city with electricity and water only in an unreliable and rationed form.

22.3 Making Our Tools: The Workshop

Complex issues need a multidisciplinary approach. In an educational program dealing with future urban environments, it is necessary to combine multiple approaches and viewpoints concurrently, in order to embrace the complexity that each specific urban landscape faces. Therefore, we see the value of using quick, intuitive ways of working to create a rich base of shared knowledge all through the use of short but intensive workshops.

22.3.1 Research by Design

All workshops are based on a research by design approach, meaning that we continuously combine various research methods with actual design proposals. Thereby, these are not seen as separate components, but as an integrated whole. In *Resources*, the concept and methodology of design is considered essentially cross-disciplinary, and in this sense we view all participants as designers, regardless of their background.

While the themes and geographies vary each year, the combination of aspects that influences the urban landscape, including climate, fossil fuel dependency, socioeconomic, and demographic patterns, and how these are all directly related to energy production and consumption are relevant for all cities and are therefore simultaneously investigated in the workshops. While we constantly strive to develop and adapt the workshops, there are a number of them that we find essential and that recur.

22.3.2 Investigating Lifestyles

22.3.2.1 Intimately Approaching the Other

How do we approach a new place? What methods can we employ to investigate and understand another place and its people? Lifestyle studies have become a key for us to try to understand an unfamiliar urban landscape by imaging how lives are lived today and will be in the near future. Although investigating a place through individual experiences may seem subjective and vague, it provides insights that enable us to get engaged in the place and fleshes out stories that statistics alone cannot provide. The lifestyle studies consist of an exploration of a family or social group. How will they live beyond peak-oil and fossil fuels? How are dwindling resources going to affect their lives? Will lifestyles become limited or will the potentials of new energy sources lead to new opportunities? Emphasis is placed on a creative speculation and on compelling images of the families' future situations, embracing everything from energy, dwelling, food, material consumption, work, mobility, public space to entertainment, and fun.

22.3.2.2 Objects and Fictions as Entry Points

The lifestyle workshops have utilized varying starting points. One entry point has been to start with an unfamiliar *tool* or an object particular to the place we are investigating. What has the tool been used for? What role has it played in society? What would the tool and its function look like in a future society of limited resources? How would this, in turn, even affect its cultural significance? The speculation on the interconnectedness between the object's function and its cultural relevance reveals the complicated nature of the issues we are dealing with. Another point of departure for an unfamiliar context has been through the investigation of *fictional material*. A third has been through *artistic works*, such as family portraits from the specific urban landscape. A text or other work of art, just as an object, is a site-specific artefact that is a distillation of inherent knowledge.

22.3.3 Envisioning Information

New images and alternative maps that enable us to read the shifting urban landscape are of fundamental importance for envisioning new realities. We continually receive requests from municipalities and state institutions to use the images created in our research to help them communicate new ideas. Apparently, there is a lack of imagery beyond the now-clichéd green-utopian type we have all become so familiar with.

Parallel to this development, we can see how the digital visualization tools of the architectural profession have become increasingly complex. Although these tools offer sophisticated possibilities for visualization, the tendency is rather that ideas become oversimplified, instead of more subtle and nuanced. By developing methods of communication, we can deliver ideas in a lucid manner, while retaining their complexity.

In association with Paolo Ciuccarelli and Luigi Farrauto from *Density Design Lab* at *Politechnico di Milano**, we have investigated the use of statistical

---

* See www.densitydesign.org.

information and its subjective nature. We combine statistical categories in unexpected ways in order to reveal new ideas graphically. For example, by looking at the overlaying of energy *and* health, food *and* employment, or transport *and* leisure, new connections can be discovered. These methods are then utilized to raise our consciousness on how we communicate to a wider audience. Significantly, the question of *what is considered to be statistics* has developed and changed during the years. *Density Design Lab* is currently looking into "subjective" statistics, such as social media flows, rumors, and chat, and how knowledge about the urban can be visualized from these sources.

22.3.4 Mapping and Governing Systems

Reaching an urbanism that actively engages in its energy needs requires the dynamic interaction of certain urban features and phenomena. *Systems thinking* is one of the key instruments for reconsidering the urban in relation to limited resources. Synergies and closed-looping are the first steps toward the efficient use of resources and energy. However, urban systems can embrace more than just the obvious and tangible flows of water, waste, and energy. Social, economic, cultural, and spatial patterns, as well as functions and exchanges within time and space, can all be integrated in technical and ecological urban systems to establish a metabolic vision of the urban. Regarding society as a network of agents, actions, and flows allows us to see urban inventions as an integral part of a system.

This workshop explores how innovations can propagate their own system and work together with others. Who will govern the advancement of this innovation? What processes need to be set in motion for this innovation to prosper? How will it generate a system in its own right, and how will this be integrated with other urban systems? System limits are investigated and the relevance of scale for these systems is defined. In order to understand ecosystems in relation to governance and innovation, we collaborate with experts from Stockholm Resilience Centre, such as political scientist Victor Galaz.*

Our notion of a systems approach has recently broadened to include an investigation of spatial and environmental justice. To move beyond using systems theory merely to achieve *green-tech* solutions, the notion of justice must be addressed as a fundamental issue when defining a given system.

22.3.5 Deploying Scenarios

We see the urgency of understanding the socioeconomic and socioecological consequences of the urban in ways beyond the classic "vision-approach" traditionally deployed by architects and planners. One possible method toward this end is *forecasting scenarios*. This method employs two fundamental and largely independent variables in order to create a field of plausible futures.

---

* See www.stockholmresilience.org.

Resources work on scenarios with systemsthinker Philippe Vandenbroeck from Belgian Shift'N* and policy strategist Karl Hallding from Stockholm Environment Institute.† Together, we have developed a methodology for urban planning through a series of four specific vantage points. These include *the contextual, the normative, the metabolic,* and *the spatial*—each one of them positioning the chosen urban situation in relation to the challenges and urgencies relevant for a certain spatial scale. The scenario framework acts as an orienting device providing us with a large canvas upon which multiple possibilities can be considered and compared—enabling us to position the various ideas from the workshop sessions in a larger, comparative context. It also creates a springboard towards a proposal. The Pune scenarios, for example, became a field for fractal interventions. By juxtaposing the variable of scale *infinite–infinitesimal* with the variable of materiality *material–immaterial*, we could describe a future for Pune that captured a multiscaled strategy that questioned given notions of development and totality (Figure 22.4).

Our way of employing scenarios has spread beyond the confines of the course. We have been invited to introduce the method and participate in a number of contexts, such as seminars for municipalities, architectural competitions and workshops as well as other academic situations.

22.3.6 Meeting the Other

While the fall term is a laboratory for facts, ideas, and fictions, the spring term begins with an onsite study and research trip. In this way, the students come to the specific urban environment with knowledge and ideas. Although these ideas are eventually questioned and rethought, it still enables them to immediately begin their research. The study trip is structured by individual research assignments conducted in groups, which creates a dynamic exchange between site work and discussions. Experts are invited to give lectures and to provide guidance and feedback. We collaborate with local organizations, NGOs, or academic institutions that act as working partners for the extent of the project.

The onsite workshop commences with a presentation of the scenario framework and selected workshop results to a local audience. The resultant discussion provides important input for the onsite research. The workshop concludes with a second public hearing to which our partners are invited. Building working relationships with experts and stakeholders enable us to establish a dialogue that continues during the remainder of the spring term and beyond.

Back in Stockholm, the findings are formulated into a framework and strategy for the proposal that guides the continued work. In order to develop the proposal, two additional workshops are held during the spring term.

---

* See www.shiftn.com.
† See www.sei-international.org.

FIGURE 22.4

The scenario diagram, which resulted from the question—how will Pune overcome the crunch of development with less resources and an emerging climate crisis? It depicts a future of innovations on multiple scales—from the infinite to the infinitesimal—that take place in either material or immaterial ways. (Image by Resources.09.)

22.3.7 Urban Tools

When investigating nonwestern urban landscapes, one is struck with how the needs of the residents are often met by tools or inventions that are smaller in scale than that of architecture, but larger than that of an object. How can we facilitate the urban inhabitant through the invention of a "middle scale" of tools?

This particular workshop has been led by different invited guests, such as Arvind Gupta, a scientist, pedagogue, and toy inventor, who uses play and "as-found" materials as a resource to spread scientific knowledge to all segments of society.* In addition, we have visited Haakon Karlsen Jr. at his

---

* See www.arvindguptatoys.com.

MIT-FabLab in Tromsø, Norway, to use this rapid-prototyping facility that enabled us to quickly test our ideas in full scale.* Results from these activities have been instrumental. Small-scale solutions are intuitive and immediately graspable with the students quickly developing their ideas. In the contemporary discourse on urban planning, small-scale solutions are often contended to be empowering, locally adaptable, and appropriate for a *self-managed urbanity*—this in contrast to top-down urban planning. In the case of the Pune project and its fractal strategy, *urban tools* became an essential part of the final proposal.

22.3.8 How Nature Does It

Biomimicry has emerged within the last 15 years as a cross-disciplinary field, combining design thinking with the knowledge of biology. Since this kind of work requires in-depth knowledge of biological functions and insight into the latest scientific research, *Resources* has investigated the use of biomimicry in urban planning together with Fredrik Höök, professor in biophysics, and researcher Patric Wallin both from Chalmers University in Gothenburg. In this workshop, we have posed the question: Could evolutionary-tested solutions from nature be applied on urban design challenges? For example, distribution of water systems has been based on the structure of leaves, evaporative cooling principles have been inspired by the dog's tongue and elephant's ear, and solutions for water harvesting and treatment have been explored in studies of various kinds of insects and lizards.

This workshop often plays an important part in influencing the final proposal. The challenge for us has been to transfer biomimetic thinking, thus far primarily utilized in the field of industrial design, to the urban planning discourse. Discussions often center on the split between "nature as an inspiration for function" versus "nature as an inspiration for form," the latter being the traditional way in which nature has inspired architects in the past. However, our experience is that these two sides are difficult to separate and that even "nature as an inspiration for organization" has been very rewarding for our studies.

22.3.9 Some Conclusions on the Method

The interdisciplinary approach of *Resources*, in which architects and planners join artists, filmmakers, journalists and other creative disciplines, creates an immediate, stimulating space for discussion. However, one should be aware of the differing working methods of these professions and the frustration that often comes when having to abandon a familiar way of doing things for a new way and then compounding that insecurity by also having to work in groups. The challenges of interdisciplinary work

---

* See www.fablab.no.

have to be addressed directly and need to be consistently and repeatedly framed as a part of the creative process.* Toward this end, we have introduced *storytelling* as an artistic method to facilitate dialogue and group work. Storytelling intertwines with the workshops, which in the end are all essentially about the same issue—telling a story about the future from different vantage points.

22.4 Three Postoil Concepts

The three years devoted to *Cities and Energy* have produced three distinct proposals, each containing a multitude of solutions, some directly related to energy transitions and others indirectly, via food production, transport, building typologies, economics, lifestyles, and consumption. These solutions are also defined in relation to their appropriate scale—that is, when is it necessary to speak of small-scale solutions for energy production, including its social implications, and in contrast, when are large-scale solutions most applicable?

22.4.1 Syncity

Our approach for Shanghai's central district of Luwan was to develop the city's urban systems and create synergies in both time and space. The scenario work explored the relationship of density and self-sufficiency—a crucial discussion for any city within the context of sustainability, but particularly relevant for Shanghai with its communist tradition of independent industrial city blocks, now being transformed into gated communities bringing with them a rapidly changing density but also the potential for greater self-sufficiency.

The proposal called *Syncity* defined itself in the future of a dense but self-sufficient urban environment, which placed specific demands on *how* density is to be achieved. Three site-specific typological phenomena were identified and provided the framework for systems thinking. The first one was the large-scale block structure of the area, at present densities each containing approximately 5000 inhabitants—an appropriate size to begin exploring the potentials of block-scale self-sufficiency. The second one was the system of elevated motorways that dominate the urban landscape and may well meet a new fate when China repeals its present policy of petrol subsidies for private citizens. How can these behemoths of the fossil era coordinate transport with public space? The third phenomenon was the heterogeneous character of the city at this specific point in time—in metamorphosis from a homogenous

---

* Editors' comment: The interdisciplinary character of other studios is discussed in depth in Chapters 20 and 21.

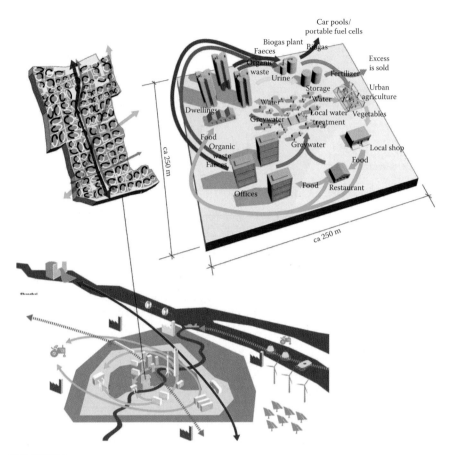

FIGURE 22.5
Different energy solutions need to be solved on differing scales in the Shanghainese society. The Syncell concept proposes a semiautonomous block structure of 5000 inhabitants that produce their own fertilizers, vegetables, fish production, and biogas for car-pooling from the domestic waste as well as solar cooling from PVCs placed on rooftops. (Image by Resources.07.)

Lilong structure (the traditional two-story building typology) to what may well become an equally homogenous landscape of high rises. In this moment of transition, the urban environment contains a unique mix of parallel histories, social strata, and living situations—a plurality that possesses urban resilient qualities (Figure 22.5).

Focus was put on developing *the Syncell* block as a self-sufficient entity organized by synergetic thinking. To retain the transitory quality of heterogeneity, a block typology was developed that retained the same built density as the 30-story high-rise buildings, but with an entirely different connection to the street, as well as providing other economic, social, and spatial qualities. This building typology was based on biomimetic principles initially developed during the earlier workshop and was reworked to include a passive cooling system to reduce the present reliance on oil-based electricity.

FIGURE 22.6
The elevated motorways are reenvisioned as a structure for new rail-bound transport systems. It also enables densification through the introduction of multilevel public spaces. (Image by Resources.07, *Syncity*, 2008. www.kkh.se/index.php/en/study-programmes/mejan-arc/architecture.)

The strategy for the elevated motorways had several dimensions. At the core however, was the reuse of these structures for different rail-bound transport modes. The invention of "food trains"—a system of mobile super-markets that connected the agricultural hinterlands to the inner-city areas also utilized the existing structures. At the height of the elevated motorways, the increased density resulting from new transport modes enabled us to reformulate a concept of multilevel public space. The group statement, *we see peak-oil as an opportunity* bears witness to the proposal's innovative and up-beat take on the grand challenges we face—a life beyond oil (Figure 22.6).

22.4.2 Fifth Ecology

The *Fifth Ecology proposal* explored the infamous L.A. River as a new public space based on immaterial consumption, bridging the east and west of L.A. The concept of *spectacular morality*, that is a morality put on show, took center stage in the city where the public realm is a theater for its vibrant citizens. In the spectacular city, the concert outplays the CD, the after-party over-shadows the show, and the entire event is valued by the stories that it emits. Ownership of object and place is only worth the experience that results from them. The abundance of resources Los Angeles enjoys from sunshine to its ethnic diversity; all are seen as essential components in the reformulation of a postmaterial urban experience. Resources are valuable to the degree that they are accessible assets, and accessibility in turn can be seen as a resilient quality for a city. In The Fifth Ecology, we investigated a scenario of more or less accessibility versus material or postmaterial consumption (Figure 22.7).

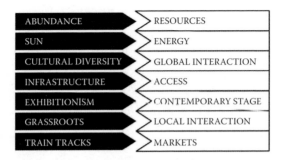

FIGURE 22.7
A diagram showing how the abundances of Los Angeles can be seen as potentials for a new sustainable L.A. urban landscape. (Image by Resources.08.)

22.4.2.1 Access and Abundances

Our strategy for the L.A. River was inspired by the charm bracelet. The chain with its lucky charms—here embodied in the proposed neighborhoods adjacent to the river—that signify important events in the life of the wearer was combined with the American strip typology to create a place filled with spectacular morality and a new notion of access. A spectacular staging of energy production and water treatment became key elements in the physical manifestation of *the Strip*.

The *Agrimids*, an example of one charm that placed food production in the heart of Downtown, managed to provide this function while simultaneously generating new public spaces. Skidrow's surplus of abandoned factories and underutilized parking lots was identified as a resource. Pyramidal structures for light-weight industrial agriculture were built on top of the existing factory structures and provided the Downtown area with a new spectacular skyline. The shaded spaces below became breathtaking semi-public interiors connected to small housing cooperatives built on top of the existing parking lots (Figure 22.8).

22.4.2.2 Walking the L.A. River Strip

The Strip as an urban typology is essentially a commercial space. At the same time, it is emblematic for the city. The Venice Beach Boardwalk and Sunset Boulevard are iconic, and both exude an aura that transcends their physical boundaries. The proposed L.A. River strip connects on an *ecology* scale two of the city's most public of spaces—the mountains and the beach. In this regard, we reconnect to the British art historian Reyner Banham who in *Los Angeles: The Architecture of Four Ecologies* (Banham 1971) described the city in terms of four distinct ecologies.

The L.A. River Strip thus became the city's fifth ecology—with its own microclimate, typology, and lifestyle, all in keeping with Banham's original

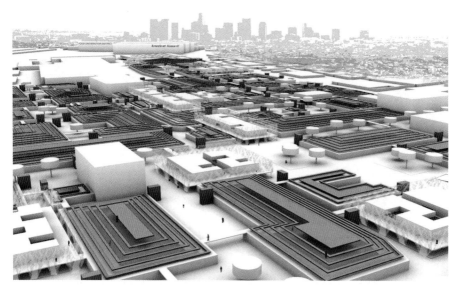

FIGURE 22.8
The Agrimids are lightweight pyramidal structures for industrial production of vegetables and placed on top of abandoned industrial structures in Skidrow. The resultant interiors provide semipublic shared spaces for new multifamily housing typologies built on top of existing parking lots. (Image by Resources.08, *Fifth Ecology*, 2009. www.kkh.se/index.php/en/study-programmes/mejan-arc/architecture.)

definition. Various methods for evaporative cooling, taking advantage of the river water itself, water treatment and wetland management, all contributed to our concept of the Fifth Ecology. The length of the L.A. River strip was shaded by *the green ray*—an urban-scaled roof composed of transparent tubes for algae production, which in turn cleaned the river water while lighting up the strip at night with a phosphorous glow from the algae. In combination with a pod-car system for waste management, *the green ray* fed directly into an abandoned industrial building that was transformed into a biogas plant (Figure 22.9).

The Fifth Ecology was conceived in the beginning of the financial crisis, in the fall of 2008. It is not without irony that the present crisis, a housing one, can be seen as stemming from the crisis preceding it in the 1930s, since we argued that The Great Depression found one of its solutions in the consumption-based pattern predicated on mass production combined with the breakthrough and proliferation of the suburban lifestyle. Paul Romer's assertion that—*A crisis is a terrible thing to waste*—became a guiding principle during the year. What could be an urban response to the current crisis?

FIGURE 22.9
The Strip—using algae production as a sheltering roof structure for a new public space as well as to clean the water of the river and to produce biofuels. Evaporative cooling from the "second river" creates its own microclimate. (Image by Resources.08, _Fifth Ecology_, 2009. www.kkh.se/index.php/en/study-programmes/mejan-arc/architecture.)

22.4.3 Pune Matters

Pune Matters investigated the fractal manner in which the Indian infrastructure is solved and built upon this "tradition" in order to develop new infrastructural concepts that could provide alternatives to western models. Scale became an important parameter explored in the Pune scenario framework and combining elements from all four scenarios became the strategy. Indian society can be understood in terms of its simultaneity, which means that solutions are not necessarily organized spatially or temporally in a hierarchic manner, but rather can occur parallel with varying degrees of intensity, permanence, and in an overlapping of their functions.

Two phenomena became central to the proposal—the Indian middle class and the use of public space. In India, public space is a contested field, which to a great degree is used by the city's poor as a space to make their livelihoods through the informal economy. In contrast, those who have access to climatized space seek to escape the city's public spaces and find refuge in their air-conditioned cars and shopping malls. When the urban environment is dysfunctional, often exhibited through segregation and injustice, then shared space is either lacking or used by limited segments of society.

Shared space in the future city will certainly be a key to addressing resource depletion, food shortage, and climate change. Energy production and consumption can in this regard act as an instrumental tool to attain spatial justice. To maintain the social importance of public space and the necessity of the middle class's participation in the urban everyday sphere were two ways of resisting the movement from a caste to a class system and to encourage another development beyond energy-dependent private consumption.

22.4.3.1 Scale Matters

Three urban strategies based on scale were introduced: *Powers of the Many*, *Urban Shift*, and *City as a Tree*. The *Powers of the Many* conveyed the notion that many small solutions can create a large effect. This strategy revisits E. F. Schumacher's *Small is Beautiful: Economics As If People Mattered* (Schumacher 1974) in which he proposed that small-scale, labor-intensive and locally adapted solutions are preferable to large-scale, rational, and general ones. One example of this thinking is seen in the proposal initially generated in the *Urban Tools workshop* for a low-tech compost bag for biogas production in the home. These bags would in turn create nutrient-rich soil that would be attended to by the waste-pickers who would collect and use them for local vegetable gardens and thereby increase their economic independence.

The *Urban Shift* was a way of doubling or tripling the infrastructure systems of the city by layering them. This may be considered the "natural" state of the Indian city but one which can be more consistently employed. For example, water is available through the public waterworks, but only for a few hours each day. The rest of the time people use the local well, and when those sources are exhausted, they continue by collecting water on their own roofs, or buy it from a water vendor. None of these systems are complete or reliable, but together they create a functioning solution. In our proposal, we continued to develop ideas around water by replacing the existing local water tanks with transparent ones filled with light-emitting algae that lit up public spaces at night. Large common water collectors became public spaces in their own right.

The *Urban Shift* questions the western precept that solutions should be all-encompassing. In a postfossil society, perhaps a totality is impossible to achieve? The *City as a Tree* demonstrates how incomplete attempts to achieve comprehensive solutions can be complemented by smaller, more flexible ones in order to create another kind of whole. In this case, we investigated a half-completed BRT system, and how this could be combined with the agile network of rickshaw taxis. These rickshaws were upgraded to a higher degree of comfort with an air-conditioned zone in the main seating area and with an added jump-seat on the back, allowing poorer citizens to hitch a ride in the direction the taxi was already going. By using their bus ticket as a transfer, money could be exchanged between the BRT and taxi systems. The taxis would run on electricity from solar energy and would be charged at newly formed hubs that would be strategically located in conjunction with the BRT's route.

22.4.3.2 Public Space Matters

The Mutha River cuts through Pune's inner city like an unfortunate afterthought—partly because of its polluted water but also because it seasonally overflows its banks, which makes the surrounding environment unreliable to use. Uncertainty is one of the important issues future cities will need to manage. To see the temporary as a necessary quality and experience is therefore essential. By changing the form of the river in section, providing a deeper central channel and more shallow banks, we could negotiate the periodic flooding while creating a productive public landscape that took advantage of this fluctuating situation. On the banks of the river, renewable energy is generated and food grown in a varied and playful environment that provided recreation space for all segments of society (Figure 22.10).

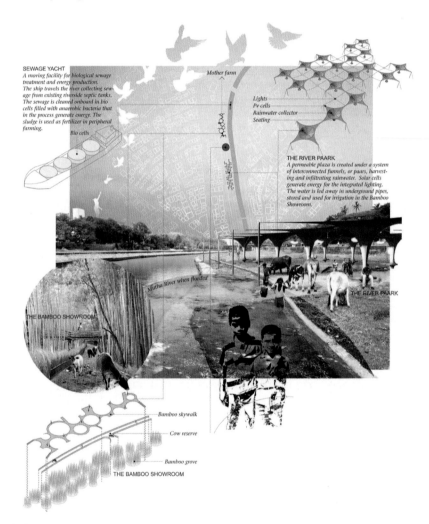

FIGURE 22.10
Temporality as a quality—the fluctuating levels of the Mutha River provide a changing place for playful energy production and a shared space for the middle class as well as for the urban farmers and their cattle that graze along the riverbed (Image by Resources.09, *Pune Matters*, 2010. www.kkh.se/index.php/en/study-programmes/mejan-arc/architecture.)

22.5 Follow-Up, Interaction, and Continuation

We use the freedom of academia to immerse ideas in the reality outside of our institutions. To reach this intersection between theory and practice, relationships need to be built on mutual interests but also on the acceptance of differences. The act of "giving back" to the context is essential for this relationship, but equally important is the attentive ear of the receiver.

Each year, the final proposal is presented in both catalogue and exhibition form at the conclusion of the spring term. A public seminar day is arranged in which the central themes of the project are discussed. The subsequent fall, we return to the city of study for a second exhibition. A smaller group of students organize financing, the installation of the exhibition, as well as the public events held during the exhibition period. In later years, we have expanded this portion of the program through the establishment of research grants for continuing research proposals based on the concepts generated in the project. This module involves the continued participation of the working partners and is conducted partly on site. The final research work is subsequently documented in a printed format.

The three projects that make up *Cities and Energy* have been featured in popular journals as well as television programs, but the most common form of communication is through invited presentations in which the students become "ambassadors" of each project. Many of the students have received job opportunities as a direct result of the work they have done, since they become recognized as experts within the sustainability discourse in Sweden. Some have returned to China, the United States, or India to work or continue their research. Such was the case with the architect Måns Tham, who continued his investigation of the L.A. freeways. By wrapping the Santa Monica freeway with a snakeskin of shimmering solar-cell scales, it was transformed into a spectacular piece of infrastructure. This project has been published widely and is currently in the next phase of development (Figure 22.11).

How we govern our cities is primary to how we think about transformation. Who initiates change and who manages these initiatives and their processes? Who ensures that the initiatives take consideration of our global challenges? In an "ideal society," politicians and civil servants would have the necessary knowledge as well as the trust of its citizens, but we all know that short-sightedness and corruption are real impediments to achieving satisfactory and creative energy solutions.

Now, emphasis should be put on testing a slew of ideas, conducting case studies and realizing pilot projects that are given the space, time, and freedom to provide long-term solutions. Toward this end, *Resources* and similar programs have an important role to play by providing a catalogue of ideas and solutions that stem from radically different urban landscapes. Each city is different and just as no solution developed in a specific urban context can

FIGURE 22.11
The Santa Monica freeway wrapt in a spectacular snakeskin of shimmering PVC solar cells.
(Image by Måns Tham.)

be directly applied on another, we can equally expect each city to generate
unique and exciting solutions—if we rely on local potentials and the inherent capacity and knowledge of each place.

Acknowledgments

Special thanks to all of the students from Resources.07, Resources.08, and
Resources.09 that have worked endlessly, produced amazing new ideas and
inspired us enormously.

Bibliography

Banham R. 1971. *Los Angeles: The Architecture of Four Ecologies.* Harper & Row, New York.
Benedetti P. 1996. *Forward through the Rearview Mirror: Reflection on and by Marshall
    McLuhan.* MIT Press, Cambridge, MA.
Benyus J.M. 2002. *Biomimicry. Innovation Inspired by Nature.* Perennial, New York.

Burdett R. 2010. *The Endless City*. Phaidon Press Ltd., London, U.K.

Burdett R. and Sudjic D. 2011. *Living in the Endless City*. Phaidon Press Ltd., London, U.K.

Correa C. 1989. *The New Landscape: Urbanisation in the Third World*. Butterworth Architecture, London, U.K.

Correa C. 2000. *Housing and Urbanisation*. Thames & Hudson, London, U.K.

Fainstein S.S. 2010. *The Just City*. Cornell University Press, Ithaca, NY.

Gottlieb R. 2007. *Reinventing Los Angeles. Nature and Community in the Global City*. The MIT Press, Cambridge, MA.

Hamidi N. 2004. *Small Change. About the Art of Practice and the Limits of Planning in Cities*. Earthscan, London, U.K.

Jencks C. 1993. *Heteropolis. Los Angeles. The Riots and the Strange Beauty of Hetero Architecture*. Academic Editions, London, U.K., Ernst & Sohn, Berlin, Germany.

Kunstler J. 2006. *The Long Emergency: Surviving the Converging Catastrophes of the 21st Century*. Atlantic Books, London, U.K.

Resources.07. 2008. *Syncity*. www.kkh.se/index.php/en/study-programmes/mejan-arc/architecture

Resources.08. 2009. *Fifth Ecology*. www.kkh.se/index.php/en/study-programmes/mejan-arc/architecture

Resources.09. 2010. *Pune Matters*. www.kkh.se/index.php/en/study-programmes/mejan-arc/architecture

Schumacher E.F. 1974. *Small Is Beautiful: A Study of Economics as If People Matter*. Abacus, London, U.K.

Soja E.W. 2010. *Seeking Spatial Justice*. University of Minnesota Press, Minneapolis, MN.

Steel C. 2009. *Hungry City. How Food Shapes Our Lives*. Vintage, London, U.K.

Walljasper J. 2010. *All That We Share. A Field Guide to the Commons*. The New Press, New York.

Part V

Epilogue

23

Conclusion

Andy van den Dobbelsteen and Sven Stremke

CONTENTS

23.1 Welcome to a New Transdisciplinary Field

Based on the chapters of this book, we hope the reader agrees with us that sustainable energy landscapes are an uplifting and inspiring subject, arousing the enthusiasm of many scholars and experts. A first interesting observation concerns the multidisciplinary background of the authors involved and the scale at which they operate. One could state that energy landscapes bring together people that are traditionally engaged in different areas. Look at us, the editors, and you will see a landscape architect (Sven) collaborating with a building engineer (Andy). This book project was not our first endeavor together; over the past 5 years, we have collaborated in many research projects such as *Synergy of Regional Planning and Exergy*. From Sven's point of view, energy was becoming an increasingly important factor in the planning and design of sustainable landscapes, whereas Andy noticed that the best improvements to the energy system could be made on a larger scale than his natural focus on buildings. And this is exactly what is happening: a new field is emerging—a field that brings together spatial planners and landscape architects, urban designers and architects, civil and building engineers, energy and exergy experts, economists, social and natural scientists, among many others. Rather than being another field for multidisciplinary research, the design of sustainable energy landscapes is becoming transdisciplinary, bringing together researchers, designers, and stakeholders with different disciplinary backgrounds. We believe that this presents a precondition for innovation in the field of energy landscapes, and sustainable innovation is what our world today needs.

23.2 Crucial Components of the New Field

In order to substantiate and advance the new transdisciplinary field of sustainable energy landscapes, a number of crucial components have been discussed in this book:

1. Understanding the relationship between energy and landscape
2. Learning how to design sustainable energy landscapes
3. Exploring alternative spatial planning approaches
4. Validating the impact through case studies
5. Conveying knowledge and experience

These will be discussed in the following five sections.

23.2.1 Understanding the Relationship between Energy and Landscape

This book provides a comprehensive overview of state-of-the-art knowledge, experience, and education with respect to sustainable energy landscape. As Pasqualetti describes accurately, we are starting to understand the consequences of different energy sources on the landscapes around us. Linking the physical landscape with technical characteristics by means of energy potential mapping, Dobbelsteen et al. define one potential role of environmental designers in the process of energy transition. In spite of the necessity to engage in energy transition, Sijmons and Van Dorst remind us how the shift from landscapes based on fossil fuels to landscapes that make use of renewable energy sources arouses a lot of opposition. This awareness—which is shared by many of us working in the field—supports the need for a better integration of aesthetics and environmental psychology into the shaping of sustainable energy landscapes. In a similar vein and based on reference studies, Schöbel et al. stress that future energy landscapes should be presented in a realistic and yet ethically correct fashion. In this way, renderings and maps can help understand the relationships between energy and landscape and, most importantly, develop alternative energy landscapes that are welcomed by inhabitants—a key objective of sustainable development.

23.2.2 Learning How to Design Sustainable Energy Landscapes

This book provides insight into the various ways of designing sustainable energy landscapes while stressing the importance of recognizing the uniqueness of each landscape—the *genius loci*. Stremke delineates the five-step approach to the planning and design of sustainable energy landscapes at the municipal and regional scale. He provides a methodological framework within which one can conduct, for example, energy potential mapping, landscape analysis, and scenario studies, all to identify innovative and yet realistic energy-conscious spatial interventions. Grêt-Regamey and Wissen-Hayek elaborate a different approach; they work with multicriteria decision analysis tools and GIS-based visualizations to integrate renewable energy technologies into the landscape. Stöglehner and Narodoslawsky propose an approach toward integrated spatial and energy planning at the community scale.

23.2.3 Exploring Alternative Spatial Planning Approaches

Understanding the potential magnitude of reintegrating renewable energy in our everyday landscapes, one may realize that the development of sustainable energy landscapes needs to be approached in a manner that differs from conventional planning. Roggema describes energy transition as a wicked problem that cannot be solved by traditional planning and design approaches. He proposes swarm planning as a fundamental paradigm shift in planning. Basta et al. too stress on the need for a different attitude toward spatial planning and call for an advanced approach referred to as individuals' active planning. Identifying the novelty of such an undertaking paves the way for new means of assessing the effectiveness of both the current and the desired future state of landscapes. We are in need of quantitative approaches that can help assessing the sustainability of energy landscapes in order to facilitate decision making. In this book, Jørgensen suggests employing eco-exergy and carbon accounting to assess the sustainability of energy landscapes.

23.2.4 Validating the Impact through Case Studies

Theoretical approaches may be promising but only have an impact when applied to and tested in case studies, either virtual or in the real world. This book presents a number of case studies conducted in different parts of the world, demonstrating that sustainable energy landscapes can be developed everywhere, yet differently in every location. Bunschoten presents two fascinating Asian projects of CHORA: the "Smart City Chengdu" and the "Taiwan Strait Smart Region." Thün and Velikov describe their transnational energy landscape in the Great Lakes megaregion in North America. Keeffe discusses a selection of research and design projects from the UK that address the critical relationships between renewable energy provision and food production in cities. Schroth et al. present various case studies on community energy planning in Canada and share their experiences from planning sustainable energy landscapes with different stakeholder groups and at various spatial scales. Schmuck et al. present how their team facilitated the development of so-called bioenergy villages in Germany—an approach that has transformed rural landscapes across the country. In their second chapter, Stöglehner and Narodoslawsky discuss how the GIS-based energy zone mapping tool and the ELAS calculator have been put to work in Austria. Jørgensen and Nielsen illustrate the application of eco-exergy to compute the degree of sustainability on Samsø—a renewable energy island in Denmark. Finally, Pulselli et al. exemplify the use of two systemic indicators in their study of the existing energy landscape in the Abruzzo region in Italy.

23.2.5 Conveying Knowledge and Experience

Another very important and not necessarily last component to the development of sustainable energy landscapes is formed by the conveyance of knowledge and experience. In the book, this was exemplified by means of graduate and postgraduate education in three universities. De Waal et al. discuss student results after having applied Stremke's five-step approach in their master's studio at the University of Wageningen, the Netherlands. Lehrman et al. present the impact of multidisciplinary and energy-conscious design courses at the University of Minnesota, USA. Last but not least, Palmer and Dudley present an overview of international, energy-related projects executed at the Royal Institute of Art in Stockholm, Sweden. We believe these educational examples are very promising in their approach and results; we highly recommend the inclusion of sustainable energy landscapes in students' curricula across the world.

23.3 And Now?

Frankly speaking, we are only at the verge of understanding the complexities of energy transition, in general, and sustainable energy landscapes, in particular. This book is one of the first attempts to bring related aspects into focus. Collectively, we should continue to study best practice and exchange experiences. In addition, a lot can be learned from other fields such as climate change adaptation. Publications and conferences can support this knowledge consolidation. Further research and design will also help establish criteria that define a sustainable energy landscape, which is a challenge since the general criteria for sustainability are still being negotiated. Nevertheless, we should get started—this book presents first aspects, definitions, frameworks, methods, and tools to be discussed in the research and design community. In the meantime, the future generation of designers and planners of sustainable energy landscapes need to be educated; energy transition offers many possibilities for our young colleagues to engage in the shaping of a sustainable future.

In our opinion, many scientists, environmental designers, and policy makers have switched too fast from mitigation of climate change to climate change adaptation. Adaptation, essentially, is a response to inevitable events, while the consequences of fossil energy use—at least over periods of generations—can still be influenced. There is no need to accept another (final) 100 years of resource depletion, exergetic destruction, and pollution caused by inefficient energy conversion and powered by low-quality coal and oil sands. We hope that this book demonstrates to you that the alternative, a sustainable and clean future, is a possibility. Since the future of a sustainable energy provision lies in the use of space, our approach to landscapes will become decisive. Sustainable energy landscapes offer many opportunities (many of these empirically proven) to farmers, for job creation, stronger regional economies, reduction of energy losses, people empowerment, etc. Next to the siting and design of renewable energy technologies, consumer behavior is an essential factor for energy transition. Can the exposure of inhabitants to artifacts of energy provision—living within an energy landscape—trigger changes in the way we consume energy? There is a need for substantial research on this decisive factor: it presents one of the key challenges (and risks) to the present-day physical environment and to inhabitants themselves.

We may conclude that with this book on sustainable energy landscapes we have touched upon a vast new field that needs to be further explored, discussed, and tested in the future. But equally important, we should start developing sustainable energy landscapes across the world. We still have time to do so, but not much. If we do not act, we will reach a tipping point where energy and material resources left are not longer sufficient to build a completely sustainable energy system, inevitably resulting in disturbances and leading to irreversible deterioration in the natural environment and society. Some scholars estimate that this tipping point, without significant changes in the energy system, lies around the year 2025. Well before that time, societal investments should be directed toward a sustainable transformation and not continued to prolong the fossil era. We hope we have made a modest contribution in the right direction.

Index